Biological Diversity Conservation and the Law

Legal Mechanisms for Conserving Species and Ecosystems

HUNTING
Aquatic Resources
Innovation Centre
York Science Park
University Road
Heslington, York, YO1 5DG, UK
Tel: ++ 44 (0) 1 904 435165
Fax: ++ 44 (0) 1 904 416611

IUCN – THE WORLD CONSERVATION UNION

IUCN – The World Conservation Union brings together States, government agencies and a diverse range of non-governmental organisations in a unique world partnership: some 770 members in all, spread across 123 countries.

As a union, IUCN exists to serve its members – to represent their views on the world stage and to provide them with the concepts, strategies and technical support they need to achieve their goals. Through its six Commissions, IUCN draws together over 5000 expert volunteers in project teams and action groups. A central secretariat coordinates the IUCN Programme and leads initiatives on the conservation and sustainable use of the world's biological diversity and the management of habitats and natural resources, as well as providing a range of services. The Union has helped many countries to prepare National Conservation Strategies, and demonstrates the application of its knowledge through the field projects it supervises. Operations are increasingly decentralised and are carried forward by an expanding network of regional and country offices, located principally in developing countries.

IUCN – The World Conservation Union seeks above all to work with its members to achieve development that is sustainable and that provides a lasting improvement in the quality of life for people all over the world.

Biological Diversity Conservation and the Law

Legal Mechanisms for Conserving Species and Ecosystems

Cyrille de Klemm
in collaboration with
Clare Shine

IUCN Environmental Policy and Law Paper No. 29

IUCN Environmental Law Centre
IUCN Biodiversity Programme

IUCN – The World Conservation Union
1993

Published by: IUCN, Gland, Switzerland and Cambridge, UK

IUCN Environmental Law Centre, Adenauerallee 214, D–5300 Bonn 1, Germany

IUCN Biodiversity Programme, Rue Mauverney 28, CH-1196, Gland, Switzerland

IUCN
The World Conservation Union

Citation: de Klemm, C. and Shine, C. (1993), *Biological Diversity Conservation and the Law*, IUCN, Gland, Switzerland and Cambridge, UK. xix + 292 pp.

ISBN: 2-8317-0192-9

Text layout by: IUCN Publications Services Unit, Cambridge, UK, on desktop publishing equipment purchased through a gift from Mrs Julia Ward

Cover photo: Ujung Kulon NP, West Java, Indonesia: WWF/Anton Fernhout

Cover design by: IUCN Publications Services Unit

Printed by: Page Brothers (Norwich) Ltd, Norwich, UK

Available from: IUCN Publications Services Unit
219c Huntingdon Road, Cambridge, CB3 0DL, UK
or
IUCN Communications and Corporate Relations Division
Rue Mauverney 28, CH-1196 Gland, Switzerland

The presentation of material in this book and the geographical designations employed do not imply the expression of any opinion whatsoever on the part of IUCN concerning the legal status of any country, territory, or area, or of its authorities, or concerning the delimitation of its frontiers or boundaries.

The views of the authors expressed in this publication do not necessarily reflect those of IUCN.

The text of this book is printed on Fineblade Cartridge 90gsm made from low chlorine pulp

Contents

Foreword

Plants and animals, evolving over hundreds of millions of years, have made our planet fit for the forms of life we know today. They help maintain the chemical balance of the Earth, stabilize climate, protect watersheds and renew soil. All societies, urban and rural, industrial and non-industrial, continue to draw on a wide array of ecosystems, species and genetic variants to meet their ever-changing needs. The diversity of nature is a source of beauty, enjoyment, understanding, and knowledge—a foundation for human creativity and a subject for study. It is the source of all biological wealth, supplying all our food, much of our raw materials, a wide range of goods and services and genetic materials for agriculture, medicine and industry worth many billions of dollars per year. People spend additional billions of dollars to appreciate nature through recreation and tourism. Biological diversity should be conserved as a matter of principle, because all species deserve respect regardless of their use to humanity, and because they are all components of our life support system. For all of these reasons, conserving biodiversity and using biological resources sustainably have become issues of major international concern.

Prudence dictates that we keep as much biodiversity as possible. But natural diversity is more threatened now than at any time since the extinction of the dinosaurs 65 million years ago. The trend is steadily downward, as more habitats are converted to exclusively human uses. While we are still uncertain about how many species now exist, some experts calculate that if present trends continue, up to 25 percent of the world's species could become extinct, or be reduced to tiny remnants, by the middle of the next century. Many more species are losing a considerable part of their genetic variation, making them increasingly vulnerable to pests, disease, and climatic change.

The conservation activities of governments, institutions, and concerned individuals are still too disparate, fragmented, and limited to bring about the fundamental changes that are required to reverse this trend. Converting depletion into sustainable use of our biological capital requires a diverse, yet coordinated and participatory programme that attacks the problem at its roots, builds support among a wide range of institutions and individuals, makes use of the best modern science, and establishes the conservation of biodiversity in its rightful place as a major development goal. Such a change needs to be supported by law, a need recognized by governments when over 150 of them signed the Convention on Biological Diversity in Rio de Janeiro on 5 June 1992. With the entry into force of the Convention at the end of 1993, governments will be challenged to ensure that national law also helps to conserve biodiversity.

The Biodiversity Convention is a landmark, for several reasons. Above all, it embraces biological diversity as such as a common concern of humankind, while at the same time recognizing the responsibility of each State to conserve it.

The Convention is the culmination of a process of evolution which took place over more than two decades: other treaties have paved the way to the Biological Diversity Convention, as has the evolution of the concept of conservation and conservation law generally in the recent past.

At the heart of the thoughts which have generated these developments is the acknowledgement that biodiversity conservation is dependent on measures to be taken at national level; that protective measures—for species and ecosystems alike—are a cornerstone of conservation; but also that protection alone is not sufficient: biodiversity conservation also necessitates the sustainable use of biological resources, and the control of processes which lead to the deterioration of the natural environment. In short, conservation can only be achieved through a cocktail of complementary measures.

Law and legal mechanisms play an important role in achieving these goals: international obligations prescribe common commitments and measures to attain them; national legislation provides a framework to regulate certain behaviour, to provide incentives to achieve certain results, and to set appropriate institutions in place.

Those who are looking for a single volume covering the full spectrum of legal mechanisms available for conserving biodiversity need look no further than this book. Part I, on species, discusses the scope of state powers to conserve wild species and habitats, the legal mechanisms for controlling taking, the procedures for listing species in need of conservation measures, the mechanisms for controlling trade, and the challenges of enforcement.

Part II covers area-based conservation at local and national level, as well as in transfrontier protected areas and areas beyond national jurisdiction. It discusses the basic instruments for protecting areas, including public ownership, voluntary agreements, and regulatory measures. It goes beyond traditional tools for protecting areas by looking at a number of innovative instruments, planning controls, incentives and disincentives.

This book, therefore, will be of great utility to those who are preparing national legislation for implementing the Convention on Biological Diversity and the many other international instruments discussed in this book. Many of these points are, of course, also highly relevant at the local level, where the role of legislation may be to establish the policies that are required to enable local communities to manage their own biological resources in ways that will provide them with a sustainable benefit.

Law, being man-made, is as imaginative as those who craft it; this is why it is so important to learn from experiences and solutions applied and continuously "invented" in all parts of the world. We hope that this book will provide food for thought to all those who seek to improve their own legal tools through comparison to the ideas of others, throughout the globe.

This book is jointly produced by the IUCN Biodiversity and Law Programmes in collaboration with the World Resources Institute and the United Nations Environment Programme, with the financial support of USAID. It is hoped that it will assist in improving the law on biodiversity conservation, and contribute to the much needed progress in this field.

Jeffrey A. McNeely
Chief Biodiversity Officer
IUCN – The World Conservation Union
1196 Gland, Switzerland

Françoise Burhenne-Guilmin
Head, IUCN Environmental Law Centre
Adenaurallee 214
D-53113 Bonn, Germany

14 November 1993

Editorial Preface

The conservation of biological diversity is increasingly becoming one of the greatest challenges of our times. It raises the fundamental question of whether we have the right to destroy species and ecosystems.

As regards both present and future generations, the conservation of biological diversity may procure a wide range of advances in science, medicine and potentially in many other fields. Particular emphasis should be placed on the enormous possibilities offered by genetic engineering, provided that the raw material, namely the genes of millions of wild species, is preserved. Wild species are also necessary for the proper functioning of ecosystems, whether natural or transformed by human action, as key species such as pollinators, seed dispersers, predators or prey play an essential role in maintaining the ecological balance.

The conservation of biological diversity is also important for moral reasons. Species have an intrinsic value, regardless of their value for humankind, and evolution will only be able to continue its natural course if the diversity of species is maintained.

The acceleration of the destruction of biological diversity over past decades has four major direct causes: overexploitation, the destruction of habitats, pollution and the introduction of exotic species. However, the principal underlying causes are population pressure, which leads to the clearing of increasingly large areas, and the requirements of development. Most importantly, indifference to the consequences of the loss of biological diversity results in a situation where there is little motivation to preserve it, little staff and, above all, very little money.

Furthermore, biological diversity has no recognised owner and has, until now, had no holders of rights over it, although the Convention on Biological Diversity is a turning point in the latter respect. This lack of ownership means that no-one exists in law to speak on behalf of biological diversity. Such representation is particularly important given that its conservation may impinge on public freedoms and more immediate human interests.

It has been repeatedly stated during the past decade that conservation and development are two sides of the same coin, and that only development will make conservation possible in the poorest countries. This is undoubtedly true of the many damaging processes that generate erosion, desertification, alteration of the water regime and depletion of natural resources of commercial value, but may not necessarily be the case for biological diversity. It is unlikely that development will in itself stop the destruction of biological diversity, because its value to the economy and to development is only potential and cannot be evaluated in monetary terms. Even in the most developed societies, the conservation of biological diversity cannot be justified on economic grounds or for its contribution to development.

As a result, the setting aside of areas which have a potential development value (such as forests and wetlands) for the purposes of biodiversity conservation is all too often perceived as an illegitimate obstacle to growth and to a whole range of other legitimate interests, both public and private, whatever the stage of development of the country concerned.

The current controversy over the preservation of ancient forests in the north-eastern region of the United States for the conservation of the spotted owl, *Strix occidentalis caurina*, is a good illustration of the difficulties of safeguarding biological diversity, even in the most developed countries.

The potential benefits of preserving natural habitats may nevertheless exceed the economic and social costs of conservation. There are many examples of this. One of the most recent is that of Taxol, a product found in the bark of a yew tree, *Taxus brevifolia*. The substance has important anti-cancer properties, but it has so far been impossible to reproduce it synthetically. However, a microscopic fungus, *Taxomyces andreana*, growing on yews and able to produce taxol, was recently discovered in an old-growth forest in the State of Montana, opening up possibilities of industrial production of the substance. [1]The value of the forest concerned was therefore immense in terms of its potential to alleviate human suffering. Had that forest not been preserved, it is by no means certain that the fungus would have been found elsewhere.

Scientific and moral arguments for the conservation of biological diversity are strong, but as long as the economic arguments remain weak, it will continue to be difficult to adopt and enforce adequate conservation measures.

Law is a reflection of the needs and demands of society. Gaps and shortfalls in present conservation legislation are the consequence of continuing indifference, if not opposition, to the fate of biological diversity amongst segments of public opinion that are still relatively large, although this varies from one country to another. At the same time, however, the very existence of legislation is proof that some consensus has developed concerning the importance of conserving species and ecosystems.

The main purpose of this paper is to review the evolution and principal features of conservation law. The paper is limited to legal instruments for the protection of species and areas, which are the major instruments used to meet threats of overexploitation and habitat destruction. Pollution will not be addressed, as it cannot really be separated from its effects on humans[2] and pollution control legislation generally makes little reference to the conservation of biological diversity.

The introduction of exotic species would merit a study in its own right, but apart from a few countries which have suffered considerably from past introductions, particularly Australia and New Zealand, existing legislation on introductions generally establishes little other than a permit requirement.

Pollution and the introduction of exotic species are processes caused by humans which affect biological diversity. There are of course many more activities in this category, such as the alteration of river flows, dredging, drainage, mining and quarrying, intensive agriculture, shifting cultivation, overgrazing, deforestation and certain outdoor recreational activities, to mention but a few. However, most of these activities are associated with the direct destruction of habitats and will therefore be considered under Part II of the paper dealing with area-based conservation.

The traditional division between species and habitats will be followed for the sake of convenience in this paper, because legislation is largely based on this distinction. It would perhaps be more rational to analyse the threats to biological diversity on the basis of potentially

---

[1] Science, 9 April 1993.

[2] Except perhaps in a few cases, such as the use of lead shot by waterfowl hunters and lead weights by fishermen.

destructive processes, as is now required by the Convention on Biological Diversity, together with the legal means, such as environmental impact assessments, to eliminate or limit their damaging effects. However, the difficulties of such a task would be considerable: in most countries, these rules are set out in a wide range of laws and regulations which are not always easily available.

Given that this paper retains the division between species and areas, there are some inevitable overlaps between the two parts, especially in the description of international conventions dealing with both aspects of conservation. Damaging processes are dealt with incidentally, especially with regard to the protection of areas.

INTRODUCTION

**THE DEVELOPMENT OF INTERNATIONAL
CONSERVATION LAW**

A. The Principle of National Sovereignty over Natural Resources

The most fundamental rule in international relations is that States are sovereign entities and that, subject to international law, they may conduct their business as they please. States exercise sovereign rights over all natural resources on their territory, which means that they may conserve, exploit or destroy them, or allow them to be destroyed as they wish.

The conservation of natural resources and habitats is, in consequence, a matter which comes under the exclusive jurisdiction of the States on whose territory they are situated. The State also has sovereignty over marine resources found in the Exclusive Economic Zone (EEZ) and on the continental shelf, as well as in its inland waters and territorial sea. In contrast, the rule applicable to the high seas, beyond the jurisdictional limits of any State, is that of the freedom of the high sea which includes the freedom of fishing. However, pursuant to the Law of the Sea Convention of 1982, the extension of the EEZ to 200 miles has in fact probably brought more than 95% of marine species under national sovereignty.[3]

The principle of national sovereignty over natural resources, which include all living resources, is restated in several international instruments relating to the conservation of species.

Principle 21 of the Stockholm Declaration, adopted by the United Nations Conference on the Human Environment which was held in Stockholm in 1972, states that

"States have, in accordance with the Charter of the United Nations and the principles of international law, the sovereign right to exploit their own resources pursuant to their own environmental policies. . ."

This principle is reproduced word for word in article 3 of the Convention on Biological Diversity of 1992. It also appears in Principle 2 of the Rio Declaration, adopted by the United Nations Conference on Environment and Development in Rio de Janeiro in June 1992 (UNCED), but has been altered to refer to "environmental and developmental policies."

The sovereign right to destroy is qualified, however, by another general rule of law, also embodied in Principle 21 of the Stockholm Declaration and in the new Convention on Biological Diversity, which is that States have

"the responsibility to ensure that activities within their jurisdiction or control do not cause damage to the environment of other States or of areas beyond the limits of national jurisdiction."

Subject to this restriction, the sovereign right to destroy nature is in theory absolute. However, as in any other area, States may of course always voluntarily accept limitations of their sovereign rights by treaty.

---

[3] The jurisdictional provisions of the Law of the Sea Convention are explained in greater detail later in the Introduction.

The Definition of Natural Resources

"Natural resources" are not defined by any of the above texts, although they clearly include wild animals and plants. Nevertheless, it is important to try to determine the scope of sovereign rights over such resources with as much precision as possible.

There is no doubt that individual animals or plants are subject to the sovereign rights of a State where they are situated on its territory or in an area under its jurisdiction.

A different problem arises with regard to sovereignty over a species as a whole, that is to say, sovereignty over the genetic characters of species which would equate to some form of proprietary interest in the capacity of genes to generate biological processes. Indeed, a species is much more than the sum of its individual members as it constitutes a unique set of animals or plants sharing the same genetic attributes. The question of whether there can be sovereignty over a species, as distinct from its individual components, is particularly acute where the range of a given species extends over the territory of more than one State, as is generally the case. Whilst each Range State has sovereignty over all members of the species which happen to be under its jurisdiction at any point in time, that State cannot have sovereignty over the whole species. On the other hand, there is no principle of joint sovereignty in international law. In consequence, real sovereignty can effectively only be exercised over species which are endemic to a single country.

International law has already progressed beyond national law, where there is still considerable confusion between species and their individual components. International law is gradually evolving towards the recognition of a legal status for species which is distinct from that of the individual animals or plants which compose them.[4]

The first step towards the formulation of this concept was the recognition that Range States could be assigned responsibilities and duties with regard to species under their jurisdiction. Much more recently, in connection with the Convention on Biological Diversity, came the realisation that Range States could also have rights over species, as distinct from their existing sovereign rights over individual components of those species.

This interesting development probably arose as a consequence of the increased emphasis on the potential value of wild species for humankind as a justification for their conservation. Although technically the right to any benefits from such species is legally independent of the duty to conserve them, it is now increasingly perceived as a fair counterpart to that duty. This linkage of a new right to a pre-existing responsibility differs from the customary evolution of legal obligations, whereby an acknowledged right is generally only later coupled with a corresponding duty.

B. The Formation of a Consensus to Conserve Natural Resources

Concern has steadily grown about the need to conserve species and natural habitats in the face of rapidly-developing threats of all kinds. There are two very different strands to this concern

---

4   Discussed at Part I, Chapter I(E), below.

about the loss of biological diversity. Firstly, the anthropocentric view is centred on a loss to science and the economy, as well as a more general loss of potential benefits for both present and future generations. Secondly, what is now referred to as the "ecocentric" view is concerned with the intrinsic value of biological diversity, which humanity may use but which it has no moral right to destroy, as well as with its fundamental role in maintaining the life-sustaining systems of the biosphere and the evolutionary potential of the Earth.

Over recent decades, the idea has taken shape that all States, and the international community in general, have at the very least an interest in the conservation of wild species and the habitats in which they live. Widespread disquiet at environmental degradation has slowly crystallised in the form of a consensus to establish rules of international law intended to achieve a better balance between sovereign rights over natural resources, including the right to destroy them, and the need to preserve wild species. The broad term, 'species', was gradually changed, firstly to 'genetic resources', and still later to 'biological diversity' which encompasses the diversity of ecosystems, species and genes still found on the planet.

The above consensus has been formalised over the years in three different spheres, namely scientific, political and legal.

The scientific community was instrumental in this process through the International Biological Programme in the 1960s, as was the International Union for the Conservation of Nature and Natural Resources (IUCN   the World Conservation Union), particularly through its Survival Service Commission (now the Species Survival Commission) and its publication of the first Red Data Books. Scientific consensus was reached at an intergovernmental conference organised by UNESCO in 1968 on the scientific bases of the wise use and conservation of biosphere resources,[5] from which the present Man and Biosphere Programme (MAB) was derived.

In 1980, a document of major importance for the development of world conservation policies was published by the IUCN in cooperation with the United Nations Environment Programme (UNEP) and the World Wide Fund for Nature (WWF), and in collaboration with FAO and UNESCO. This document, entitled the World Conservation Strategy (WCS), laid down three major conservation goals. These were respectively:

- the maintenance of essential ecological processes;
- the preservation of genetic diversity;
- the sustainable use of species and ecosystems.

The WCS has had considerable influence on the development of national conservation policies and legislation, as well as on certain treaties, such as the ASEAN treaty,[6] which were concluded subsequently. The three objectives of the WCS are now often included in the Preambles or basic provisions of nature conservation laws throughout the world, and even sometimes in national constitutions, such as the recently-adopted Constitution of Namibia.

---

[5] The biosphere is that part of the globe where life is possible.

[6] The Agreement on the Conservation of Nature and Natural Resources, concluded in Kuala Lumpur on 9 July 1985 between the six States of the Association of South-East Asian Nations (ASEAN), comprised of Brunei, Indonesia, Malaysia, the Philippines, Singapore and Thailand.

Following the establishment of a consensus on the principles for conserving biological diversity, the next step was to develop a consensus on the objectives. To this end, the World Conservation Strategy was further developed by the publication of a follow-up document in October 1991, entitled "Caring for the Earth —a Strategy for Sustainable Living". In addition to laying down basic principles, this document describes the kind of corresponding actions which should be taken and offers guidelines to adapt the WCS to the user's needs and capabilities. The document places particular emphasis on the need to improve the conservation of wild flora and fauna through a combination of in situ and ex situ methods, to harvest wildlife resources sustainably and to increase incentives to conserve biological diversity (Action Points 4.9–4.14).

Political agreement on the need to conserve biological diversity was achieved at the United Nations Conference on the Human Environment in 1972, which resulted in the creation of UNEP and the adoption of the Stockholm Declaration, discussed below. This consensus formed the foundation for the development of a number of international instruments laying down certain general conservation objectives and sometimes very specific conservation rules. The very fact that these instruments could be adopted confirmed the evolution of the principle of national sovereignty over natural resources, as well as the recognition that the absolute character of this principle must be tempered because the international community was acknowledged to have a real interest in the conservation of certain of these resources.

Few, if any, of these international instruments could have been developed if there had not already been a large number of national laws relating to protected species and areas in existence. In turn, internationally agreed principles and rules constitute a basis upon which national legislation may be further developed or improved. There is therefore a process of dynamic interaction between national and international legislation which is naturally conducive to gradual changes towards better conservation standards and norms.

States voluntarily accept limitations on their sovereignty by agreeing to international obligations to conserve certain of their natural resources, which then generally helps them to justify the taking and enforcement of domestic conservation measures. Furthermore, by setting objectives and minimum standards, international instruments provide a normative basis from which national legislation may be developed. In parallel, as awareness of conservation needs increases, national standards are often raised correspondingly and new laws are enacted. This responsiveness to new scientific information means that in some countries, a third generation of conservation statutes are now already in force. These national measures may, in turn, lead to the development of more exacting international instruments with higher conservation objectives and standards.

The political consensus materialised in the adoption of a number of important non-binding soft law instruments which formed the basis for the development of binding legal rules. This emerging legal consensus crystallised in the conclusion of a certain number of treaties, culminating in the adoption of the Convention on Biological Diversity in 1992 at UNCED. The legitimacy of these legal instruments is founded on the growing recognition of the incalculable value that biological diversity possesses for humankind, both now and in the future, making its conservation a common concern of the world community.

International conservation instruments may take the form of soft or hard law.

C. The Main Soft Law Instruments

Soft law instruments, such as declarations of principles, charters or resolutions of international organisations, are not binding in law. In other words, they do not contain legal obligations that could be enforced in a court of law in the event of non-compliance. The moral value of such instruments may be very high, however, especially where they can be considered as the manifestation of a broad consensus on the part of the world community.

The Stockholm Declaration of 1972 sets out 26 Principles from which a body of international environmental law has since been developed. The Declaration places great emphasis on the need to protect both species and their habitats, particularly at Principles 2 and 4 which read as follows:

"The natural resources of the earth including the air, water, land, flora and fauna and especially representative samples of natural ecosystems must be safeguarded for the benefit of present and future generations through careful planning or management, as appropriate." (Principle 2)

"Man has a special responsibility to safeguard and wisely manage the heritage of wildlife and its habitat which are now gravely imperilled by a combination of adverse factors. Nature conservation including wildlife must therefore receive importance in planning for economic development." (Principle 4)

Ten years later, on 28 October 1982, the United Nations General Assembly adopted and solemnly proclaimed a World Charter for Nature, which had been prepared by IUCN at the request of the President of Zaire. The Charter proclaims principles of conservation "by which all human conduct affecting nature is to be guided and judged", and incorporates not only the Stockholm Principles but also the three objectives of the World Conservation Strategy, mentioned above.

The Charter states in its Preamble that

"every form of life is unique, warranting respect regardless of its worth to man."

This represents the formal recognition by the UN General Assembly of the intrinsic value of species, which should entail, at the very least, a corresponding moral duty to preserve them.

In the Charter's operative part, the following "General Principles" are set out:

"The genetic viability on the Earth shall not be compromised, the population levels of all life forms, wild and domesticated, must be at least sufficient for their survival and to this end, necessary habitats shall be safeguarded." (Principle 2)

"All areas of the earth, both land and sea, shall be subject to these principles of conservation; special protection shall be given to unique areas, to representative samples of all different types of ecosystems and to the habitats of rare and endangered species" (Principle 3).

"Ecosystems and organisms as well as the land, marine and atmospheric resources that are utilised by man, shall be managed to achieve and maintain optimum sustainable productivity but not in such a way as to endanger the integrity of those other ecosystems with which they co-exist." (Principle 4)

The World Commission on Environment and Development (also called the Brundtland Commission after its chairman, now Prime Minister of Norway) also placed considerable emphasis on the need to preserve biological diversity and to abide by the principle of optimum sustainable yield in the use of natural animal and plant resources. The Conclusions of the Brundtland Report were adopted by the UN General Assembly in 1987 as a framework for future cooperation in the field of environment and development.

The precautionary principle is increasingly seen to be of great importance to the conservation of biological diversity, as has now been formally recognised in both hard and soft law. For example, Principle 2 of the Rio Declaration of 1992 places emphasis on this principle, although the Declaration is otherwise silent on biological diversity, perhaps because this was not felt to be necessary in view of the adoption of the Convention on Biological Diversity at the same time.

Finally, Agenda 21, the Action Plan drawn up by the UNCED in June 1992 has devoted chapter 15 to the conservation of biological diversity. The content adds relatively little to the new Convention which is discussed at the end of this Introduction, but is nonetheless important as a framework for further action.

D. Treaties and Other Binding Instruments

Treaties are of course essential where the species or population to be preserved or harvested is international, namely shared between two or more countries or found in the high seas. Natural resources of this international nature include animal populations which are shared in space (extending on both sides of a border, such as in a boundary river or lake, or stocks found in adjoining territorial seas or EEZs) or in time. The latter category includes migratory species, the individual members of which accordingly come under the jurisdiction of different States at different times during their migration.

Treaties are also needed where measures to control international trade in wild flora and fauna are required: illegal traffic in these species can only be prevented by close cooperation formally established under a treaty. This is the justification for the Convention on International Trade in Endangered Species, generally known as CITES, which was signed in Washington on 3 March 1973.

It is usually necessary to conclude treaties or bilateral agreements where it is sought, *inter alia*, to conserve transboundary protected areas effectively and to create a framework within which joint measures can be developed and enforced in respect of the whole ecological unit. Such instruments may also establish requirements to carry out environmental impact assessments in respect of activities proposed in one State which are capable of having significant adverse effects on the environment of a neighbouring State.[7]

Furthermore, it was recognised early on that treaties relating to the conservation of national flora and fauna of the Parties may serve a very useful purpose, even where the absence of an international element means that they are not essential in law. Such treaties are valuable because:

- they establish uniform conservation rules;
- they permit conservation priorities to be identified and decided jointly;

---

[7] The Convention on Environmental Impact Assessment in a Transboundary Context, signed in Espoo, Finland on 25 February 1991, is discussed in Part II of this paper.

- they organise international cooperation;
- they serve more generally as an expression of the commitment of the Parties to conserve certain species.

In the case of federal or regionalised States, treaties may confer powers on central Government which would otherwise be constitutionally lacking. Once the treaty has been concluded, federal Government is then empowered to legislate in the area concerned. One example is the treaty between the United States and Canada on migratory birds, concluded in 1916.

Before giving an overview of the main international conservation instruments, it should be emphasised that many of them include measures relating to the protection of both species and areas. This Introduction will therefore be restricted to a description of the background and general scope of these instruments, whilst more detailed analysis of their species—or area-based provisions will be set out in Parts I and II of this paper respectively.

1. Early Treaties

The recognition of the need for concerted action to protect wildlife came as early as 1900, when the London Convention on the Protection of Wild Fauna in Africa was concluded on 19 May of that year, although it was never ratified.

On 19 March 1902, the Convention for the Protection of Birds Useful to Agriculture was signed in Paris, and is still in force although it has been completely superseded. This treaty was effectively regional as well as sectoral, since all the Parties to the Bird Treaty of 1902 were in fact European.

On 8 November 1933, the London Convention Relative to the Preservation of Fauna and Flora in their Natural State was signed. This Convention had similar objectives to the aborted 1900 Convention, namely to protect species of value as hunting trophies and to create protected areas in Africa. It remained in force until 1968 when, as a result of the independence of the majority of African States, it was replaced by the new African Convention on the Conservation of Nature and Natural Resources, signed in Algiers on 15 September 1968.

The last Convention to be signed before the war, in Washington on 12 October 1940, was the Convention on Nature Protection and Wildlife Preservation in the Western Hemisphere, which remains in force to this day although it is of relatively little influence. It was based on the London Convention on Africa of 1933 and was intended to serve as its counterpart for the American continent. It too combines measures for the protection of both species and areas, its broad objective being the protection and preservation in their natural habitats of all species of native fauna and flora.

The early trend in international conservation law was therefore either regional, dealing with specific geographic areas, or sectoral, dealing with particular subjects of species.

The idea of a World Treaty on Conservation was originally launched before the First World War. A meeting in Berne in 1913 agreed on a text establishing a Standing Committee of Government representatives, which was given the task of collecting and publishing data on international nature protection and of issuing propaganda for that purpose. However, nothing further happened after the outbreak of war.

The concept of a World Treaty was subsequently revived at an IUCN meeting in Fontainebleau in 1948 and discussed at the first UNESCO-IUCN International Technical Conference on Nature Protection at Lake Success (USA) in 1949. Many delegations felt that disparities between the conditions and legislations of countries would make such a treaty almost unworkable. The feeling of the meeting was that it would be better to continue to work through regional and sectoral treaties. It was decided that IUCN should undertake further work on a world conservation treaty in the future, but at a date which could not yet be set.

In the event, the momentum to develop a global treaty was only renewed in the early 1980s, culminating in the conclusion of the Convention on Biological Diversity in 1992 which is discussed in detail in section (E) below.

2. Regional Treaties and Other Instruments

a. Africa

The African Convention on the Conservation of Nature and Natural Resources was signed in Algiers on 15 September 1968. By 1991, 30 African States were Parties.

The Convention starts by establishing fundamental principles, set out in article 2:

"The Contracting States shall undertake to adopt the measures to ensure conservation, utilisation, development of soil, water, flora and faunal resources in accordance with scientific principles and with due regard to the best interests of the people."

Each of these elements is then dealt with individually in a specific article, although only in the form of very general obligations.

Amendments to widen the scope of the Convention and to strengthen some of its provisions have been prepared by IUCN at the invitation of the Organisation of African Unity (OAU). However, these amendments have not so far been officially adopted.

b. America

In addition to the Western Hemisphere Convention of 1940, six central American States[8] signed a Convention for the Conservation of Biological Diversity and the Protection of Priority Wild Areas in Central America on 5 June 1992, International Environmental Day.

c. Europe

i. The Berne Convention

The regional conservation treaty for Europe, the Convention on the Conservation of European Wildlife and Natural Habitats, was signed in Berne on 19 September 1979, under the auspices of the Council of Europe. Membership is not, however, limited to the Member States of that organisation. The Convention is open to any non-Member State which is invited to accede by the Council of Europe's Committee of Ministers. In consequence, Burkina Faso and Senegal are Parties, as is Estonia since 1992. Morocco, Tunisia and several eastern European countries,

---

[8] Costa Rica, El Salvador, Guatemala, Honduras, Nicaragua and Panama.

including the Russian Federation, have also received such invitations to accede. There are regular annual meetings of the Parties within a Standing Committee. The Secretariat is provided by the Council of Europe.

The aim of the Convention is

"to conserve wild flora and fauna and their natural habitats, particular emphasis being given to endangered and vulnerable species."

Its measures to protect species and natural habitats will be discussed in Parts I and II respectively of this paper.

ii. The Benelux Conventions

The Benelux Convention on Hunting and the Protection of Birds of 1970 has as its main purpose the harmonisation of the legislation in this area by the three countries concerned, Belgium, Luxembourg and the Netherlands. These countries also signed a Convention on Nature Conservation and Landscape Protection in 1982.

iii. European Community Legislation

The European Community initially had no jurisdiction over environmental matters as the Treaty of Rome contained no provisions dealing with this matter. However, the Treaty allows the Council of the Community to enact appropriate provisions where Community action appears necessary to achieve one of the aims of the Community, in cases where the Treaty has not provided for the requisite powers of action (article 235). The Community used article 235 as the legal basis for developing Environmental Action Programmes and has adopted Community legislation to implement these Programmes. The first such Programme was adopted in 1973. The EC Birds Directive (79/409) was adopted under article 235, based on the fact that birds occurring in the Community crossed intra-Community borders and therefore constituted a common heritage and that effective bird protection was typically a trans-frontier environmental problem entailing common responsibilities.

Since 1974, the Community has ensured that provisions on ratification or accession of economic integration organisations are included in the text of conservation treaties and has, in its own right, systematically joined such treaties open to any or all of its Member States where the subject-matter was within EC competence.

Participation by the Community in a treaty enables it to adopt legally-binding Community instruments, known as Directives or Regulations, for the implementation of the treaty throughout the Community. Regulations and Directives are adopted by the Council or the Commission. Directives are binding upon Member States as to the results to be achieved, but leaves the choice of form and method to the national authorities. Regulations are directly applicable in Member States and do not require any transposition into national legislation.

Until the adoption of the European Single Act on 27 February 1986, competence in environmental matters was very limited. The Act gives the EC very broad jurisdiction in the field of the environment, and these powers have been extended still further by the Maastricht Treaty of 1991.

In the exercise of these wider powers, the Council adopted Directive 92/43 on the Conservation of Natural Habitats and of Wild Fauna and Flora of 21 May 1992. The main aim of the Directive is to promote the maintenance of biodiversity. Measures taken pursuant to the Directive must be designed to maintain or restore at favourable conservation status natural habitats and species of wild fauna and flora of Community interest. The Directive therefore provides for two sets of measures covering the protection of species and species' habitats and the protection of habitat types, discussed in Parts I and II of this paper respectively.

The Community has also adopted a Regulation dated 3 December 1982 for the uniform implementation of CITES by its Member States.

iv. The Alpine Convention

The Convention on the Protection of the Alps, which was signed in Salzburg, Austria on 7 November 1991, has not yet entered into force. All Alpine countries (France, Italy, Switzerland, Liechtenstein, Germany and Austria) have signed: the only exception is Slovenia which is expected to sign shortly. It is the first time in the world that a Convention has been concluded in respect of a complete terrestrial ecological unit, whereas the the Regional Seas Conventions have long been in existence in respect of certain marine areas.

The treaty is a framework Convention which only lays down obligations of a general kind, covering the whole range of environmental problems, namely: population and culture; physical planning; air quality; soil protection; water regime; conservation of nature and maintenance of landscape; mountain agriculture; mountain forests; tourism and recreation; energy; and waste.

The Convention will operate through an Alpine Conference and a Standing Committee. The Conference may make recommendations to the Parties with regard to the achievement of the objectives of the Convention. It may also adopt Protocols to the Convention, subject to their ratification by the Parties. The Standing Committee essentially prepares the work of the Conference. The Convention provides that a Secretariat may be established if the Conference so decides.

d. Asia

The Agreement on the Conservation of Nature and Natural Resources was concluded in Kuala Lumpur on 9 July 1985 between the six States of the Association of South-East Asian Nations (ASEAN), comprised of Brunei, Indonesia, Malaysia, the Philippines, Singapore and Thailand.

The purpose of this very modern treaty is to achieve global environmental protection, based on the objectives of the World Conservation Strategy, by means of specific measures concerning each element of the environment: air, water, soil, plant cover, forests, fauna, flora and ecological processes. The Agreement places great emphasis on the need to prevent, reduce and control the deterioration of the natural environment and also to reduce pollution.

The Agreement, which provides for meetings of Contracting Parties every three years and for the establishment of a Secretariat, may rightly be considered as the most comprehensive, detailed and forward looking of all conservation treaties for many reasons, in particular because of its provisions on protected areas which are discussed in Part II, Chapter I(A) below.

e. The Pacific

The Convention on the Conservation of Nature in the South Pacific was concluded at Apia in Samoa on 12 June 1976 and entered into force on 26 June 1990. It adopts a similar approach to the African and Western Hemisphere Convention, in providing for the creation of protected areas and the protection of species, a list of which is to be drawn up by each Party. No meetings of Contracting Parties are provided for, although the Convention refers to the possibility of consultations to give effect to its provisions.

f.   Regional Seas

The United Nations Environment Programme (UNEP) has established a massive marine conservation programme which currently covers eleven seas or coastal regions and may soon be extended to the North-West Pacific and the Black Sea. The procedure in each case is for UNEP to prepare an Action Plan upon which a Framework Convention is then based. A certain number of Protocols are then adopted to amplify each of the Framework Convention's basic provisions.

These Regional Seas instruments are designed to achieve a global approach by the riparian States of a given sea to its particular environmental problems. Such problems largely concern pollution, but most of the Regional Seas Conventions emphasise the need to preserve natural habitats. The legal basis on which the Protocols to these Conventions were developed to deal with protected marine and coastal areas is to be found at article 194.5 of the United Nations Convention on the Law of the Sea of 1982 (still not in force), which creates a general obligation to protect

"rare or fragile ecosystems as well as the habitat of depleted, threatened or endangered species and other forms of marine life".

Protocols relating to the conservation of natural areas have already been concluded in respect of four regional seas: the Mediterranean (Geneva, 1982), East Africa (Nairobi, 1985), the South-East Pacific (Paipa, Colombia, 1989) and the Caribbean region (Kingston, Jamaica, 1990).

g.   Areas beyond National Jurisdiction: the Antarctic

The Antarctic region is protected by a system which includes the following five instruments:

- the Antarctic Treaty, signed in Washington on 1 December 1959, and its Protocol signed in Madrid on 4 October 1991;

- the Convention for the Conservation of Antarctic Seals, concluded in London on 1 June 1972;

- the Convention on the Conservation of Antarctic Marine Living Resources, signed in Canberra on 20 May 1980;

- the Convention on the Regulation of Antarctic Mineral Resource Activities, signed in Wellington on 2 June 1986 but never ratified, now superseded by the Madrid Protocol of 1991.

The Antarctic Treaty applies only to the Antarctic continent and to the ice-shelves south of the 60th degree of latitude South. It is primarily a political treaty which was concluded principally to freeze all claims to national sovereignty in the region, to ban nuclear tests, to authorise only peaceful activities and to reinforce cooperation between the Parties in the field of scientific research, the conduct of which is proclaimed to be completely free throughout the region.

Although the Treaty has no provisions on environmental protection other than a ban on the disposal of radioactive waste, it authorises the Contracting Parties during their meetings to consider, formulate and recommend to their Governments measures "in furtherance of the principles and objectives of the treaty, including measures regarding [. . .] preservation and conservation of living resources in Antarctica" (Article IX–1).

Pursuant to this provision, the Parties adopted a Recommendation at the third Consultative Meeting of the Parties in Brussels in 1964, entitled "Agreed Measures for the Conservation of Antarctic Fauna and Flora". The Measures, which are in effect a small Treaty within a Treaty,[9] are discussed in Part I, Chapter I(f) below.

The Madrid Protocol to the Treaty, adopted in 1991, creates a Committee for Environmental Protection whose functions are to provide advice and formulate recommendations to the Parties on matters pertaining to its implementation, including the operation and further elaboration of the Antarctic Protected Area System.

3. Sectoral Treaties

Sectoral conservation treaties deal either with certain species or with certain types of natural habitat or protected area, although species—and area-based measures may of course frequently be combined. Such treaties may be either global or regional.

a. Treaties dealing with Species

Instruments dealing primarily with the protection of wild species fall into three broad groups.

i. Species whose Range is Shared by Several States

Where the range of a given species covers several neighbouring States, effective measures for its protection are usually dependent upon the conclusion of international agreements for the taking of joint conservation and management measures to conserve stocks, control trade and preserve the natural habitat of the species in the region concerned.

Such agreements have for the most part been concluded in respect of species which have a very high commercial value and which may therefore become endangered if their taking and trade are not regulated. These include the vicuna, which is only found in certain countries in South America and, as regards marine species, the stocks of certain species of seal which have long been protected in international law. The polar bear and its natural habitat is now protected by an Agreement on the Conservation of Polar Bears, concluded between the five circumpolar States on 15 November 1973.

ii. Migratory Species

Migratory species are increasingly seen as an international resource, given that different States have jurisdiction at different points along their migration routes. Such species may be terrestrial, as in the case of the caribou herds protected by a bilateral agreement between Canada and the United States, or marine. The Law of the Sea Convention of 1982 contains provisions specifically dealing with certain migratory marine species.[10]

---

[9] Under Artical IX–4 of the Treaty, measures adopted by meetings of the founding Contracting Parties become effective once approved by all these Contracting Parties. From that moment, therefore, they are legally binding.

[10] See Part I, Chapter I(B) (4) below.

The majority of international agreements are limited to the protection of migratory birds, dating back to the early conservation treaties described earlier in this Introduction. These were largely restricted to Europe. More recent instruments include a series of bilateral agreements concluded in the 1970s and 1980s which only provide coverage for North America, the Pacific and parts of Asia. In addition, an unofficial agreement between the United States, Canada and Mexico, the North American Waterfowl Management Plan, was signed in 1985 to protect migratory species of waterfowl which are hunted as game and appears to have been highly effective.

Global measures to protect all migratory species were first proposed by Recommendation 32 of the 1972 Stockholm Conference Action Plan, which urged Governments to consider

"the need to enact international conventions and treaties to protect species inhabiting international waters or those which migrate from one country to another."

The Convention on the Conservation of Migratory Species of Wild Animals was subsequently adopted in Bonn, Germany, on 23 June 1979. It entered into force on 1 November 1983 after the deposit of the fifteenth instrument of ratification. Its provisions are discussed in detail in Part I, Chapter I(A)(e) below.

By the end of 1992, there were only 39 Contracting Parties, of which 17 were in Europe (including the European Community), 13 in Africa, 6 in Asia and only 3 in America. This means that countries of major importance for migratory birds, such as the Russian Federation where the great majority of European and Asian waterfowl nest, the United States, Canada, most of the Latin American countries, China, Japan, and the South-East Asian countries are all still outside the Convention. The effectiveness of the Convention has suffered from the lack of sufficient Parties to cover the majority of species included in the Appendices and their migration routes.

As regards the machinery for its implementation, the Convention establishes a Conference of the Parties, empowered inter alia to amend its Appendices and to adopt the necessary financial provisions, as well as a purely advisory Scientific Council to be set up by the Conference at its first meeting. The Conference also determines the exact functions of the Council, which is to be composed of experts nominated by the Parties together with any additional experts which the Conference may choose to nominate. Finally, the Convention establishes a Secretariat, provided by the United Nations Environment Programme, which is currently located in Bonn, Germany.

iii. Treaties regulating the Trade in Wild Species

The global Convention on International Trade in Endangered Species of Wild Fauna and Flora (CITES) was signed in Washington on 3 March 1973 and entered into force on 1 July 1975. Its principles have been widely accepted by a large number of States and implemented and enforced with considerable success. The regulatory, scientific and management provisions of CITES are discussed in detail in Part I, Chapter V(C).

CITES is a world convention to which a large number of States are now Parties. It was therefore felt that there was no longer any need to cover international trade in wild species in other conventions, such as the Berne Convention and ASEAN Agreement.

iv. Treaties regulating the Exploitation of Wild Species

Certain species-based treaties may be classified as 'exploitation treaties' where their primary aim is the conservation not of biological diversity but of the basis for an economic activity. In order to prevent over-exploitation of the natural resource in question, it is necessary for joint regulatory measures to be adopted and implemented and to share the results of scientific research into the population and management of stocks.

Major agreements currently in force include the International Convention for the Regulation of Whaling, signed in Washington on 2 December 1946, and the Convention on the Conservation of Antarctic Marine Living Resources (CCAMLR), signed in Canberra on 20 May 1980. Sealing is also regulated by a series of different regional conventions.

Most recently, the growing use of large drift nets has become regulated in order to prevent the incidental taking of dolphins in significant numbers. Following bi- or trilateral agreements covering the North Pacific, and the conclusion in 1989 of a convention for the South Pacific, measures were finally adopted for the high seas as a whole by a Resolution of the United Nations General Assembly on 15 December 1989 (Res.44(225)).

b. Area-Based Conservation Treaties

The two global treaties dealing with the conservation of ecosystems are the Convention on Wetlands of International Importance especially as Waterfowl Habitat, signed in Ramsar, Iran on 2 February 1971, and the Convention concerning the Protection of the World Cultural and Natural Heritage, signed at UNESCO in Paris on 16 November 1972.

Both treaties establish an international site-specific conservation system based on official lists of protected areas.[11] However, the extent of their coverage is limited. The Ramsar Convention exclusively concerns wetlands, albeit broadly defined, whilst the World Heritage Convention confers protection on a relatively small number of areas of outstanding universal value from the point of view of science, conservation or natural beauty.

The Ramsar Convention is primarily concerned with the conservation and management of wetlands included in the List of Wetlands of International Importance. Parties are also required to promote the "wise use" of wetlands on their territory and to take measures for the conservation of wetlands and waterfowl by establishing nature reserves on wetlands, whether they are included in the List or not.

By the end of 1992, the Ramsar Convention had 71 Contracting Parties. This number will shortly increase to 76, as a further five States were in the process of acceding, but had not yet designated a wetland for inclusion on the List. There is no obligation on States to establish protected areas, but they cannot be considered as Parties until at least one site has been included on the List. There are at present 582 sites on the List, covering more than 36 million hectares.

The World Heritage Convention was adopted, *inter alia*, to conserve and transmit to future generations the natural and cultural heritage situated on their territory. By the end of 1992, there were 129 Parties to the Convention. The inclusion of a site on the World Heritage List requires the approval of the World Heritage Committee: the List now contains a total of 378 sites, of which 276 are cultural, 87 natural and 15 "mixed" cultural and natural.

Interestingly, the World Heritage Convention embodies the two principles of law necessary for the common concern for the protection of biological diversity to be translated into effective action in the field. Firstly, the Convention considers that there should be a legally recognised duty on the part of all States to conserve their natural and cultural heritage. Secondly, the uneven distribution of this heritage which, in the case of biological diversity, is richest in many of the

---

[11] These provisions are discussed in Part II, Chapter I(B) below.

poorest countries in the world, means that there should also be a corresponding duty on the part of the world community to contribute financially to the conservation of that natural heritage in those countries.

The World Heritage Convention accordingly established a special financial mechanism to assist Parties to discharge their obligations in respect of listed sites. The World Heritage Fund has been extremely successful, initiating a system whereby the world community acknowledged its duty to contribute financially to the conservation of biological diversity in developing countries. In 1990, the Parties to the Ramsar Convention set up a Wetland Fund along on similar lines to contribute to the preservation of Ramsar sites. As discussed in section (5)(e) below, the Convention on Biological Diversity has endorsed and extended these principles.

4. Law of the Sea

The international law of the sea has emerged over centuries, during which time the sea has acquired increasing importance for trade, transport and the supply of marine resources, particularly fish. Traditionally, the principle which applied to all parts of the sea other than the territorial sea or internal waters was that of the freedom of the high seas, including the freedom of fishing mentioned above. This tenet was laid down as early as 1609 by Grotius, a Dutch lawyer, in his famous book, Mare Liberum.

As it now stands, the law of the sea is the result of an unprecedented effort of codification of the rules relating to all aspects of the use of the marine environment by man. The First United Nations Conference on the Law of the Sea was held in 1958, and led to the adoption of no less than four conventions, respectively dealing with the High Seas; Fishing and Conservation of the Living Resources of the High Seas; the Continental Shelf; and the Territorial Sea and Contiguous Zone.

The present Law of the Sea Convention was finally signed in Montego Bay in December 1982 after some ten years of negotiations. The Convention, which is still not in force and has not yet been signed by some of the wealthier maritime States, is exceptionally broad in scope. It defines, *inter alia*, the boundaries of each part of the sea and the continental shelf and the legal regime applicable therein; the rights and conditions of passage for shipping through other State's waters; jurisdiction over ships on the high seas and exceptions to the principle of the freedom of the high seas; and the legal and management regime for the exploitation of mineral resources on the deep sea-bed and ocean floor beyond national jurisdiction.

In addition, the Convention establishes detailed provisions concerning the prevention of pollution. States are under a fundamental obligation "to protect and preserve the marine environment" (article 192). They are required, separately or jointly with other States, to take all necessary measures compatible with the Convention to prevent, reduce and control pollution of the marine environment from whatever source. The very wide definition of pollution includes not only pollution caused by the use of technology (which is not specified), but also the deliberate or accidental introduction of alien or exotic species which are capable of causing significant and harmful changes to the marine environment (article 196).

This paper will concentrate principally upon those parts of the Convention that deal with jurisdiction, the conservation and exploitation of marine species and the possibility of establishing protected areas within the marine environment. The latter two subjects will be addressed in the appropriate chapters of Parts I and II of this paper respectively. For ease of reference, the jurisdictional provisions of the Convention are set out here. These are now almost universally applied, despite the Convention not yet having entered into force.

Under the Convention, States are fully sovereign over their "internal waters". This is the first zone to be crossed when travelling from the shore towards the high sea and includes all waters on the landward side of the baseline, usually as far as the high water mark or the salinity limit of the river mouth. The normal baseline is the low water mark along the coast. Where the coastline is deeply indented or if there is a string of islands along the coast, straight baselines from one headland to another are generally used. There is no difference in legal status between internal waters and land areas.

Next comes the "territorial sea" which may extend to a distance of twelve nautical miles from the baseline. The legal status of the territorial sea is very similar to that of internal waters, as this part of the sea is also considered to form part of the territory of the State concerned. There is nevertheless a legal restriction on sovereignty within the territorial sea. Coastal States cannot prohibit the "innocent passage"[12] of foreign ships through these waters, although they can regulate it, for instance by the establishment of sea lanes.

Coastal States are entitled under the Convention to claim an Exclusive Economic Zone (EEZ), the last zone before the high sea. The EEZ extends beyond the limit of the territorial sea to a maximum of 200 nautical miles from the baseline. Under no circumstances may the combined breadth of the territorial sea and the EEZ exceed 200 nautical miles. Within the EEZ, the coastal State has sovereign rights for the purpose of exploring, exploiting, conserving and managing the natural resources found therein, including fisheries, as well as the responsibility for preserving the marine environment.

Coastal States also have sovereign rights over the continental shelf and its natural resources, as far as the outer limit of the shelf. The shelf is defined as comprising the sea-bed and sub-soil of the submarine areas that extend beyond the territorial sea, as far as the edge of the continental margin. In some cases, the shelf may therefore go far beyond the outer limit of the EEZ. However, where the edge of the margin is within the 200 mile limit, the shelf is deemed nonetheless to extend as far as that limit. These sovereign rights do not affect the rights of other States in or on the superjacent waters.

The last jurisdictional zone in the ocean is the high sea beyond the 200-mile outer limit of the EEZ. No State has jurisdiction over the waters of this zone. Any limitation on the taking of species must necessarily be the subject of an international agreement.

Under the Convention, the sea-bed and ocean floor and subsoil thereof beyond the limits of national jurisdiction ("the Area"), together with the mineral resources found therein, are declared to constitute the "common heritage of mankind" (article 136). No State may claim sovereignty or sovereign rights over any part of the Area or its resources. Once the Convention comes into force, the new International Sea-Bed Authority shall exclusively organise and control exploration and exploitation in the Area for the benefit of mankind as a whole. The Authority shall provide for the equitable sharing of economic benefits derived from activities in the area, and shall show special consideration for the interests of developing States.

---

[12] Passage is "innocent" so long as it is not prejudicial to the peace, good order or security of the coastal State and is in conformity with the applicable rules of international law.

5.   The Convention on Biological Diversity

a.   The Background to the Adoption of the Convention

As the rate of destruction of biological diversity has increased over past decades, there has been a growing recognition that the world community should take concerted action to ensure the conservation of species and ecosystems.

Until the adoption of the Convention on Biological Diversity, the sectoral and regional nature of international instruments for the protection of species and ecosystems resulted in considerable gaps in coverage in both cases. In consequence, the priority of the new Convention was to extend the scope of conservation obligations to a much larger range of situations than those presently covered by the body of international conservation law in force, including global instruments.

Regional conservation conventions are limited to certain parts of the world, leaving many regions unprotected by such treaties, including the Arctic, the Middle-East and most of Asia, and they vary widely in the substance of their obligations. Some recent conventions, such as the ASEAN Agreement and the Kingston Protocol, are very elaborate and deal with many different issues, whereas earlier conventions tend to be limited to the protection of certain species and the establishment of protected areas.

Conventions also differ greatly in the effectiveness of the mechanisms and institutions that they establish. Earlier instruments such as the Western Hemisphere and African Conventions have become sleeping conventions with few practical achievements in the field, whereas CITES probably has the best-developed implementation provisions, although it could still be improved.

Moreover, the regional conservation conventions concluded in the late 1970s and during the 1980s do not provide adequate financial and technical assistance for the purpose of preserving biological diversity, particularly through the establishment of well-secured protected areas. If adequate financial means cannot be found, the purpose of these newer treaties may well be defeated in practice. The only recent international instrument[13] to combine comprehensive conservation rules and a financial mechanism to assist in their implementation is the EC Habitats Directive of 21 May 1992.

Finally, no mechanism existed to coordinate action taken under existing conventions. Each convention has its own Parties, which are not always the same even under the global conventions, as well as its own governing body and Secretariat which are based at different locations.

The concept of a World Convention was put forward not to replace these existing conventions, but rather to establish general obligations for the preservation of biological diversity and to provide a coherent framework for action in the future. The idea was initially launched by the IUCN at its 15th General Assembly in Christchurch in 1981, which instructed the Secretariat of the Union to carry out a preliminary study of the matter.

The next IUCN General Assembly, held in Madrid in 1984, requested the IUCN Secretariat to develop a number of principles to serve as a basis for a preliminary draft of a global instrument on the conservation of the world's genetic resources: at its 17th Session in San José (Costa Rica),

---

[13]   As mentioned above, financial assistance for conservation is also available under the World Heritage and Ramsar Conventions.

the Assembly decided that the first draft which had been produced should be further developed. After intensive consultations within IUCN, a final draft was completed in June 1989.

The IUCN proposals centred on three main points:

- a general obligation for all States to conserve biological diversity;
- the principle of freedom of access to wild genetic resources; and
- the principle that the cost of conservation should be shared equitably by all nations.

In parallel, the World Commission on Environment and Development recommended in its report, "Our Common Future", that Governments should

"investigate the prospect of agreeing to a 'Species Convention' similar in spirit and scope to the Law of the Sea Treaty and other international conventions reflecting the principles of 'universal resources'. A Species Convention... should articulate the concept of species and genetic variability as a common heritage."

At its 14th Session in 1987, the Governing Council of UNEP requested the Executive Director of that organisation to establish an *ad hoc* Working Group of Experts

"to investigate the desirability and possible form of an umbrella convention to rationalise current activities in this field, and to address other areas which might fall under such a convention."

The Ad Hoc Working Group first met in November 1988. Draft provisions respectively prepared by IUCN and FAO, and "Comments for a Draft" jointly prepared by UNEP, FAO, UNESCO and IUCN,[14] were made available to the Group as background documents. At its second meeting in February 1990, the Working Group agreed that there was a need for a legally binding framework instrument

"to engage concrete and action-oriented measures for the conservation and sustainable utilisation of biological diversity".

There was also general agreement that an international instrument would be meaningless without firm commitments to finance the additional costs of conservation resulting from the obligations under the new convention, and that in a spirit of common responsibility, those costs should not fall disproportionately upon countries with significant biological diversity.

The Ad Hoc Working Group met again in July and November 1990 and also requested the UNEP Secretariat to prepare a first draft of the convention for its consideration. This draft, which was made with the assistance of a small group of legal experts, was submitted to the Working Group at its following session in February 1991. The Group was then officially renamed Intergovernmental Negotiating Committee for a Convention on Biological Diversity and the formal negotiation process started.

The negotiations met with considerable difficulty as developing countries rightly emphasised the need to counterbalance the conservation obligations they would undertake under the Convention with the recognition of rights over the genetic material of animals and plants under their jurisdiction. For the same reason, they also sought the recognition of the right to a fair and equitable share of the benefits resulting from the utilisation of such material and the right

---

[14] These organisations together comprise a coordinating body called the Ecosystems Conservation Group.

to the transfer by developed nations of the financial means which are necessary to meet the costs of conservation.

After more than three years of hard bargaining, the Convention was eventually adopted and opened for signature on 5 June 1992 at the United Nations Conference on Environment and Development in Rio de Janeiro. The Convention has been signed by 158 countries since that date, and some ratifications have already taken place. Thirty ratifications are needed for the Convention to come into force. It is expected that this figure will be reached in 1993. If this expectation is met, the speed with which the Convention has come into force will represent a significant achievement in international environmental law.

b. The Principles of the Convention

The Preamble recognises that biological diversity is disappearing fast and that it should be conserved for both ecocentric and anthropocentric reasons.

Interestingly, the ecocentric reasons are put first. The very first words of the Preamble are that biological diversity has an "intrinsic value", which implies that it must be protected for its own sake. The Preamble also states that it should be preserved for the continuation of evolution and the maintenance of the life-supporting systems of the biosphere. Moreover, biodiversity must be conserved for humankind because of its ecological, genetic, social, economic, scientific, educational, cultural, recreational and aesthetic values for present and future generations.

For all these reasons, the Preamble "affirms" that the conservation of biological diversity is a common concern of humankind.

The logical consequence of this is that although States have sovereign rights over their biological resources under international law, they also have the responsibility for conserving their biological diversity and for using their biological resources in a sustainable manner. This is "reaffirmed" by the Preamble.

The Preamble also stresses the importance of preventive measures and of attacking the causes of biodiversity loss at source. It recognises that information and knowledge regarding biological diversity is generally lacking and that it is therefore urgent to develop scientific, technical and institutional capacities to provide the basic understanding upon which to plan and implement appropriate measures. In the meantime, lack of scientific certainty shall not be used as a reason for postponing conservation measures, consistent with the precautionary principle.

The conservation of biological diversity and the sustainable use of its components are two of the major objectives of the Convention, as provided by article 1.

The third major objective, also laid down in that article, is the equitable sharing of the benefits arising out of the utilisation of genetic resources. This right to an equitable share follows on logically from the principle of sovereignty over natural resources and is a fair counterpart to the obligation to conserve biological diversity.

This right can be exercised in several different ways.

Access to genetic resources is now a matter which may be regulated by national legislation (article 15.1). Sovereign States have always had that right, of course, although they seldom exercised it, but this is the first time that it has been affirmed by a treaty. Access to genetic resources shall be subject to prior informed consent of the Contracting Party providing the resource, unless that Party decides otherwise (article 15.5). In consequence, the collection of genetic material will generally now be subject to a permit, and the conditions of access will have to be mutually agreed. This implies that access permits will probably be accompanied by contracts, in which the conditions for access, including any financial conditions, will be stipulated. These will normally include an access fee.

In addition, article 15.7 of the Convention requires that the results of research and development, as well as the benefits arising from the commercial and other utilisation of genetic resources, must be shared in a fair and equitable way with the Contracting Party providing such resources, again upon mutually agreed terms. This applies, in particular, to the results and benefits arising from biotechnologies based upon genetic resources (article 19.2). These requirements will probably be routinely included in the access contract as a condition for access. However, the amounts that will have to be paid cannot be set unilaterally by the State providing the genetic material, but will instead have to be freely negotiated in each individual case.

Finally, the Convention establishes a general obligation for all Parties to provide access to and transfer two particular kinds of technology to other Parties. The first includes those technologies which are relevant to the conservation of genetic resources, such as in gene banks, and which may make it easier for the country of origin of those resources to implement the Convention's conservation obligations. The second includes technologies which make use of genetic resources and do not cause significant damage to the environment. With regard to the latter category, this obligation is additional to the requirement of fair and equitable sharing of the benefits with countries that have provided the genetic resources.

Moreover, article 16.1 provides that in respect of developing countries, access and transfer of these two categories of technologies must be provided or facilitated on "fair and most favourable terms", including on concessional and preferential terms where mutually agreed.

There are interesting implications of these provisions, which were not necessarily in the minds of the negotiators when they agreed that both conservation technologies and technologies based on the utilisation of genetic resources must be made widely available to all Parties. The first is that the availability of conservation technologies is a means for Parties not only to discharge their responsibilities under the Convention, but also to keep control over genetic material originating from their territory and thereby to strengthen their proprietary rights. The second implication is that in requiring access to technologies using genetic resources to be facilitated for all countries, the Convention effectively treats such resources as if they were indeed the common heritage that the negotiators had refused to recognise.

All these obligations are subject to an important proviso, however, as they do not affect intellectual property rights, such as patents. Were the utilisation of genetic resources no longer to be the subject of intellectual property rights, there would be little incentive to research the potential uses of such resources, few benefits to countries providing them and even fewer incentives to conserve them. On the other hand, it is increasingly perceived as illegitimate for intellectual property rights to be used to prevent the transfer at a reasonable price of technologies making use of genetic resources, especially biotechnologies, to developing countries, especially those which provided the resources from which these rights were derived.

.The Convention therefore endeavours to strike a balance between these two opposing views. On the one hand, the transfer of technology must be provided on terms which recognise and are consistent with the adequate and effective protection of intellectual property rights (article 16.2).

On the other hand, Contracting Parties must cooperate to ensure that such rights are supportive of the obligations of the Convention and do not run counter to its objectives. This may therefore be a matter for negotiation between Parties and possibly a subject for a Protocol under the Convention. In addition, article 16.3 requires Parties to take legislative or other measures to facilitate the access of Parties which provide genetic resources, particularly those which are developing countries, to technologies making use of these resources, including those which are protected by patents and other intellectual property rights.

The Convention is silent on the methods which are to be used for that purpose. However, it is clear that a forced transfer of technology which disregarded intellectual property rights would not be consistent with article 16.2. It is therefore incumbent on the Parties from whom this transfer is required to develop the necessary incentives, whether financial, fiscal or other, in order to make technology transfer attractive to patent holders.

c. The Scope of the Convention and the Main Conservation Obligations

The Convention provides a comprehensive definition of biological diversity in line with the most modern scientific thinking:

"Biological diversity means the variability among living organisms from all sources including, *inter alia*, terrestrial, marine and other aquatic ecosystems and the ecological complexes of which they are part; this includes diversity within species, between species and of ecosystems."

The Convention also generally applies to

"all processes and activities which have or are likely to have significant impacts on the conservation and sustainable use of biological diversity."

The Convention therefore adopts a broad approach to conservation. It requires Parties to adopt national strategies, plans or programmes for the conservation and sustainable use of biological diversity, and to integrate the conservation and sustainable use of biological diversity into relevant sectoral or cross-sectoral plans, programmes and policies (article 6).

Parties must identify the components of biodiversity important for its conservation and sustainable use, monitor such components where so identified, and identify processes and categories of activities which are likely to have significant adverse impacts on biodiversity (article 7).

Particular emphasis is placed on *in situ* conservation and the maintenance of viable populations of species in their natural surroundings. Processes and categories of activities which may be destructive of biological diversity must be regulated and managed. The introduction of exotic species which threaten ecosystems, habitats or species must be prevented and species already introduced must be eradicated or controlled. The risks associated with the use and release of living genetically modified organisms must be regulated, managed or controlled.

Parties also have the obligation to prepare environmental assessments of proposed projects that are likely to have significant adverse effects on biological diversity, with a view to avoiding or minimising such impacts, and also to ensure that the environmental consequences of their programmes and policies which are likely to have such impacts are duly taken into consideration. Curiously, there is no corresponding provision relating to plans, such as land-use plans or other physical planning instruments, although these may clearly also result in significant adverse impacts on biological diversity.

Measures required for the conservation of protected and other areas are discussed in greater detail in Part II, Chapter I of this paper, whilst measures for the protection of species will be considered in Part I, Chapter I(D).

None of the conservation obligations set out in the Convention are absolute. Instead, they are all qualified by the phrase, "as far as possible and as appropriate". This wording makes allowance for the difficulties faced by most countries in fulfilling the performance obligations laid down by the Convention. On the other hand, this systematic qualification of the obligations weakens them considerably, as it becomes a matter of judgment as to whether conservation is possible and appropriate in the face of other pressing needs.

Whether the Convention will ever amount to more than a statement of good intention and become an effective tool in the conservation of biological diversity will depend on the political will of all the Parties. Equally critical to its success will be the effectiveness of the mechanisms established by the Convention to oversee its implementation and to further its purposes, and the financial means provided by developed countries to that end.

d. The Mechanisms of the Convention

The Conference of the Parties is required to keep the implementation of the Convention under review. It may consider and undertake any action that may be required for the achievement of the objectives of the Convention in the light of experience gained in its operations. These powers are very broad but will again depend on political will for their effectiveness.

Parties are required to submit reports to the Conference on the measures taken to implement the Convention and on their effectiveness in meeting its objectives. Non-governmental organisations (NGOs) may be admitted as observers to the meetings of the Conference.

The Conference may establish subsidiary bodies. One has already been established by the Convention itself, entitled the Subsidiary Body on Scientific, Technical and Technological Advice (article 25). This is to be a multidisciplinary body with some interesting functions, which include the requirements to:

- provide scientific and technical assessments of the status of biological diversity;
- prepare scientific and technical assessments of the effects of types of measures taken in accordance with the provisions of the Convention; and
- identify innovative, efficient and state-of-the-art technologies and know-how relating to the conservation and sustainable use of biological diversity and advise on the ways and means of promoting development and/or transferring such technologies.

This subsidiary body therefore has the capacity to be a powerful driving force behind the implementation of the Convention. Its potential weakness is that it is to be entirely composed of Government representatives and cannot in consequence be considered as an independent scientific authority. The Conference of the Parties will of course be free to designate independent subsidiary bodies if it so wishes.

An important function of the Conference of the Parties is the adoption (and amendment) of Protocols and Annexes. The Convention does not specify the subjects which may be dealt with by Protocols, which means that any matter covered by the Convention may be the subject of a Protocol. Once adopted by the Conference, Protocols must of course be ratified before they can enter into force.

In contrast, Annexes to the Convention or to any Protocol must be restricted to procedural, scientific, technical and administrative matters, whether it is a question of a new Annex or the amendment of an existing Annex. Once adopted, an Annex forms an integral part of the Convention or the Protocol, as the case may be, and may be amended by a simplified procedure which does not require ratification by the Parties. Annexes enter into force one year after their adoption, except for those Parties which have made a Declaration of Objection during that period of time.

The Convention also establishes a Secretariat to service the meetings of the Conference, to prepare reports on the execution of its functions and to perform any functions as may be determined by the Conference of the Parties or assigned to it by any Protocol. At its first meeting, the Conference of the Parties shall designate the Secretariat from amongst those existing international organisations which are competent in this field and which have signified their willingness to carry out these functions.

e. The Provision of Financial Resources under the Convention

Various financial mechanisms for the Convention were suggested during the negotiations, although the problem was not studied in any depth until a relatively late stage.

The proposal originally put forward by IUCN, that conservation actions should be financed through levies on the commercial applications of wild products, which would be paid into an international conservation fund established under the convention, was not retained. Similarly, the possibility of creating an "enterprise" for the commercial exploitation of genetic material, of which States would be shareholders, was not seriously investigated.

It was therefore concluded that any financial mechanism would have to be exclusively based on compulsory or voluntary contributions by Contracting Parties. This immediately raised doubts as to whether such contributions could ever be sufficient to meet global conservation requirements under the Convention.

In parallel, it was recognised from the very beginning of the negotiations that substantial financial means would need to be transferred to developing countries for conservation purposes if the Convention was to have any chance of being effective. Clearly, many of the countries which are the richest in biological diversity are also amongst the poorest in the world. As a result, they would be completely unable to meet their obligations under the Convention without such financial transfers. This partly explains the constant use of the expression, "as far as possible and as appropriate", throughout the Convention.

Moreover, these financial transfers would have to be additional to any other development assistance that might be required, as such transfers should be exclusively devoted to meeting the "incremental costs" which developing countries would incur in fulfilling their obligations under the Convention.

The Convention therefore establishes a legal interrelation between the conservation obligations of developing countries and the obligation on the part of developed country Parties to provide the former with new and additional financial resources. Without the provision of such resources, developing countries are considered by the Convention as no longer bound by their conservation obligations.

The Convention provides that the level of such contributions will be governed by the incremental costs for developing countries incurred as a result of implementing measures which fulfil their obligations. For each developing country Party, however, these costs must first be mutually agreed upon by that Party and an special institutional structure to be established by the Conference at its first meeting. This institution will be created specifically to develop and implement an appropriate mechanism for the provision of financial resources to developing country Parties.

The mechanism thus established shall function under the authority and guidance of the Conference and shall be accountable to that body. The Conference shall determine the policy, strategy, programme priorities and eligibility criteria governing access to these resources and their utilisation.

For the first time and as a result of the above provisions, a global financial mechanism for the conservation of biological diversity is at last in view, with the potential to attack priority problems. Its effectiveness will obviously depend on the level of contributions, which again comes down to a question of political will.

Much will also depend on the nature of the institutional structure that is to be established. The Convention provides that the mechanism shall operate within a democratic and transparent system of governance. There was considerable opposition between developed and developing countries during the negotiations as to the institutional structure that would be chosen to implement the financial mechanism. The former favoured the Global Environmental Facility (GEF), already instituted jointly by the United Nations Development Programme, UNEP and the International Bank for Reconstruction and Development. The developing countries felt that this institution lacks democracy and transparency and that a separate structure should instead be instituted under the Convention, which would be under the direct control of the Conference of the Parties.

For the time being and until the Conference of the Parties decides otherwise, the Convention provides that the GEF will serve as the required institutional structure on an interim basis, "provided it has been fully restructured" in accordance with the requirements that the Convention lays down with regard to democracy and transparency.[15]

f. Relationship with other Conventions

The Convention is not an umbrella convention encompassing the conservation conventions already in force. It functions instead as a framework convention for future action, which lays down general obligations, rules and principles and also provides for the adoption of protocols and annexes which will treat specific issues in greater detail.

There are in fact no legal or practical means to encompass existing conventions in this way. In law, each treaty stands on its own, with its own Parties, institutional mechanisms and 'sponsoring' international organisation, unless it has been specifically drawn up as a protocol to another convention. It would therefore have been legally impossible to bring existing conservation treaties under the umbrella of the Convention without amending them for that specific purpose. This would have taken many years, as such amendments would have had to be ratified by a majority of the Parties to each treaty before they could come into force.

Even to assign a coordination role to the Conference of the Parties to the Convention on Biological Diversity would have met with strong resistance.

Existing conventions will accordingly continue to be independent.

To this end, article 22 of the Convention on Biological Diversity provides that the provisions of the Convention shall not affect the rights and obligations of the Parties deriving from any existing international agreement. This means that no changes have been made with regard to any of the other conservation conventions.

---

[15] Article 39, dealing with Financial Interim Arrangements.

Nevertheless, there is an interesting exception which may well affect other agreements in practice. Article 22 makes an exception

"where the exercise of those rights and obligations would cause a serious damage or threat to biological diversity".[16]

In other words, where there is a conflict between the Biological Diversity Convention and any other agreement with regard to the conservation of biological diversity, the provisions of the former will prevail.

Strangely, this provision does not seem to apply to the Law of the Sea treaty, as the Biological Diversity Convention states that Parties must implement the Convention consistently with the rights and obligations of States under the law of the sea.

Despite the above provisions, it is essential that institutional relations and a minimum degree of coordination be established between conservation conventions if the world conservation system is to operate satisfactorily. A certain degree of cooperation with other conservation conventions is therefore provided for by the Convention.

The Conference of the Parties must contact, through the Secretariat, the executive bodies of other conventions dealing with biological diversity with a view to establishing appropriate forms of cooperation with them. The Secretariat must coordinate with other relevant international bodies (these are not limited to those providing Secretariats for other conventions, and could include FAO, UNESCO and IUCN, for example) and may, in particular, enter into such administrative and contractual arrangements as may be required for the effective discharge of its functions.

Finally, no time has been lost since the Convention was first opened for signature. Although it has still not entered into force, meetings have already been held on a number of subjects. An interim Secretariat has been provided by UNEP, in accordance with the Convention. This will function until the first meeting of the Conference of the Parties, at which time the permanent Secretariat will be designated.

---

[16] This provision setting out the application of the Convention to other treaties is rather extraordinary. The Convention prevails if any other treaty (not necessarily a conservation treaty) seriously affects biological diversity. Theoretically, at least, the Convention would therefore prevail over GATT.

PART I

SPECIES-BASED CONSERVATION
AND THE LAW

CHAPTER I
THE INTERNATIONAL LAW OF SPECIES

Treaties on species may concern the protection of individual species for nature conservation purposes, or may have as their objective the organisation of the rational exploitation of species harvested by more than one country. As the purposes and mechanisms of such treaties are different, they will be considered separately in the sections dealing respectively with conservation and exploitation treaties. The species conservation provisions of the Convention on Biological Diversity will be considered at the end of this chapter.

Although CITES is essentially a conservation treaty, it also deals with exploitation and will be discussed in Chapter V on Trade below.

A. Species Conservation Treaties

1. Early Treaties

The London Convention on the Protection of Wild Fauna in Africa, concluded in 1900 but never ratified, was the first instrument which took as its purpose the conservation of species as such. The Convention was intended to prevent the "uncontrolled massacre" and ensure the conservation of African wild animal species which are useful to man or inoffensive. There was a list of protected species as well as of species the shooting of which had to be regulated.

The Paris Convention for the Protection of Birds Useful to Agriculture of 1902, which is still in force although superseded, prohibits the taking of and trade in certain listed species, which are essentially small passerines and owls. It also prohibits the use of mass destruction or capturing devices, and lists noxious species, the destruction of which is encouraged. These include diurnal birds of prey, in particular, as well as fish-eating birds such as herons, pelicans and divers.

The 'next' African Convention, the London Convention Relative to the Preservation of Fauna and Flora in their Natural State, was in force from 1933 to 1968. It set out a list of species which had to be strictly protected (including a plant, the spectacular Welwitschia bainesii of the Namib Desert), and a list of partially protected species, the taking of which was to be strictly regulated. The Convention also imposed domestic and international trade restrictions and prohibitions on certain methods of hunting.

The Convention on Nature Protection and Wildlife Preservation in the Western Hemisphere of 1940 is intended to protect and preserve all species of native fauna and flora in their natural habitats, but fails to provide effective means to achieve this objective with regard to species. The Convention does not incorporate lists of protected species which all Parties are committed to protect. Instead, it simply contains sets of national lists, which do not form part of the Convention, which list the species which each Party intends to protect. Parties are free to change these lists at any time. Accordingly, there are no binding obligations under the Convention in respect of the protection of species.

2. Regional Treaties and Other Instruments

a. Africa

With respect to faunal resources, the African Convention on the Conservation of Nature and Natural Resources of 1968 requires Contracting States to ensure their conservation, wise use and development within the framework of land-use planning and economic and social development. Outside protected areas, Contracting States shall manage exploitable wildlife populations for an optimum sustainable yield.

There are also some substantive provisions on protected species, hunting and protected areas, which broadly reproduce those of the London Convention of 1933. Class A of the Annex to the Convention sets out a list of strictly protected species, including three plants; Class B contains a list of species which may only be hunted, killed, captured or collected with a special permit. The lists are longer than those of the London Convention of 1933. National legislation must regulate the issue and use of permits and prohibit certain hunting methods. All methods liable to cause mass destruction of wild animals are prohibited by the Convention.

The Convention contains two new and interesting provisions. The habitat necessary to the survival of species threatened with extinction must be accorded special protection. Where such a species is endemic to the country of only one Contracting State, that State has particular responsibility for its protection. There are also provisions on traffic in specimens and trophies. Parties have the obligation to regulate trade in and transport of all specimens and trophies, whether or not they are obtained from protected species. Special permits are necessary for the import, export or transit of specimens and trophies of species belonging to Classes A and B. However, this latter provision has of course lost much of its importance, now that a large number of African countries have become Parties to CITES.

b America

The only treaties in force are the Western Hemisphere Convention of 1940 and the Convention for the Conservation of Biological Diversity and the Protection of Priority Wild Areas in Central America of 1992. The latter contains general provisions for the protection of species, which include the obligation for each Party to encourage the development of national legislation for the conservation and sustainable use of the components of biological diversity; to promote species recovery plans; to establish machinery to strengthen controls on illegal traffic in specimens of wild fauna and flora between the countries of the region; to control the collection of biological resources in natural habitats; and to regulate domestic trade in such resources by national legislation. There is no list of protected species under the Convention.

c. Europe

i. The Berne Convention of 1979

Under the Convention on the Conservation of European Wildlife and Natural Habitats of 1979, plant and animal species listed in Appendices I and II respectively must be strictly protected by the Parties. The deliberate picking, collecting, cutting or uprooting of plants thus protected must therefore be prohibited by the Parties, together with the deliberate killing, capturing and keeping of protected animals. The Convention also prohibits deliberate damage to or destruction of the breeding or nesting sites of Appendix II animal species, as well as deliberate disturbance to and the destruction, taking or keeping of their eggs. Contracting Parties are also encouraged to prohibit the possession and sale of strictly protected plants and animals.

Appendix III lists animal species whose exploitation must be regulated in order to keep their populations out of danger. Appendix IV contains a list of methods of hunting and other forms of exploitation which are prohibited for mammals and birds.

There are 517 species of plants listed in Appendix I and most of the European mammals (except rodents), birds, reptiles and amphibians are listed in Appendices II or III, together with some freshwater fishes and invertebrates.

One of the important provisions of the Convention is that Parties must also protect the habitats of protected species. The Standing Committee has decided to draw up a list of species the habitats of which require priority attention for conservation by the Parties, as not all species listed in the Appendices are in need of such measures. Another particularly useful provision is the obligation to preserve endangered habitat types, which is discussed in Part II of this paper.

ii. European Community legislation

The first Community instrument on the protection of wild species was Directive 79/409 on the Conservation of Wild Birds, adopted on 2 April 1979, which will be discussed in greater detail in the section on Sectoral Treaties below.

The Birds Directive implements the Berne Convention within the Community for birds, as does the Directive on the Conservation of Natural Habitats and of Wild Fauna and Flora of 21 May 1992 (92/43) for other species. As regards other treaties dealing with species, the EC is also party to the Bonn Convention, the Regional Seas Conventions for East Africa and the Caribbean and their Protocols,[1] and some fisheries treaties, such as the Convention on the Conservation of Antarctic Marine Living Resources of 1980.

The EC Regulation of 3 December 1987 deals with the implementation of CITES in the Community. The EC is not a Party to CITES as such, but the CITES Convention has been amended to allow "organisations of economic integration" to accede. Although this amendment was adopted in 1983, it is still awaiting ratification by a number of CITES Parties and is not in force. Nevertheless, the Regulation of 1987 was necessary to oblige Member States to apply uniform measures for CITES implementation throughout the Community. A new and more comprehensive Regulation has now been prepared, but has not yet been adopted by the EC Council.

Turning to the Habitats Directive, Annex IV sets out a list of Animal and Plant Species of Community interest in Need of Strict Protection, including a fairly long list of invertebrates. This list includes species protected under the Berne Convention, but excludes birds as these continue to be covered by the Birds Directive.

The above species are protected by prohibitions which are essentially the same as those under the Berne Convention and the Birds Directive. In addition, the Habitats Directive establishes a system to monitor the incidental capture and killing of listed animal species. In the light of information thereby gathered, Member States must take further measures as required to

---

[1] Certain Member States, such as France, have overseas territories which are covered by the Regional Seas Protocols.

ensure that incidental taking does not have a significant negative impact on the species concerned.

Member States also come under an obligation to prohibit the keeping, transport, sale or exchange of specimens taken from the wild after the Directive has been implemented.

In addition, Annex V sets out a list of Animal and Plant Species of Community Interest whose Taking in the Wild and Exploitation may be Subject to Management Measures. These comprise a relatively small number of species of mammals, amphibians and freshwater fish, as well as a few invertebrates and plants.

Member States are only required to impose restrictions on the taking of Annex V species, if they "deem it necessary". This contradicts the provisions of the Berne Convention in respect of those species which are listed under both Annex V of the Directive and Appendix III of the Convention, as the Berne Convention imposes mandatory restrictions on species listed under Appendix III.

Where Member States do deem such measures necessary, they also have a choice of measures which may include:

- temporary or local prohibitions on taking;
- the regulation of periods and/or methods of taking;
- the application of hunting and fishing rules which take into account the conservation of certain populations;
- the establishment of licences or quotas;
- the regulation of the purchase or sale of specimens;
- the breeding in captivity or artificial propagation under strictly controlled conditions, with a view to reducing the taking of specimens in the wild,
- assessment of the effectiveness of the measures adopted.

It should be noted that the Directive contains no restrictions on the import or export of listed species. It is expected that the new Regulation for the implementation of CITES will deal with all aspects of international trade in species covered by the Birds and the Habitats Directives.

With regard to species' habitats, Annex II lists Animal and Plant Species of Community Interest whose Conservation requires the Designation of Special Areas of Conservation. Some 160 plant species and 22 animal species from this long list are marked as "priority species", whose host sites are automatically deemed to be Sites of Community Importance.

Many of the animal species listed in Annex II are also listed as specially protected species in Annex IV, although a few are in Annex V. However, some animal species in Annex II are not listed in these other Annexes, which means that no prohibitions or restrictions on their taking are required, although measures for the protection of their habitats are obligatory. Such species include freshwater fish, in particular, and some invertebrates.

Conversely, some species in Appendix IV and most of those in Appendix V are not listed in Appendix II. In this situation, no specific habitat conservation measures need be taken, but prohibitions or restrictions on taking are required.

In contrast, all plant species listed under Annex II as requiring habitat conservation measures are also listed in Annex IV. Annex IV also contains a number of species whose taking is prohibited, but for which no specific habitat protection measures are required. No specific habitat conservation measures are required for Annex V plants.

Procedures for the selection and financing of these Special Areas of Conservation are discussed in Part II of this paper.[2] By way of summary, the Directive aims to establish "a coherent European ecological network" called Natura 2000, which will comprise the above-mentioned Special Areas of Conservation as well as the Special Protection Areas established by Member States under the Birds Directive of 1979.

Article 19 sets out the procedure for amending the various Annexes in line with technical and scientific progress. Non-priority species and habitat types may be recharacterised as priority species or habitats, and additional species or habitats may be listed for the first time by the Council acting by qualified majority on a proposal from the Commission. However, amendments to Annex IV on Animal and Plant Species of Community Interest in Need of Strict Protection must be adopted unanimously. The latter provision reflects the opposition of hunters during the preparation of the Directive, who were afraid that game species might become fully protected against the will of their Governments.

iii. The Alpine Convention

As mentioned in the Introduction, the Convention on the Protection of the Alps of 1991 (not yet in force) is a framework Convention covering the whole range of environmental problems in very general terms.

A draft Protocol on nature conservation has been prepared but not yet adopted. There will not be an Annex with a list of protected species, as it was felt that this would be better achieved through the Berne Convention, if necessary by amending the Appendices to that Convention.

d. Asia

The only treaty covering the protection of species in Asia is the ASEAN Agreement on the Conservation of Nature and Natural Resources of 1985. Parties are required to give special protection to threatened and endemic species and to preserve those areas which constitute the critical habitats of endangered or rare species, of species that are endemic to a small area and of migratory species.

In addition, an Appendix listing endangered species deserving special attention will be adopted by a meeting of the Contracting Parties. The taking of and trade in these species is prohibited. A draft Schedule, prepared in 1987 for submission to the Parties, will be adopted when the Convention comes into force.

e. Regional Seas

Only two Regional Seas Conventions, those applicable to East Africa and the Caribbean, have so far been followed by Protocols specifically dealing with the protection of species.

---

2   Provision is made for the possibility of co-financing from the new Community Environmental Fund (LIFE) which was established by Regulation 1973/92 of 21 May 1992.

The Protocol on Protected Areas and on Wild Fauna and Flora was adopted in Nairobi on 21 June 1985, the same day as the Convention for the Protection, Management and Development of the Marine Environment and the Coastal Areas of the East African Region.[3] Article 10 of that Convention requires Parties

"individually or jointly , [to] take all appropriate measures to protect and preserve rare or fragile ecosystems as well as rare, depleted, threatened or endangered species of wild fauna and flora and their habitats in the Convention area. To this end, the Contracting Parties shall ... establish protected areas, such as parks and reserves, and shall regulate and, where required and subject to the rules of international law, prohibit any activity likely to have adverse effects on the species, ecosystems or biological processes that such areas are established to protect."

The Protocol applies to all the marine areas of the Indian Ocean which come under the jurisdiction of the Parties, namely their territorial sea and EEZ, as well as to their coastal areas and their internal waters related to the marine and coastal environment (article 1).

The General Undertaking contained in Article 2 of the Protocol is inspired by the World Conservation Strategy since it obliges Parties to take

"all appropriate measures to maintain essential ecological processes and support systems, to preserve genetic diversity, and to ensure the sustainable utilisation of harvested natural resources under their jurisdiction."

Although Article 2 of the Protocol then follows the terms of Article 10 of the Convention, it weakens the latter's binding obligation into a non-binding requirement that the Contracting Parties simply "endeavour to protect and preserve rare or fragile ecosystems". Article 2 also requires Parties to "develop national conservation strategies and coordinate, if appropriate, such strategies within the framework of regional conservation activities."

More specifically, provision is made for the conservation of certain plant and animal species listed in four Appendices. Appendix I lists just eleven protected plants, some of which are very rare insular endemic forms. Article 3 requires the Parties to take all appropriate measures to ensure the protection of these species by prohibiting, as appropriate, activities having adverse effects on their habitats as well as picking, collecting, cutting or uprooting, and possession or sale.

Appendix II lists numerous animal species in need of special protection, which include ten mammals, ninety birds, eleven reptiles, six marine molluscs, the coconut crab, corals, and two insects. The list of birds is directly drawn from the IUCN Red Data Book of Endangered Species. Ironically, for all that the Protocol deals with a marine area, very few marine species are actually listed!

Parties must take all appropriate measures to ensure the strictest protection of Appendix II species, including the regulation or prohibition of activities having adverse effects on the habitats of such species. Where necessary, prohibitions must be imposed on all forms of capture, keeping or killing; the deterioration or destruction of critical habitats; the disturbance of fauna, particularly during the period of breeding, rearing and hibernation; the destruction, taking or

---

[3] The East African Region includes the following States: Comoros, France, Kenya, Madagascar, Mauritius, Mozambique, Seychelles, Somalia and Tanzania. All these States participated in the negotiations on the Convention and on its Protocol.

keeping of eggs; possession of or internal trade (external trade is dealt with by CITES) in animals, whether dead or alive, and any readily recognisable part or derivative thereof (Article 4).

Appendix III lists exploitable species for which protection measures are necessary. Very few are marine species: the list includes seventeen mammals (such as elephants and zebra) as well as rock lobsters and two species of marine turtle. It should be emphasised that this characterisation of marine turtles as "exploitable" is inconsistent with their totally protected status under both the Algiers and Bonn Conventions and their inclusion in Appendix I of CITES. A possible explanation is that some of the Parties to the Nairobi Convention are not Parties to the Algiers or Bonn Conventions.

The exploitation of Appendix III species must be regulated in order to "restore and maintain the populations at optimum levels". Each Party must develop management plans for the exploitation of these species, which may include the prohibition of indiscriminate means of capture and killing; closed seasons; the temporary or local prohibition of exploitation; the regulation of possession, transport and sale; the safeguarding of breeding stocks of these species and their critical habitats in areas specially protected to this end; and exploitation in captivity.

Appendix IV deals with a few migratory species, including two whales,[4] the dugong and five species of marine turtle occurring along the African coast. The Parties are required to coordinate their protection efforts in respect of these species (Article 6).

Article 7 of the Protocol deals with the introduction of exotic species, and corresponds to Article 196 of the Convention on the Law of the Sea 1982. Parties are required to take

"all appropriate measures to prohibit the intentional or accidental introduction of alien or new species which may cause significant or harmful changes to the region".

The Convention for the Protection and Development of the Marine Environment of the Wider Caribbean Region was adopted in 1983. Under this Convention, a Protocol concerning Specially Protected Areas in the Wider Caribbean Region, more comprehensive than the Nairobi Protocol, was signed in Kingston, Jamaica on 17 January 1990. The Convention and the Protocol apply to the territorial sea, the Exclusive Economic Zone and the internal waters of the Parties, as well as to any related territorial areas, including watersheds, that may be designated by the Parties having sovereignty or jurisdiction over them. The application of the Protocol to territorial areas therefore depends on the decision of each individual Party.

Contracting Parties are required to identify endangered or threatened species of wild flora and fauna under their jurisdiction and to accord them protected status. The Parties are also under a performance obligation to take appropriate measures to prevent species from becoming endangered or threatened, and are required to establish cooperation programmes to assist with the management and conservation of protected species and to develop regional recovery programmes.

---

[4] Already strictly protected under the International Whaling Convention, discussed under Exploitation Treaties in section (C) below.

Parties must also take cooperative measures to ensure the protection and recovery of species listed in three Annexes, which were adopted in June 1991. Annex I lists plant species to be fully protected by the Parties, Annex II lists animals to be fully protected by the Parties, and Annex III lists species of flora and fauna whose use must be regulated in accordance with management plans which the Parties must adopt.

The Protocol also regulates the introduction of non-indigenous or genetically altered species in the whole of the geographical area concerned, rather than just in areas protected under the Convention.[5] Parties are required to prohibit the

"intentional or accidental introduction of non-indigenous or genetically altered species to the wild that may cause harmful impacts to the natural flora, fauna or other features of the Wider Caribbean Region."

The Protocol also provides for the preparation of environmental impact assessments in respect of activities liable to have adverse effects on protected areas. Institutional arrangements include regular meetings of the Parties to review and direct the implementation of the Protocol. The Protocol establishes a Scientific and Technical Advisory Committee which advises the Parties inter alia on the listing of protected species and the formulation of common guidelines and criteria. The Secretariat is provided by UNEP.

Finally, mention should also be made of the Convention on the Protection of the Natural Resources and Environment of the South Pacific (the SPREP Convention), signed in Noumea on 24 November 1986. Article 14 of the Convention deals with specially protected areas and the protection of wild flora and fauna, and could accordingly form the basis to a Protocol. However, such a Protocol has not yet been adopted to this Convention.

f. Areas beyond National Jurisdiction: the Antarctic

The Agreed Measures for the Conservation of Antarctic Fauna and Flora, adopted in 1964 by the Parties to the Antarctic Treaty of 1959, apply to mammals other than whales, birds and Antarctic vegetation as a whole. All species are in principle protected. Exceptions are possible in certain cases, apart from specially protected species listed in an Annex, for which permits for hunting or capture shall be delivered only upon good scientific reasons and provided that this entails no danger to the natural ecosystem or to the survival of the species itself. Activities which are harmful to the normal living conditions of mammals and birds must be reduced to a minimum. The unauthorised introduction of animal or plants species into the Treaty area is prohibited.

The Treaty also provides for the designation of Specially Protected Areas and Areas of Outstanding Scientific Interest, which are discussed in Part II, Chapter I of this Paper below.

In October 1991, a Protocol to the Antarctic Treaty was adopted in Madrid for the purpose of developing a comprehensive regime for the protection of the Antarctic environment and its dependent and associated ecosystems in the interest of mankind as a whole. To this end, Antarctica is designated as a "natural reserve devoted to peace and science". Activities in the Treaty Area must be planned and conducted so as to avoid detrimental changes in the distribution, abundance or productivity of species or populations of flora and fauna or further jeopardy to endangered species or populations.

---

[5] As is the case for the Protocol applicable to the Mediterranean, adopted under the Geneva Convention of 1982, which is discussed in Part II below.

Annex II of the Protocol deals specifically with the conservation of Antarctic fauna and flora. The taking of indigenous species is regulated by the general rule that the taking of or harmful interference with fauna and flora shall be prohibited without a permit, which may only be issued in special circumstances and under strict conditions. The issue of permits must be limited so as to ensure the maintenance of the diversity of species, as well as the habitats essential to their existence, and the balance of the ecological systems existing within the Antarctic Treaty Area.

Specially protected species are listed separately and a permit for their taking may only be granted for scientific reasons.

Annex II also regulates the introduction of non-native species, parasites and diseases: the introduction of such species requires a permit.

The above provisions are broadly similar to those set out in the Agreed Measures. In addition, the Protocol requires that all activities which may have more than a minor or transitory impact on the environment must be subject to an environmental impact assessment (article 8). Furthermore, any activity relating to mineral resources, other than scientific research, is prohibited. In effect, the latter provision means that the Treaty of Wellington on the extraction of minerals has now been superseded.

Once the Protocol comes into force, the Agreed Measures of 1964 will cease to have effect.

3. Sectoral Treaties: Migratory Species

The only sectoral treaties dealing with wildlife are those dealing with migratory species or with species between one or more country. Fisheries treaties will be dealt with separately in section (D) below on Exploitation Treaties, as will those agreements that have been concluded for the protection of other marine species such as whales and seals.

Migration may be defined as a cyclical, and therefore predictable, phenomenon whereby certain animals perform periodic movements between two separate geographic areas, one usually being where they breed. Migratory species may be terrestrial, fresh water or marine. Treaties are obviously essential for the protection of migratory species, as it would otherwise be impossible to take joint conservation measures along the migration routes of the species to be conserved.

Some of the regional treaties examined above have provisions on migratory species. These include the Convention on Nature Protection and Wildlife Preservation in the Western Hemisphere of 1940, which requires Contracting Parties to adopt appropriate measures to protect migratory birds of economic value or aesthetic interest, or to prevent the extinction of a given species which is under threat. Under Article 10 of the Berne Convention, the Parties undertake to coordinate their efforts for the protection of migratory species. The ASEAN Agreement of 1985 also establishes an obligation to cooperate in the conservation, management and, where appropriate, the regulation of the taking of the international resources which these species constitute. However, all the above provisions are very general and do not establish any specific mechanisms for cooperation or joint management.

a. Species other than Birds

Some treaties have been concluded to protect particular species other than birds. Amongst treaties for the protection of species found on land, three international conventions[6] have been signed in the last 25 years to protect the vicuna. This is a high-mountain South American camelid with the finest and most expensive wool in the world, which lives in frontier regions and is liable to move from one country to another. The vicuna is also listed in Appendix I of CITES.

A Bilateral Agreement on the Protection of the Porcupine Caribou Herd was concluded between the USA and Canada in Ottawa on 17 July 1987, to protect a large herd of caribou which regularly migrates between the Canadian Far North and Alaska.

Polar bears straddle the border between land and marine species as they spend so much time on the pack-ice. They are protected by an Agreement on the Conservation of Polar Bears which was signed in Oslo on 15 November 1973 by the five circumpolar States, namely Canada, Denmark (for Greenland), Norway, the USA and the former USSR).

The Agreement has three main objectives: to encourage the Parties to coordinate their research programmes; to restrict the killing and capture of polar bears; and to protect the ecosystems of which polar bears are a part. The taking of polar bears is prohibited with a certain number of exceptions, and domestic and international trade in polar bear skins is also prohibited. Parties must protect the ecosystems hosting polar bears, especially their denning and feeding sites, together with the bears' migratory patterns. Polar bears are also protected by the Berne Convention and are listed in Appendix II to CITES.

b. Early Bird Treaties

Birds have traditionally formed the subject of international conventions. As mentioned earlier, the Convention for the Protection of Birds Useful to Agriculture of 1902 was essentially limited to the protection of small passerines and nocturnal raptors. The Paris Convention of 1950 established the principle that all birds, with a small number of exceptions, must be protected. Parties were required to ban hunting during the period when birds return to their nesting sites, as well as certain hunting methods, and were encouraged to create nature reserves.

Neither of the Paris Conventions is limited to migratory birds, but it is doubtful whether they would have been concluded if migration had not been a major issue. The Conventions were ratified by only a small number of countries and although they are still technically in force, they have in practice been superseded. Moreover, they lack any machinery for joint cooperation and management measures and their effectiveness has therefore been extremely limited.

c. The EC Birds Directive

EC Directive 79/409 on the Conservation of Wild Birds of 2 April 1979 is compulsory for all Member States. The objective of the Directive is to maintain the populations of all bird species

---

[6] The Convention for the Conservation of Vicuna, concluded in La Paz in 1969 to which Argentina, Bolivia, Chile, Peru and Ecuador are Parties; the Convention for the Conservation and Management of Vicuna, signed in Lima on 16 October 1979, to which Bolivia, Chile, Peru and Ecuador are Parties; and the bilateral Agreement between the Bolivian and Argentinean Governments for the Protection and Conservation of Vicuna, signed in Buenos Aires on 16 February 1981.

in the EC at a level which corresponds to ecological, scientific and cultural requirements. To this end, the Directive requires Member States to establish a general system of protection for all birds, prohibiting in particular their deliberate killing, capture or disturbance, the deliberate destruction of or damage to their nests and eggs, the taking or removal of eggs and nests, the keeping of eggs and the keeping or sale of live or dead birds or of readily recognisable parts or derivatives thereof. Exceptions from these rules are made for a relatively small number of listed game species. There is also a list of prohibited hunting methods.

Member States are also required to establish Special Protection Areas to protect the habitats of all species listed in Annex I, many of which are migratory, as well as the habitats of all migratory species, whether listed or not. These obligations are considered in greater detail in Part II of this paper.

d. Bilateral Treaties

In practice, the above instruments are only implemented in Europe although in law they are more widely applicable. Elsewhere in the world, a number of bilateral agreements have been concluded on the conservation of migratory birds.

The oldest of these agreements is a Convention between the United States and Great Britain (in the name of Canada which had not then acceded to independence) concluded in 1916. A similar treaty was concluded in 1936 between the United States and Mexico, a supplementary Protocol to which was signed in 1972.

No further bilateral treaties on this subject were signed until the 1970s and 1980s. These are respectively the treaties between United States—Japan, 1972; Japan—the former USSR, 1976; China—Japan, 1981; India—the former USSR, 1984; and Australia—China, 1986. It will be seen from this list that the migratory birds of the world outside Europe are unprotected in international law, other than in the North American and Pacific regions and in Asia to a limited extent.

These bilateral treaties have several common features. In general, they forbid the taking of species listed on an appendix, except if an open season has been declared by the Party concerned. Discretion as to which species may be harvested is therefore left to each Party. The hunting of game birds is usually permitted except in the breeding season. The Parties are often encouraged to establish reserves for migratory birds.[7] There is usually a provision controlling the introduction of exotic species.

On the other hand, such bilateral treaties do not cover the whole range of species or the entire length of the migration routes of the species in question. They do not establish any cooperation machinery to facilitate the implementation of the instrument in question or to ensure coordination between conventions dealing with the same species in the same geographical region. They are no substitute for a world convention designed to address the conservation problems of migratory species globally.

---

[7] The United States—former USSR Convention is the only treaty to require its Parties to identify breeding, wintering, feeding and moulting areas of particular importance for migratory birds and to protect them against pollution and any other deterioration of the environment.

e. The Global Treaty on Migratory Species: the Bonn Convention of 1979

The Convention on the Conservation of Migratory Species of Wild Animals was adopted in Bonn, Germany, on 23 June 1979.

Main Obligations

The principal obligations of the Convention are to protect certain endangered species listed in Appendix I of the Convention (Article III) and to

"endeavour to conclude agreements for the protection and management of migratory species whose conservation status is unfavourable and of those whose conservation status would substantially benefit from the international cooperation deriving from an agreement." (such species are listed in Annex II)

Since a species may often be both endangered and likely to benefit from international cooperation in favour of its conservation, the Convention rightly specifies that if the circumstances so warrant, a migratory species may be listed both in Appendix I and II (Article IV-2).

Species Covered

The Convention defines migratory species as those which cyclically and periodically cross one or more jurisdictional boundaries. The definition therefore includes marine species which migrate between adjacent EEZs or between an area under the jurisdiction of a coastal State and the high seas. In the latter case, States exploiting a migratory species on the high seas are considered by the Convention to be Range States of the species concerned. The definition obviously excludes species which are found exclusively in the high seas.

At present, Appendix I only includes 51 severely-threatened species. The list is comprised of 18 mammals, including four species of whale, the monk seal, several antelopes and the mountain gorilla; 24 birds; 8 reptiles, including 6 of the 7 species of marine turtle; and one fish.

The Appendix does not claim to be an exhaustive list of all endangered migratory species, but rather a representative sample of the most threatened species. The list may be progressively amended by the Conference of the Parties, in the light of new scientific data. The first meeting of the Conference in 1985 added a further eleven species to the Appendix, including four of those species of marine turtle.

In contrast, Appendix II lists many species, including some 2,000 species of birds or nearly a quarter of existing species. Whole families of bird species are included, such as the Anatidae (geese and ducks), all the wading birds belonging to the Charadriidae and Scolopacidae families, the entire Muscicapidae family of passerines, and all the diurnal raptors apart from the non-migratory secretary-bird. The Appendix also lists some mammals such as certain populations of seals and several species of small cetacean, the African elephant, the dugong, the vicuna, two African antelopes, all the marine turtles, two other reptiles, a fish and the Monarch butterfly.

The Conference of the Parties has already amended Appendix II twice to include the European populations of bats, small cetaceans, certain populations of seals and most cranes. Some Appendix I species, such as the monk seal, several species of crane and four marine turtles, also feature in Appendix II.

The two Appendices also list some species which only migrate locally, albeit across borders, and whose conservation status is precarious. Appendix I species in this category include the mountain gorilla, which is only found on the border between Rwanda and Zaire; Grevy's zebra (Ethiopia and Kenya) and the Barbary stag (Algeria and Tunisia).

A simplified procedure is established for amendments to both Appendices, which can be adopted by the vote of a two-thirds majority of the Parties present and voting. Amendments enter into force after ninety days and bind all Parties that have not made a reservation on the amendment during this period (Article IX–4–6).

Obligations under Appendix I

Parties must prohibit the taking of animals listed in Appendix I, whether or not such species are also listed on Appendix II and are covered by agreements concluded under the Convention for their conservation and management. Exceptions may be made in certain cases, such as for scientific purposes or for the needs of traditional subsistence hunting, provided that they are precise as to content, limited in space and time and do not operate to the disadvantage of the species concerned (Article III–5).

Parties should also endeavour to conserve and, where feasible, restore the important habitats of Appendix I species; to prevent, remove, compensate for or minimise the adverse effects of activities or obstacles that seriously impede or prevent migration; and finally to prevent, reduce or control factors which endanger or are likely to endanger these species, including strictly controlling the introduction of exotic species and controlling, limiting or eliminating those exotic species which have already been introduced (Article III–4).

Obligations under Appendix II

The only obligation imposed under the Convention in respect of Appendix II species is to endeavour to conclude Agreements.

Annex V sets out very detailed guidelines on the content of such Agreements. Although these are obviously not binding, they provide a very full catalogue of the type of measure which should be taken to ensure the conservation and management of these migratory species concerned.

Such measures include the periodic review of the conservation status of the species covered by an Agreement; coordinated conservation and management plans; research into the ecology and population dynamics of these species; conservation and, where necessary and feasible, restoration of important habitats; control of exotic species; maintenance of a network of suitable habitats appropriately located in relation to the migration routes; elimination of activities and obstacles which hinder or impede migration; prevention, reduction or control of the release of polluting substances into the habitats of the species concerned; and measures based on ecological principles to manage the taking of the species.

Appraisal of the present status of the Convention

As mentioned in the Introduction, many countries of major importance for migratory birds are still outside the Convention, which means that there are insufficient Parties to cover the majority of species included in the Appendices and their migration routes. As a result, the inclusion of many species in Appendix I is largely symbolic since their Range States are not Parties to the Convention. This also makes it difficult to start negotiating Agreements in respect of most of the Appendix II species.

Nevertheless, a small number of Agreements have been concluded, invariably where a large number of the Range States of the species concerned are already Parties to the Convention and an Agreement is therefore workable. At its first meeting in Bonn in October 1985, the Conference of the Parties decided that Agreements should be drawn up for European bats, the white stork, two small North Sea and Baltic cetaceans and the geese and ducks of the Western Palaearctic region (i.e. Europe as far as the Urals and North Africa).

The Agreements on bats and cetaceans were formally adopted in 1991, although they are not yet in force. In addition, the three countries bordering the Waddensee, a large area of shallow waters, sand-banks and mudflats along the coasts of Denmark, Germany and the Netherlands, have also signed an Agreement for the protection of the populations of seals living in that area.

The agreement on the white stork is still under discussion. Similarly, a preliminary draft has been prepared on Western Palaearctic waterbirds, but detailed negotiations have not yet begun. It is doubtful whether the number of Range States which are Parties to the Convention is sufficient for an effective Agreement in respect of these waterbirds. The absence of the Russian Federation (nesting sites) and African wintering countries on the Atlantic flyway that the scope of any Agreement eventually concluded would inevitably be incomplete, unless major Range States which are not Parties to the Convention can be persuaded to join the Agreement.

Conclusions

Progress is being made, albeit slowly, towards the more effective application of the Bonn Convention. The number of Parties, though still inadequate, is increasing, as are the Secretariat's very limited resources, and the Scientific Council is very active. The Contracting Parties only meet once every three years and have so far not been able to deal with problems of substance.

The situation is not helped by the Convention's two-tier system, by which Agreements must be separately negotiated, signed and ratified by the Contracting Parties concerned. On the other hand, this system is highly flexible in keeping with the enormous range of situations and migratory species covered by the Convention. The Convention does at least provide a framework and incentives for the conclusion of such Agreements, which would almost certainly not be negotiated if the Convention did not exist.

The fact that each Agreement generally needs to be formally ratified by each Party's legislature entails further delay, as well as uncertainty as to how urgent Parliaments will consider the need to improve the conservation status of, say, bats. One option might be for the Government departments responsible for wildlife protection in the States concerned to conclude non-binding simplified agreements as an interim step to the conclusion of official Agreements. Interim agreements would at least establish a basis for inter-State cooperation for the conservation of migratory species, which hardly exists at present.

This simplified interim procedure would not, of course, be suited to the negotiation of financial obligations: commitments of this kind could only be undertaken within a formally ratified international Agreement. Nevertheless, it is conceivable that part of the Convention's Budget might be used to finance interim agreements, and that the States concerned might make voluntary contributions to the Budget for this purpose.

f. The North American System

One example of an informal arrangement which does work in practice is the system developed in North America, where an unofficial agreement between the Canadian and American Government departments responsible for fauna has contributed significantly to the conservation of ducks and geese.

The North American Waterfowl Management Plan covers 37 species, mostly hunted as game. Signed in 1985, this agreement establishes a framework for long-term planning and sets targets for the restoration of the bird populations concerned, essentially through protection, restoration and management of their habitats. The legal basis for this agreement is supplied by the Treaties of 1916 and 1936, concluded by the United States on the one hand and Canada and Mexico on the other for the protection of migratory birds.[8]

The Plan was extended in 1988 to Mexico by the conclusion of a "Memorandum of Understanding" between the Government services responsible for wild fauna in the three countries. The Memorandum is intended to prepare and implement a global conservation strategy for migratory birds and their habitats, in application of the two treaties of 1916 and 1936. A tripartite Commission will be established to draw up this strategy, together with coordinated management plans, where appropriate.

In the United States, the North American Wetlands Conservation Act was passed on 13 December 1989 specifically to implement this tripartite agreement and to enable the federal US Government to finance conservation projects in Canada and Mexico. The necessary funds are raised by a tax on hunting weapons and ammunition purchased in the United States, and through a budgetary contribution approved by Congress. The most innovative aspect of the Act is that 50% to 70% of the available funds must imperatively be devoted to financing projects in Canada and Mexico, according to priorities determined in the light of the birds' needs.

The North American system works because of the existence of the two conventions on migratory birds, the close relations which have evolved between the relevant Government departments, especially in Canada and the United States, and the political will to put a stop to the constant reduction in the numbers of certain species of ducks and geese.

One of the reasons why Canada and the United States have not ratified the Bonn Convention seems to be that the necessary legal framework is already available to them to conserve migratory birds. However, although the present system does appear sufficient for ducks and geese whose wintering range rarely extends beyond Mexico, it remains totally inadequate for the very many other species of birds which migrate much further south.

The coordination of the necessary conservation measures is obviously much easier between three countries than within the Western Palaearctic region, where dozens of countries are concerned. The proposed Agreement on Western Palaearctic waterbirds would seem to be essential if the numbers of ducks and geese are at least to be stabilised at their current levels.

Any Western Palaearctic Agreement should follow the broad lines of the North American system. In view of the economic and social importance of waterfowl hunting in most European countries, it would seem only fair that hunters in these countries should contribute to the conservation of these species' habitats in the countries where the birds winter and possibly nest,

---

[8] In contrast to these early treaties which only regulate the taking of birds, the Plan pays particular attention to habitat conservation.

as hunters in the United States now do. Such an Agreement will probably fail if it does not provide for some transfer of resources from northern to southern countries for the conservation of these birds' habitats.

Indeed, the transfer of financial resources may be equally pertinent to many Agreements concluded under the Bonn Convention at some time in the future, whether these are concerned with conservation alone or with the exploitation of species as well. The negotiators of the Convention ensured that it was legally possible to conclude treaties for the exploitation and management of migratory species of commercial importance as Agreements under the Convention, despite some opposition during negotiations from those responsible for marine fisheries.

The practice is, of course, another matter at present. However, if the principle of international co-financing for necessary conservation measures were to be recognised by the Conference of the Parties as an essential precondition to the preparation of Agreements, it is likely that the number of Parties to the Convention would increase substantially.

B. The Mechanisms of Species Conservation Treaties

1. The Technique of Species Conservation Treaties

Most treaties dealing with the protection of species impose obligations on Contracting Parties as to the means of such protection. Species are generally listed in appendices, a separate appendix being used for each different legal category (e.g. "fully protected" and "partly protected") attributed to the species in question.

Since the number of species in need of protection increases constantly as time goes by, there is a danger of the lists of species contained in appendices becoming ever longer and more obscure. The technique of the 'negative' or 'inverted' list is therefore becoming more common, whereby a whole group of plants or animals are listed as protected, with a certain number of specifically named exceptions for those species which may be hunted or destroyed.

Conservation conventions often contain detailed rules for the protection of species, especially with respect to prohibited methods of taking and the terms and conditions of permissible taking. Taking and trade restrictions are applied with varying degrees of stringency to each of the appendices: taking may be banned completely for the species on one appendix, whilst being minimised for species on another appendix by means of controls on their hunting, fishing and collection. The Berne Convention and the EC Directives have separate Appendices specifying prohibited methods of taking.

Conventions almost always provide for a simplified procedure for the amendment of appendices. This is most important for the practical effectiveness of a convention. If normal ratification procedures had to be followed, it would take a very long time before any amendment to an appendix could enter into force, as each amendment would first have to be individually ratified by a majority of Parties. In practice, the need to list a species as "protected" is often urgent, and is moreover a minor matter which does not constitutionally require ratification.

This simplified procedure makes tacit consent sufficient. Once the amendment has been adopted by the Parties at one of their regular meetings by a majority laid down by the convention in question (usually a qualified majority of two-thirds), the Parties have a certain period, generally three or six months, within which to make objections or reservations. The latter terms are the same but the terminology used varies from one convention to another.

The amendment to the appendix enters into force for all Parties, and is therefore binding upon them at the expiry of this period, including those Parties which were not present at the meeting or who abstained or voted against the amendment. The only exception concerns those Parties which have entered formal objections or reservations during the prescribed time period.

CITES permits amendments to appendices to be made by postal vote. This is extremely useful as it allows for quick action in the case of an emergency, without waiting for the next Conference of the Parties which may not be due for another two years.

Conservation conventions generally leave States free to adopt measures which are stricter than those which they have undertaken to apply. Such a provision is necessary in order to prevent those States which are most advanced in terms of the conservation of biological diversity being obliged to fall into line with the less advanced States. Where this kind of clause exists, the convention must be considered as a set of minimum rules that all Parties have agreed to apply, instead of a maximum number of rules which cannot be exceeded.

Interestingly, this type of provision is found in the EC Birds Directive and in the EC Regulation implementing CITES, whereas in most other spheres, one of the fundamental goals of the Community is to bring national legislations closer together.

2. The Implementation of Species Conservation Treaties

Nature conservation treaties are law-making treaties, setting forth rules of general application, unlike the contract-making treaties, such as trade agreements, which contain reciprocal obligations. Whereas non-compliance by one of the Parties to a contract-treaty may be sanctioned by the withdrawal of an advantage conceded by another Party, no such sanctions exist under law-making treaties. If one Party allows a protected species to be destroyed, other Parties cannot obtain satisfaction by doing the same thing.

Conventions for the conservation of biological diversity, like other law-making treaties such as those dealing with the protection of human rights, are therefore difficult to enforce in the absence of the possibility of effective retaliatory measures under the treaty itself. Where retaliation is attempted by other means, this raises difficult issues and may even be of questionable legality, particularly with reference to the GATT.

In a recent and well-publicised case, the United States imposed restrictions upon imports of tuna from Mexico, in a move intended to stop the incidental taking and drowning of dolphins in the nets of Mexican tuna fishing boats. Mexico complained to GATT on the grounds that this constituted an unpermissible trade barrier and this position was upheld by the GATT Panel to which the case was submitted. If trade restrictions in the interest of the conservation of biological diversity are to be made possible, the GATT rules would have to be amended accordingly.

The question does not, however, seem to have arisen for CITES, the legality of which in relation to GATT has not so far been challenged. Although CITES itself does not provide for any possibilities of retaliation, Parties have in practice on several occasions been asked not to accept imports of specimens from countries not complying with the Convention. There are no powers under the Convention to force Parties to do so, but these import prohibitions have nevertheless been quite effective.

Conservation conventions are little more than the sum of unilateral commitments, the non-performance of which causes no damage to other Parties. Enforcement is therefore difficult. As all or most Parties are usually guilty of breaching the provisions of such conventions, they are accordingly reluctant to exercise pressure on other Parties to obtain compliance. Furthermore, no inspection system exists as this would of course impinge on national sovereignty. Parties are excessively hesitant about accepting any system for the verification of the implementation of a convention within their territory. The rare treaties which do provide for some form of inspection all concern marine areas, such as the Whaling Convention and some fishing conventions, as well as the Convention on the Conservation of Antarctic Marine Living Resources and the Antarctic Treaty.

Given these difficulties, it does seem that the only way in which a conservation convention can be reasonably effective is to provide it with bodies that have powers to follow its application. In most cases, this means a Conference of the Parties or a Standing Committee that meets regularly and is entrusted at each of its sessions with reviewing the working and implementation of the convention, amending its appendices if appropriate and making to one or more Parties any recommendation it may deem useful. A Secretariat is also necessary to service the Conferences and to establish the essential links between Parties in the periods between sessions. Finally, the smooth working of these bodies requires a budget which must be sufficient to enable them to discharge the functions which have been devolved to them by the convention.

Where there is a regular Conference of the Parties, a commonly-used method is to ask the Parties to submit periodical reports, to which wide publicity can be given, and to have these reports discussed by the Conference. Non-compliance with the obligations under these conventions will thus generally emerge in debate and the Conference may then adopt recommendations to the effect that the Parties concerned should remedy the situation.

Certain recent conventions also provide that representatives of non-governmental organisations—such as nature conservation organisations—may be admitted as observers to the Conferences of the Parties under certain conditions. This is indisputably a victory for democracy. Since debates are public, any breaches will become public knowledge. States may hesitate to sully their image when they know that the press and public opinion could ask them to give an account of themselves.

However, all these measures to facilitate the application of conventions are only found in some of the most recent ones, which are also generally the ones which work best. Those conventions which do not have governing bodies, a Secretariat and a budget, remain 'sleeping conventions', almost totally devoid of practical effectiveness. These include the Convention of Nature Protection and Wildlife Preservation in the Western Hemisphere of 1940 and the African Convention on the Conservation of Nature and Natural Resources of 1968.

Secretariats are generally provided by an existing organisation, which not only saves money but also simplifies administration. The Secretariats of the major conventions are respectively provided by UNESCO (the World Heritage Convention), IUCN (Ramsar), the Council of Europe (the Berne Convention) and UNEP (CITES, the Bonn Convention, the Regional Seas Conventions and the Biodiversity Convention).

It remains the case that conservation conventions must survive on little money and skeleton staffs, as the Parties are generally unwilling to contribute to more than a bare-bones budget.

C. Exploitation Treaties

Exploitation treaties have quite different objectives from those of conservation treaties and have accordingly not evolved over the years to remotely the same extent.

The purpose of exploitation treaties is the conservation not of biological diversity but of the basis for an economic activity. Where this activity concerns shared resources, a treaty is obviously essential if joint regulatory measures are to be adopted and implemented. In the absence of such measures, it is impossible to prevent over-exploitation of the resource in question.

Some exploitation treaties deal with fishing in boundary waters or international rivers, but most are concerned with fisheries at sea. Sea fisheries treaties were essential before the extension of the EEZ, and therefore of national fishery limits, to 200 miles, but have lost much of their importance since then. However, by way of exception, treaties remain most important for those species that are taken in the high seas, such as tuna and whales.

The implementation of fisheries treaties is always overseen by a Commission, which is usually set up as a small independent international organisation with its own Secretariat financed by the Parties. The Secretariat is therefore not generally provided by an existing international organisation.

Such a Commission usually has important powers. It can adopt regulatory measures which become binding on the Parties after a time period during which they can make objections or reservations. The system is the same as that applicable to the amendment of lists of species under conservation conventions, but it applies here to the whole range of regulatory measures that the Commission is empowered to take.

These regulatory measures may vary from one agreement to another. They usually include the establishment of closed areas and seasons, gear restrictions and catch limitations, including total protection for certain species. There are hardly ever any habitat protection measures.

Some of the treaties also provide for inspection procedures or the presence of independent observers on board fishing vessels, as in the Whaling Convention and the Convention on the Conservation of Antarctic Marine Living Resources which will now be examined in greater detail.

1. The International Convention for the Regulation of Whaling

The International Convention for the Regulation of Whaling, signed in Washington on 2 December 1946, applies to "all waters where whaling is prosecuted", which accordingly includes waters under national jurisdiction, including the territorial sea and internal waters. No definition of whales is given. The Convention has been interpreted by the Parties as covering only baleen and sperm whales and a few others: most small cetaceans are therefore excluded.

The objectives of the Convention are

"to provide for the proper conservation of whale stocks and thus make possible the orderly development of the whaling industry."

In earlier years, however, the Commission set quotas which were too high, which meant that it failed to prevent the near-extinction of most species. It has therefore been obliged to extend complete protection to the species concerned, one after another, which has resulted in the demise of the very industry that the treaty purported to develop!

In consequence, the International Whaling Convention has gradually been transformed from a commercial exploitation treaty into a conservation treaty.

Conservation measures are in practice facilitated by a feature unusual in a exploitation treaty. The Convention is open to all States and is therefore not restricted to whaling nations. This provision does appear to have been included deliberately, as the Preamble states that signatories recognise

"the interest of the nations of the world in safeguarding for future generations the great natural resources represented by the whale stocks."

As stocks continued to be depleted and the plight of the great whales aroused increasing public concern throughout the world, many non-whaling nations became Parties, including landlocked States such as Switzerland. The three-quarters majority required in the Commission for the adoption of regulatory measures was therefore rapidly converted into a majority for conservation.

As a result of this change in composition, the majority of non-whaling States were able to impose their will on the very small minority of States that continued to whale. Of course, the latter States could always lodge objections to decisions taken by the Commission, which meant that they were not thereafter bound by such decisions: this course was often followed. However, such tactics damaged the public image of those States as their objections were widely publicised. Iceland has now denounced the treaty and is therefore no longer bound by its obligations, and others may follow.

2. The Convention on the Conservation of Antarctic Marine Living Resources (CCAMLR)

The CCAMLR treaty, signed in Canberra on 20 May 1980, is only open to States

"interested in research or harvesting activities in relation to the marine living resources to which the Convention applies."

These resources include all living organisms found south of the Antarctic Convergence.[9] Whales and seals are excluded, however, as these species are covered by the Whaling and Antarctic Seals Conventions respectively.

The Commission established by the Convention has considerable powers. It may designate protected species and the quantity of any species that may be harvested, specify open and closed seasons and areas and also regulate fishing activities and methods of harvesting. These types of measures are more or less standard in fisheries conventions, as indicated earlier. In addition, the

---

[9] The Antarctic Convergence is the meeting-place of the cold waters of the Antarctic and the warm waters from the north. It acts literally as a biological barrier separating two different ecosystems inhabited by different species. The Convergence moves slightly from year to year, but the CCAMLR treaty delimits it by fixed geographical coordinates.

Commission may take other conservation measures that it considers necessary for the fulfilment of the objectives of the Convention. These may include measures related to the effects of harvesting and associated activities on the components of the marine ecosystem other than the harvested populations.

Uniquely amongst fisheries treaties, the Convention also lays down general principles of conservation which must govern any harvesting and associated activities in the area covered by the Convention (article II(3)). These include:

- the prevention of the decrease in the size of any harvested population to levels below those which ensure its stable recruitment;

- the maintenance of the ecological relationships between harvested, dependent and related populations of Antarctic marine resources;

- the prevention or minimisation of the risk of changes in the marine ecosystem which are not potentially reversible over two or three decades.

The global objective is to make possible the sustained conservation of these resources.

CCAMLR is therefore the first and only fisheries treaty explicitly intended to function as an ecological treaty as well. It is accordingly much more than a commercial exploitation treaty and constitutes a milestone in this field.

Unfortunately, decisions of the Commission must be taken by consensus and this does not facilitate the taking of sound conservation-based decisions.

3. Incidental Taking by Large Drift Nets

Recent technological developments have made it possible to make very long drift nets of up to 50 to 60 km in length. These nets catch large numbers of non-target species, including sea birds and cetaceans.

There is nothing in existing international legal instruments to prohibit the use of these nets in the high seas, where the principle of the freedom of fishing continues to apply. However, widespread public opposition, together with fears of the considerable over-exploitation of certain marine resources as a result of this method of fishing, have now produced some results. Many States have introduced national legislation to prohibit the use of large drift nets by their nationals within their EEZ.

Nevertheless, the problem remains acute in the high sea, particularly in the Pacific. As regards the North Pacific, the situation seems to have been resolved by a trilateral agreement between Canada, Japan and the United States, and by bilateral agreements concluded between the United States and Korea and the United States and Taiwan respectively.

In the South Pacific, a Convention concluded in Wellington on 24 November 1989 requires its Parties to prohibit the use of drift nets longer than 2.5 km in areas under their jurisdiction. Parties may also prohibit the landing, processing and importing of fish taken with drift nets anywhere, as well as the possession on board any fishing vessel of such nets within these areas. Doubts have nevertheless been raised as to the compatibility of these provisions with the Convention on the Law of the Sea and with GATT.

There remained the high seas which could not be covered by these instruments. A number of States therefore decided to put the matter to the United Nations General Assembly. It is an unusual procedure for a technical question of this kind to be submitted to that political body, but it was successful.

Following long negotiations, a Resolution was finally adopted by consensus on 15 December 1989 (Res.44(225)) on the basis of a compromise. This Resolution recommends a moratorium on the use of large drift nets in the high seas as from 1 July 1991. It is understood that the prohibition may not be imposed or may be lifted in a given region, if effective conservation and management measures have been taken on the basis of a statistical analysis carried out jointly by the members of the international community having fishing interests in the region.

The Resolution has been implemented by most countries whose fishing fleets use large drift nets. It is of course not binding, unlike a treaty proper, but a Resolution of the U.N. General Assembly does have considerable moral force and in this case would appear to have been effective. In contrast, a treaty would have taken years to be negotiated, signed and ratified.

This new approach to the solution of a problem affecting shared natural resources therefore constitutes a very useful precedent. It should enable a consensus to be reached in the political arena, especially where a quick solution to the problem is urgently required.

4. Conventions for the Regulation of Sealing

As most species of seals inhabit coastal areas, only a small number of species benefit from international protection measures. Certain migratory species of seals are protected under four specific conventions.

In the Atlantic Ocean, the Agreement on Measures to Regulate Sealing and to Protect Seal Stocks in the Northeastern Part of the Atlantic Ocean of 1957 and the Agreement on Sealing and the Conservation of Seal Stocks in the Northwest Atlantic of 1971 have lost much of their importance since the extension of national jurisdictional boundaries to 200 miles. In the Pacific Ocean, the Interim Convention on the Conservation of North Pacific Fur Seals was concluded in 1957 and periodically renewed until 1984, but has now expired.

In the Antarctic, the Parties to the Antarctic Treaty of 1959 signed a Convention for the Conservation of Antarctic Seals in London in 1972. The objectives of this Convention are the protection, scientific study and rational use of all species of seal present in the region and the maintenance of a satisfactory balance within the ecological system.

Certain other seal species and their breeding grounds are also protected under both regional and global conservation conventions, including CITES, Berne and Bonn.

D. The Convention on Biological Diversity

It must be stressed that under the Convention on Biological Diversity, very few provisions are devoted to the protection of species as such. This is the consequence of the emphasis being firmly placed on the conservation of ecosystems as part of a global approach to conservation. More specifically, the Convention contains no list of species to be protected. However, there is of course nothing to prevent the adoption of Protocols or Annexes listing protected species at a later date, should the need arise and the Parties agree to do so.

Other than the above-mentioned requirement to maintain viable populations of species in their natural surroundings, the Convention establishes the following obligations with regard to the protection of species:

- the identification and monitoring of species considered to be important elements of biological diversity.

An indicative list of categories of species which may meet that criterion is contained in Annex I. These species may be considered as priority species for conservation. The categories of species include threatened species; wild relatives of domesticated or cultivated species; species of medicinal, agricultural or other economic value; species of social, scientific or cultural importance or of importance to research into the conservation and sustainable use of biodiversity, such as indicator species; and within species, described genomes and genes of social, scientific and economic importance.

This list is not much of a guide, however, as many species will meet at least one of these criteria or could at least be considered to meet them potentially.

- the development or maintenance of necessary legislation for the protection of threatened species and populations. This is the only substantive obligation with respect to the protection of such species.

- the adoption of measures for the *ex situ* conservation of components of biological diversity, preferably in their countries of origin. It is made clear that this obligation is primarily for the purpose of complementing the *in situ* measures, referred to above.

- general provisions are also laid down for the sustainable use of harvested species: Parties should in particular "adopt measures relating to the use of biological resources to avoid or minimize adverse impacts on biological diversity".

This provision should be understood in a very broad sense, covering the over-exploitation of target species, the effect of the taking of target species upon non-target species and the ecosystem as a whole, and the problem of incidental taking.

It should be noted that this provision could be used as the world-wide legal basis for the prohibition of large drift nets.

Other provisions of the Convention, such as those dealing with access to genetic resources, have already been discussed in section (D)(5)(b) of the Introduction. The fact that access may no longer be free of charge should constitute an incentive to conserve biological diversity, particularly since an equitable share of the benefits derived from the use of genetic resources must now go back to the country which has provided the resource.

E. Rights and Responsibilities of States for the Conservation of Species

1. The Foundations and Definition of Responsibility

The responsibility accepted by one State under a treaty may be towards other States, towards the Range States of certain species, such as migratory species, or towards future generations. Such responsibility is clearly always of an international nature.

The foundations of such responsibility are established by the preambles to conventions which list the general values of wild fauna and flora which are considered to justify action for their conservation. These include not only utilitarian values, but also scientific, cultural, recreational and economic values.

In a few cases, such as the Berne and Biological Diversity Conventions, the intrinsic value of wild fauna and flora has now been explicitly acknowledged. This affirmation of intrinsic value may be interpreted as the formal recognition by States of the right of species to exist, quite independently of their usefulness to humankind.[10]

The recognition of responsibility under a treaty automatically entails some limitation of sovereign rights. However, it is relatively rare for instruments to refer explicitly to responsibility or to define its scope. More usually, such responsibility must be inferred from the species' values which are listed in the preamble and from the substantive obligations set out in the articles of the convention.

Nevertheless, certain conventions, all concerning migratory species, do make use of the term. It appears, for instance, at article 2.6 of the Ramsar Convention in connection with the migratory nature of waterfowl:

"Each Contracting Party shall consider its international responsibilities for the conservation, management and wise use of migratory stocks of waterfowl, both when designating entries for the List and when exercising its right to change entries in the List relating to wetlands within its territory."

The Preamble to the Agreement on the Conservation of Polar Bears of 1973 recognises "the special responsibilities and special interests of the States of the Arctic Region in relation to the protection of the fauna and flora of the Arctic Region". The EC Birds Directive states that the effective protection of migratory bird species is typically a transfrontier environmental problem entailing common responsibilities.

In respect of marine species, the Preamble to the Convention on the Conservation of Antarctic Marine Living Resources of 1980 recognises

"the prime responsibilities of the Antarctic Treaty Consultative Parties for the protection and preservation of the Antarctic environment and, in particular, their responsibilities under Article IX, paragraph l(f) of the Antarctic Treaty in respect of the preservation and conservation of living resources in Antarctica."

In all the above cases, the limitation of sovereignty implied by this recognition of responsibility proceeds from the fact that other States along the migration route also possess sovereign rights over the same species.

This concept of international responsibility can be stretched to shared species. The sovereignty of each Range State is limited *de facto*, because the action or inaction of any one Range State is liable to affect the right of the other Range States to the continued survival of the species concerned. This right arises from their own sovereignty.

---

[10] As mentioned in the Introduction, this principle was incorporated into soft law in 1982 by the World Charter for Nature, which states that "every form of life is unique, warranting respect regardless of its worth for man."

However, this reasoning is not applicable to endemic species, because no other individual State has a legal interest in the their survival. The only possible international interest is that of the world community as a whole in the avoidance of the extinction of such species.

This interest, the common interest of humankind in the survival of species, is gradually being formally recognised. This is confirmed by the fact that at least three treaties now establish special responsibilities for the preservation of endemic species.

The African Convention on the Conservation of Nature and Natural Resources of 1968 provides that where animal and plant species are threatened with extinction, and are

"represented only in the territory of one Contracting State, that State has a particular responsibility for its protection" (Article VIII).

The Berne Convention of 1979 requires Parties to pay particular attention to endangered and vulnerable species, especially endemic ones (article 3). The ASEAN Agreement requires Contracting Parties to recognise their special responsibility in respect of species that are endemic to areas under their jurisdiction.

Finally, the Preamble to the Convention on Biological Diversity provides that States are responsible for conserving their biological diversity and for using their biological resources in a sustainable manner. This means that responsibility for all species, including endemics, has now been recognised in a world treaty, notwithstanding the principle of national sovereignty.

2. Problems of Developing an International Status for Species

The acceptance of responsibility is one thing, but the recognition of a special international status for species is quite another. The acceptance of responsibility is presently construed as a voluntary limitation upon national sovereignty, arising on a case by case basis as a result of freely negotiated treaties. In contrast, the recognition of an international status for species could mean that future treaties ceased to view the limitation of sovereignty as voluntary. Such limitation would instead be imposed upon all nations by the world community.

In consequence, the recognition even of migratory species as common resources has posed great difficulties, as such a step was felt to impinge directly on sovereignty.

Nevertheless, some treaties or other instruments have now recognised the concept of common resources, although their use of the term varies widely. In the Preamble to the Ramsar Convention, for example, the Parties recognise that

"waterfowl in their seasonal migrations may transcend frontiers and so should be regarded as an international resource."

The EC Birds Directive states that migratory species constitute a "common heritage". The Bonn Convention included the concept of "common resources" in preliminary drafts, but the negotiators disagreed and the term was finally removed from the text.

It is noteworthy that the preceding examples are restricted to migratory species. With regard to species in general, the furthest reaching text is the International Undertaking on Plant Genetic Resources, which was adopted by the FAO Conference in November 1983. This document proclaims that it "is based on the universally accepted principles that plant genetic resources are a heritage of mankind and consequently should be available without restriction."

The document also provides that

"appropriate legislative and other measures will be maintained and, where necessary, developed and adopted to protect and preserve the plant genetic resources of plants growing in areas of their natural habitat in the major centres of genetic diversity."

This text is, however, nothing more than a Resolution of the FAO Conference and is accordingly not legally binding. Nonetheless, it is of great interest, as it is unique in combining the recognition of genetic resources as the heritage of mankind with provision for freedom of access and the obligation to conserve. The Undertaking is of course limited to plants.

The new Convention on Biological Diversity leans much more to the side of national sovereignty. Biological diversity is thus only stated to be the "common concern of humankind", and the proposal to use the term, "common heritage", was rejected outright.

3. Performance Obligations arising from State Responsibility

Even where treaties do not state this explicitly, responsibility is implied to entail performance obligations to avoid the extinction of species or to maintain their populations at favourable conservation levels. These fundamental requirements appear in most conservation conventions.

An early example is the Washington Convention of 1940 on Nature Protection and Wildlife Preservation in the Western Hemisphere. Although there is no substantive obligation in the text, the Preamble sets out a clear declaration of policy:

"The Governments of the American Republics, wishing to protect and preserve in their natural habitat representatives of all species and genera of their native flora and fauna, including migratory birds, in sufficient numbers and over areas extensive enough to assure them from becoming extinct through any agency within man's control."

One of the most recent conventions, the ASEAN Agreement, states that:

"the Contracting Parties shall, wherever possible, maintain maximum genetic diversity by taking action aimed at ensuring the survival and promoting the conservation of all species under their jurisdiction or control."

It is interesting to note that although there are 45 years separating these two conventions, there is very little difference between the two texts.

Performance obligations are important not only because they flow directly from a State's recognition of responsibility, whether this is explicitly stated or only implied, but also because they provide a clear legal basis for the development of concrete measures for the conservation of species. It is of course in this latter, more practical area of conservation methods that early and more recent conservation conventions differ significantly.

This evolution in international conservation law is also illustrated by the way in which treaties increasingly deal with both conservation and exploitation issues. The effective transformation of the International Whaling Convention from a commercial exploitation treaty to a conservation treaty has already been mentioned in section (C) above. More recently, the Convention on Biological Diversity combines provisions for the conservation of biological diversity with measures for the sustainable use of its components. "Sustainable use" is defined by the Convention as

> "the use of components of biological diversity in a way and at a rate that does not lead to the long-term decline of biological diversity, thereby maintaining its potential to meet the needs and aspirations of present and future generations."

This definition, together with the substantive provisions of the Convention, imposes an unambiguous performance obligation upon Parties to ensure that measures regulating the use of wild species and natural habitats are designed to prevent such a decline of biological diversity.

4.   The Special Case of Migratory Species

The rules of sovereignty apply normally to both sedentary and migratory species, unless a treaty is concluded to the opposite effect. Migrants therefore come under the successive sovereignty of all States along the migration route. The effective conservation of migratory species therefore requires the conclusion of treaties to secure international cooperation between Range States.

The conservation of migratory species may be particularly problematic where some countries within the migratory range implement conservation measures, often at great cost, whilst others allow species to be exploited unsustainably or important habitats to be destroyed. The situation is most acute as regards anadromous species, namely species which breed in inland waters yet spend their adult life in the high sea beyond the limits of national jurisdiction.

Until recently, the law of the sea permitted these species to be harvested in the high sea without restriction, unless a treaty provided for the contrary. Any such treaty would, in any event, only have been binding on its Parties. Non-Parties accordingly remained free to proceed to any form of exploitation. Unsurprisingly, disputes arose including with regard to salmon in the North Atlantic and the North Pacific, as the countries of origin strongly objected to the uncontrolled fishing of salmon in the high sea. These disputes were settled by agreement, but the principle of the freedom of fishing remained intact. Fisheries resources ceased to be national as soon as the limits of national jurisdiction were crossed.

However, this principle was abandoned in the Law of the Sea Convention of 1982, which provides that

> "States in whose waters anadromous stocks originate shall have the primary interest in and responsibility for such stocks."

The fishing of these species in the high sea is prohibited for all States, including those in which the stocks originate. Their fishing is henceforth only permitted in the EEZ.

The Law of the Sea Convention also extends these rules to catadromous species, namely species which perform the reverse migration, such as eels. They breed in the high seas and spend their adult life in inland waters.

5.   The Rights of States over Species

For as long as sovereignty was limited to individual animals and plants, it did not interfere with the right to utilise genetic resources for scientific research or commercial or other applications. Such resources were considered to be the free products of nature, incapable of being owned or patented. Only those processes derived from genetic characteristics could be patented.

Admittedly, States have always had the right to control the collection of specimens on their territory or to charge levies upon such collection. Some countries have even prohibited the export of certain resources, such as Brazil in respect of the seeds of the rubber tree. However, once an illegal export had occurred, the State had no claim against the producers or users of the resource outside its national boundaries, because the State had no recognised right over that resource.

The result of this approach was that, in the absence of any law governing this matter, genetic processes were considered in practice to constitute a common heritage. Moreover, no obligations were imposed on the user of such processes. In practice, freely collected material was often kept in gene banks in other countries, which meant it was no longer available to countries of origin. Huge profits might sometimes be made from the sale of products derived from such material, without any returns to the countries where the material had originally been collected.

An alternative system was proposed by IUCN in the draft convention on biological diversity which it prepared prior to the opening of the negotiations on the new Convention on Biological Diversity. This system would have guaranteed freedom of access to genetic resources, although not necessarily free of charge, whilst users would in return have had to pay a fee, at least in respect of the commercial use of any products, such as pharmaceuticals, derived from the natural resources in question. The amounts raised could have been paid to an international fund to finance conservation action where this was most urgently needed.

However, the negotiators of the Convention opted instead to assert the sovereignty of States over the genetic resources of the species inhabiting their territory. This is tantamount to establishing national sovereignty over species, as distinct from the individual animals and plants which compose the species.

As mentioned in the Introduction, the right of access to genetic resources, governed by article 15 of the Convention on Biological Diversity, has been made subject to the consent of the State concerned. "The results of research and development, and the benefits arising from the commercial and other utilization of genetic resources" must be shared in a fair and equitable way with the Contracting Party providing such resources.

In other words, States are now deemed by the Convention to be the holders of rights over the species that live within their boundaries. States have the right to refuse other countries access to the genetic resources of those species. Although this right already existed, as mentioned above, it had never been set out explicitly in a legally binding instrument. Its inclusion in the new Convention means that the non-binding FAO International Undertaking on Plant Genetic Resources has now been eclipsed.

The Convention affirms the right of States to require payment for the commercial application of genetic resources obtained from their territory, which is very new. This provision is the logical extension of the acceptance that countries of origin of genetic material hold rights over the genes and processes derived from such material, especially where these are used in commercial applications.

The formalisation of such rights, equitable as it may seem, may nevertheless give rise to considerable practical difficulties, particularly where the genetic material used in a commercial application is found in a species which is present in several countries, as is more than likely.

Many legal and practical problems may arise. Is it fair, in particular, only to make payments to the country which has actually provided the resource, as the Convention requires, and not also to those countries where the same resource is found *in situ*? How will it be possible to determine the true country of origin in the case of a dispute between Range States over the actual provenance of the same genetic resources? Can a country where a species has been introduced into the wild be fairly considered as a country of origin under the Convention? What sort of recourse will a country of origin have if the genetic material has been smuggled out of the country? What happens where the country of origin is unknown?

The Convention does not answer most of these questions. Disputes are therefore bound to occur, as to the determination of the actual holders of rights over the resources and difficult problems of proof may arise. In this context, it should be mentioned that article 27 of the Convention does provide for a dispute settlement procedure.

CHAPTER II
THE SCOPE OF STATE POWERS TO CONSERVE WILD SPECIES

A. The Legal Basis for Legislation to Conserve Species

1. The Legal Status of Wild Animals and Plants

Many traditional societies considered that the natural resources they exploited for their subsistence were their collective property. As a result, such societies did not allow outsiders to trespass into their hunting or fishing grounds, and they developed their own rules for the rational exploitation and sharing of these resources among their members. Breaches of these rules were generally punished severely.

In contrast, wild animals were considered under ancient Roman law to be res nullius, meaning things which belonged to no-one. As a result, wild animals could be freely appropriated by any person who was able to take them into his possession. As many if not most existing legal systems have their roots in Roman law, the res nullius concept has gradually gained almost universal recognition. The concept of collective ownership of harvestable resources has therefore almost completely disappeared as traditional societies become increasingly disrupted and the closed systems in which they operated are opened up to outsiders.

The res nullius concept obviously had its own logic. Animals being essentially mobile, they cannot be considered as attached to the land on which they happen to be at any given moment in time. It follows that they cannot be owned by the owner of that land or anyone else.

As wild flora and fauna is increasingly recognised as a valuable part of the national heritage, their legal status has in some cases been changed accordingly, with the res nullius concept being replaced by the classification of State ownership. This was and will most probably remain the case in China and in the former Socialist countries. In China, for instance, State ownership of wildlife is clearly established by article 3 of the Law on Wild Animals of 8 November 1988, which states that "wild animal resources belong to the nation".

Similar provisions appear in the laws of several African countries, such as Benin, Congo, Zaïre and Zambia, as well as in the laws of Brazil, Colombia and Mexico in Latin America, and in the Italian Hunting Act 1977 (now replaced by the Act of 31 January 1991). The laws of several Australian States (for example, New South Wales and Queensland), some American States (for instance, Florida, Ohio and Texas) and of certain Canadian Provinces (such as British Colombia) also proclaim the public ownership of wild fauna. The constitutionality of the State laws in question may be dubious, however, as the Supreme Court has on several occasions decided that neither the individual State nor the United States as such is the owner of wild animals.[11]

Nevertheless, some of the laws asserting State ownership of wild fauna are limited in their scope as they only refer to certain categories of animals. For example, the Fauna (Protection and Control) Act of Papua New Guinea limits State ownership to protected animals, whereas the Wildlife Protection Act 1987 of Saint Vincent and the Grenadines refers only to mammals, birds, reptiles, amphibians, fishes, and crustaceans. All other species, presumably, remain res nullius. Benin's Law on Nature Conservation and Hunting 1980 defines fauna as including mammals (except bats and rodents), birds, chelonians, pythons, monitor lizards, crocodiles and fishes. Only animals which belong to these groups are characterised as State property.

State ownership has, at least in theory, several advantages. Illegal taking counts as a theft of government property, and the State should normally have standing to claim civil damages from offenders in addition to any penalty that may be imposed for violations of the law. However, State ownership also has the major disadvantage, again in theory, that the State should be liable in civil law for the damages caused to human beings and property by the animals it owns. This disadvantage may be easily remedied by a disclaimer of liability in the legislation itself. Few jurisdictions have done this, however, as the problem of damage caused by wild animals is generally covered by specific rules. Nevertheless, the law of British Columbia specifically states that "no right of action lies, and no right of compensation exists, against the Crown in right of the Province for death, personal injury or property damage caused by wildlife" (Wildlife Act of 23 July 1982, section 2).

The advantages of the characterisation of State ownership are more apparent than real. The State is always entitled to impose penalties for the unlawful taking of wild animals, whether or not this is considered to be a theft, and to empower the appropriate agency to claim damages in the Courts for injury to the national heritage. All that is necessary is a legislative provision to that effect, rather than a change in the legal status of wildlife. Finally, as regards the psychological value of State ownership versus the res nullius concept, the very least one can say is that it is doubtful whether State ownership provides better protection for wild animals. Government property is all too likely to be perceived as the property of anyone and everyone, making the net result the same as if the wildlife had no owner at all.

The status of wild flora is very different in law from that of wild animals since plants, being rooted in the soil, are the property of the owner, whether public or private, of the land on which they grow. Unless plants are otherwise protected by legislation, landowners are therefore free to preserve or destroy plants growing on their land as they wish. On the other hand, the fact that plants are owned protects them, in principle, from being collected or destroyed by outsiders, except with the owner's agreement.

In practice, however, there are often customary rules which allow certain species or forest products to be picked, such as fungi, berries, or medicinal plants. Moreover, even though such picking technically counts as theft, it would be difficult in most cases to prosecute successfully the picker of wild plants of little or no commercial value, because this is regarded by the general public as part of the natural enjoyment of the countryside. The only exception would be if the case involved large quantities and a substantial profit for the collector. As a result, for all practical purposes wild plants are de facto also res nullius unless, like trees, they have a recognised commercial value.

---

[11] See for instance *Douglas v. Seacoast Products Inc.* (431 US 265).

2. The Regulatory Powers of the State

Whatever the legal status of wild animals and plants, the State is always entitled, in the general interest, to exercise its police powers by enacting legislation to prohibit or control the exploitation of any species of fauna or flora.

Such conservation measures are usually embodied in an Act of Parliament or equivalent legislative instrument, which specifies the categories of species or taxonomic groups to which it potentially applies and lays down rules for the conservation of those species which are expressly listed for that purpose. Such a List may be appended to the Act or established subsequently by means of a statutory instrument or other secondary legislation.

Whichever the case, the Act will specifically empower the Executive branch of the Government or a designated Minister or Authority to amend the List. Such powers are obviously limited since they can only be exercised in respect of the species which belong to the categories or taxonomic groups covered by the Act. It follows that other species cannot be protected unless the Act itself is amended to make this possible.

On the other hand, the fact that a species is covered by the Act does not mean that it has to be listed, but merely gives the appropriate authority the right to list the species should the need arise. In the same way, the authority concerned is obviously entitled to remove a species from the List, if the authority considers that such protection is no longer justified, or to downgrade the species to a lower level of protection.

Whatever the field in which the police powers of the State are exercised, their legitimacy is always based on the preservation of the general interest. As regards wildlife conservation, such powers were initially used (and to a considerable extent continue to be used) to preserve valuable economic or recreational resources against uncontrolled exploitation. Such exploitation, if left unregulated, would soon result in the depletion of the species concerned, inevitably followed by the decline or even the complete stoppage of important subsistence, economic or social activities. The legitimacy of species preservation was therefore seen only as incidental to the preservation of legitimate human pursuits.

This is the basis upon which hunting and fishing legislation was developed over the years throughout the world. Moreover, a limited number of countries have used the same basis for the adoption of legislation regulating the collection of certain plant resources such as medicinal plants or particular species of trees.

a. Hunting Legislation

Legislation regulating hunting was initially applicable only to those species which were considered to be game. These were usually limited to certain mammals and birds. No other species were covered by the law and could therefore not be protected.

A first step towards the inclusion of non-game species in legislation was the adoption in 1902 of the International Convention for the Protection of Birds Useful to Agriculture. The Convention protected about 150 species which were listed in an Annex, mostly passerines and owls. Their protection at national level, in the few countries which actually enacted legislation to implement the Convention, was achieved either by listing them as protected game species or by the adoption of special legislation. For example, Belgium and the Netherlands have specific Acts dealing with the protection of non-game birds which are still in force. In the United States and Canada, following the conclusion of the Convention for the Protection of Migratory Birds in 1916 between these two States, a large number of non-game bird species were also brought under protection.

As part of the gradual evolution of hunting legislation, many game species acquired greater protection through the establishment, by means of regulations, of permanent close seasons in respect of those species. In some cases, non-game species were added to the lists of protected species adopted under the hunting legislation. By way of example, bats, snakes, lizards and amphibians were already protected in Spain under the Hunting Act before the adoption of a Nature Conservation Act in 1989. Nevertheless, the legality of using legislation aimed at specific resources or activities to protect unconnected species is dubious, and could easily be challenged in the Courts.

The potential scope of hunting legislation is further limited because the legislation does not cover the taking of many animals for such purposes as private collection or the pet trade unless it contains an extensive definition of hunting and game species. Moreover, such legislation is obviously not applicable to species covered by fishing legislation or to plants.

b. Fishing Legislation

The scope of fishing legislation is generally broader than that of hunting legislation because animals belonging to many different taxonomic groups may constitute exploitable resources. Many fishing laws are therefore potentially applicable to all kinds of aquatic organisms. In practice, however, Fisheries Departments have only tended to regulate the exploitation of species of commercial or recreational importance.

c. Legislation protecting Plants

The protection of plants was traditionally possible only by means of forestry legislation, which usually provides for the collection of forest products in public forests to be regulated. This made it impossible to protect many plant species unless special legislation was enacted for that specific purpose. This seldom occurred. Examples of such special laws include the Italian laws on the collection of truffles and of medicinal plants (the latter dating back to 1931) and the Native Plants Protection Act 1939 of South Australia (now replaced by provisions of the National Parks and Wildlife Act 1972).

d. Nature Conservation Legislation

As an ever-increasing number of non-game and plant species were threatened with extinction or considerable depletion and the need to safeguard biological diversity was widely acknowledged to be a duty towards future generations, it became apparent in the late 1960s and early 1970s that existing sectoral legislation was insufficient to provide the protective measures that were then urgently required. What was needed was legislation that made any species eligible for protection, should the need arise.

This objective has generally been achieved either through the enactment of specific nature conservation legislation, which empowers the appropriate minister to list protected species, or by the amendment of existing hunting legislation in order to broaden the range of species eligible for protection.

The legitimacy of this new approach to the exercise of the State's police powers is no longer linked to safeguarding certain resources and activities, but is instead derived from the increasingly recognized need to preserve species and biological diversity for their intrinsic value and for the use of future generations.

Most of the laws which provide for the possibility of protecting species other than game are recent. In Europe, for instance, such laws have been enacted in Sweden (1964), Switzerland (1966), Norway (1970), Belgium (1973), France, Germany and Ireland (1976), Great Britain (1981), Portugal (1987), Spain (1989) and Italy (1992). In the United States, the federal Endangered Species Act was adopted in 1973, and State legislation protecting non-game species was mostly enacted in the 1970s and 1980s. Elsewhere in the world, Australian States adopted their nature conservation legislation for the most part in the 1970s, India and Malaysia their respective Wildlife Acts in 1972, Peru in 1977 and Colombia in 1981.

There are, of course, other countries which have enacted laws which provide for the possibility of protecting non-game species, but most of these are of limited scope, covering only mammals, birds, reptiles and, in some cases, amphibians. Fish are usually excluded, as they are covered by fishing legislation, and invertebrates continue to be generally ignored.

The extent to which wildlife legislation is potentially applicable to groups other than mammals and birds does not seem to obey any logic and may be only a reflection of the biases of the draughtsmen or of the legislative bodies concerned. To quote the example of a few States within the United States, the Law of Florida is only applicable to mammals, birds, reptiles and amphibians, those of Arizona, New Mexico and Nebraska to all vertebrates, molluscs and crustaceans, that of Wisconsin to all vertebrates, molluscs, arthropods and crustaceans (sic), whereas those of Hawaii, Michigan, Rhode Island and South Dakota apply to all members of the animal kingdom.

With regard to plants, comprehensive nature conservation or wildlife legislation often also empowers the Government to designate protected plants. The protection of wild plants, however, may give rise to problems as its legitimacy may be more difficult to establish.

Animals, being either res nullius or State property, can be protected against taking (although not against the destruction of their habitat) without any infringement of private property rights. In contrast, as wild plants are the property of the owner of the land on which they grow, their protection necessarily implies the imposition of restrictions on landowners' freedom to dispose of their property as they please. As a result, the legitimacy of regulatory measures taken for the protection of wild plants on private property meets with the obstacle of equally legitimate private property rights. This is certainly one of the reasons why legislation protecting wild plants has been slow to develop and why it is often less stringent than the legislation applicable to animals.

Nevertheless, the growing recognition that the public interest in the conservation of biological diversity should prevail over private interests has enabled these legal difficulties to be overcome gradually. There are now more and more countries, particularly in Europe, whose conservation legislation provides for the conservation of wild plants, even against the will of the landowners on whose land they grow. However, a large number of countries still have no legislation providing for the protection of wild plants anywhere other than in public forests.

In spite of considerable progress, mainly achieved during the past two decades, in the development of comprehensive wildlife legislation allowing for the protection of wild animal or plant species, many countries still continue to rely on their hunting, fishing and forestry legislation for such protection, with all the limitations that this approach usually entails with regard to species coverage.

Furthermore, jurisdictional problems may often constitute a considerable obstacle to the efficient and integrated conservation of wild species, as discussed below.

B. Jurisdiction over Species

The powers to designate protected species and to establish prohibitions or restrictions on the taking of, or trade in, species may be contained in separate enactments and entrusted to different Government agencies. The allocation of such powers may be based on the type of environment in which the species live, whether terrestrial, freshwater or marine, or on the legal status of the species, namely game or non-game species.

This separation of jurisdiction is the result of long-standing tradition. The protection of species only came about as the by-product of the regulation of specific activities, particularly hunting and fishing, which were regulated by separate laws almost everywhere.

The position is even more complicated in federal or regionalised States, where certain species may come under federal jurisdiction whilst others are under the jurisdiction of the federated or regional entities. In some countries, such as Italy, there may be concurrent jurisdiction of the two levels of Government. In other countries such as Austria and Australia, however, the federal Government may have virtually no jurisdiction at all over species.

This fragmentation of responsibility results in a highly complex situation, which varies from one country to another and which may lead to considerable difficulties in the conservation and management of wild species. It may also encourage many inconsistencies in the formulation and implementation of species protection legislation.

A general review of the most commonly-encountered problems is set out below. Subject-matter or functional jurisdiction and territorial jurisdiction will be examined in turn, before consideration is given to the question of jurisdiction over marine species.

1. Subject-Matter Jurisdiction

a. Jurisdiction Based on the Type of Environment

There has long been a jurisdictional split between terrestrial and marine species, which have traditionally been dealt with by hunting and marine fisheries legislation respectively. Freshwater species are in an intermediate position. Fisheries legislation may sometimes apply to fishing in any waters, or else a distinction may be made between fishing in inland waters and marine fisheries.

Even where one Ministry is responsible for both hunting and fisheries, including marine fisheries, the relevant laws are frequently administered by different divisions within that Ministry.

This jurisdictional system worked reasonably well when its main purpose was simply to regulate hunting and fishing. However, problems have now arisen with regard to the conservation of species as required under nature conservation legislation. Where aquatic species are covered by separate legislation, it is often impossible to protect them under conservation legislation. This is particularly true in the case of marine species.

In practice, there are often separate, and different, conservation rules applicable to terrestrial, freshwater and marine species. Problems of jurisdictional overlap also arise in connection with species which move from one type of environment to another, which may be especially difficult to resolve. The resistance of the relevant administrative bodies may also impede the process of change.

The principal difficulties inherent in this separation of jurisdiction will now be examined in greater detail.

i. The Inadequate Definition of Species

There is frequently no clear definition of the species covered by the legislation applicable to the terrestrial, freshwater or marine environments. This imprecision may either cause inter-departmental conflicts or may lead to the complete neglect of certain species. In the latter situation, each of the departments potentially concerned believes that it is the other department's responsibility to deal with the species in question.

The criteria set out in legal instruments to define the species to which they apply are often unclear and artificial. For example, many laws merely state that aquatic species come under the jurisdiction of the Ministry of Fisheries. However, this leaves a jurisdictional void around those species which spend part of their life cycle on land, unless they are expressly included in the definition of "aquatic" species.

Some laws are more specific. For example, the Decreto Supremo of 1 April 1977 in Peru confers jurisdiction on the Ministry of Agriculture over "all wild animals which reproduce themselves on land". This provision is fortunately clarified by a list of species thereby covered, which includes all mammals, except for cetaceans and sirenians (so seals and others do not come under the jurisdiction of the Fisheries Department); all birds (which therefore includes sea birds); and all reptiles, except for sea snakes (including all sea and freshwater turtles). However, although the foregoing is reasonably clear, the same cannot be said for the next item on the list in the Decreto, which concerns those amphibians which reproduce themselves on land. As the vast majority of amphibians lay their eggs in the water, it is unclear whether "reproduction" denotes the place where mating actually occurs.

ii. Jurisdictional Splits based on the Movements of Species

Jurisdiction may sometimes be split in accordance with the movements of a species from water to land and vice versa during its life cycle. Sea turtles are one such species. In the United States of America, the Department of the Interior has jurisdiction over sea turtles when on land, where they lay their eggs, whereas the Department of Commerce (National Marine Fisheries Service) assumes responsibility when the sea turtles are in the sea. The coordination of species protection and management may not be easy where this formal separation of jurisdiction exists.

A further example of species which move from one environment to another are anadromous species, such as salmon. These species may come under the administration responsible for inland fisheries when in rivers, and under the administration in charge of marine fisheries when at sea. This jurisdictional split occurs very frequently, making the unitary conservation and management of such species impossible or very difficult.[12]

iii. Administrative Resistance to Jurisdictional Changes

Where provision for species protection is made under nature conservation legislation, it is often very difficult, if not impossible, to remove species from the jurisdiction of the Fisheries Department in order to bring them within the scope of the conservation legislation.

---

[12] The treatment of migratory marine species under the Law of the Sea Convention is discussed at Chapter I(B)(4) above.

This difficulty is gradually being resolved in respect of freshwater species, which have traditionally occupied an intermediate position as mentioned above. Many, though by no means all, countries now list protected freshwater species, especially fish, under their nature conservation legislation.

However, it is still virtually impossible to transfer jurisdiction over marine species in this way, other than for those species which are particularly popular in the eyes of the public, such as cetaceans, seals and sea turtles. Even in the case of these species, such transfers are by no means the universal rule.

Marine fishes and invertebrates very rarely appear in lists of protected species, even those which are annexed to international conventions such as the Berne Convention of 1979 or CITES. Indeed, at the 8th Meeting of the Conference of the Parties to CITES in Kyoto in 1992, there was such resistance to the proposal to include the bluefin tuna (Thunnus thynnus) in Appendix II that the proposal was eventually withdrawn.

Further examples of such restrictive listing are provided by the two Protocols to the Regional Seas Conventions which provide for the protection of species. These are respectively the Protocol concerning Protected Areas and Wild Fauna and Flora in the Eastern African Region (adopted in Nairobi on 21 June 1985) and the Protocol concerning Specially Protected Areas and Wildlife to the Convention on the Protection and Development of the Marine Environment of the Wider Caribbean Region (adopted in Kingston on 17 January 1990).

The overwhelming majority of species protected under these Protocols are terrestrial. The only marine species listed are cetaceans and sea turtles, to which have been added sirenians and a small number of marine invertebrates, particularly molluscs and coral. There is not a single species of marine fish listed. In fact, a group of scientists had proposed a fairly large number of fish species for inclusion in the Annexes to the Kingston Protocol, but not one species was eventually retained.

The fundamental reason for this systematic exclusion of many marine species is the separation of authority between fisheries and conservation administrations, which are responsible for the implementation of completely different laws.

b. Jurisdiction over Game and Non-Game species

As mentioned earlier, the protection of animal species was initially effected through hunting legislation. In many countries, the wildlife conservation legislation that was enacted did not take the form of comprehensive legislation covering both the protection of species and the imposition of regulatory measures. Wildlife conservation measures were only enacted in respect of non-game species, whilst hunting legislation remained in force for game species.

This approach has usually resulted in two separate laws, often administered by separate departments. For example, the conservation of non-game species is most commonly the responsibility of the Ministry of the Environment, whilst the regulation of game species is carried out by the Ministry of Agriculture or the Forest Department, which often continue to have jurisdiction over hunting.

Hunting legislation was often enacted a long time ago, and these ancient laws are frequently still in force, albeit with some amendments. The species covered by such legislation are consequently those which could lawfully be hunted at the time when the relevant law was enacted, which were of course very numerous.

Protection has gradually been extended to many of these game species by the simple means of not declaring open seasons, the only period in which the hunting of such species would be permissible. As regards most of these species, it is unlikely that their taking will ever be authorised again.

Nevertheless, from the legal point of view, these species are still classified as game species. This means in practice that their taking may be authorised at any time by Order of the Minister responsible for hunting, who is under no obligation to consult with the relevant nature conservation authority.

Moreover, none of the rules of conservation legislation are applicable to hunting legislation. The latter has its own rules with regard to possession, taxidermy and the domestic trade in game species, which may be much less strict than those enacted under conservation legislation. Furthermore, hunting legislation seldom if ever deals with imports and exports.

Hunting and conservation legislation co-exist in this way in Austria, Belgium, Denmark, Germany, Luxembourg, the Netherlands and Switzerland amongst other countries. This dual system can only evolve very slowly, with the piecemeal transfer to conservation legislation of those game species for which hunting has not been permitted for a long time.

In 1976, many species in Germany were deleted from the list of game species set out in the Hunting Act of 29 November 1952, which entailed their automatic transfer to a list of species protected under the Nature Conservation Act 1976. Species transferred in this way included all herons (except for the grey heron (Ardea cinerea)), white stork (Ciconia ciconia) spoonbill (Platalea lencorodia), glossy ibis (Plegadis falcinellus), whooper swan (Cygnus cygnus), osprey (Pandion haliaetus), crane (Grus grus), all Alcids, beaver (Castor fiber), European mink (Mustela lutreola) and many others.

In Belgium, it was only in 1987 and 1992, in the Flemish and Walloon Regions respectively, that many species for which no open season had been declared for a long time were eventually removed from the list of game species.

c. Wild Plants

Pursuant to forestry legislation, forestry departments usually have jurisdiction over all forest products, including not only trees but also herbaceous plants, ferns, mosses, fungi and so on in State-owned forests. Forestry departments are also often empowered to designate protected tree species on private land.

Where conservation legislation does apply to plants, which is still not very frequent, the power to list protected species or to grant collection permits may be split between conservation and forestry departments, according to the species or area concerned.

Difficulties arose at the above-mentioned CITES conference in Kyoto 1992 with regard to the proposal to list certain tropical forest trees in different Appendices. The hostile reaction to these proposals was certainly due, at least in part, to jurisdictional problems. In essence, forest departments were unwilling to accept decisions affecting their activities which had been taken by the representatives of conservation departments.

d. Other Jurisdictional Splits

Japan provides an interesting example of an unusual jurisdictional split.

The Act on the Protection of Cultural Property of 29 August 1950 provides for animals and plants which have a high scientific value in and for the country to be listed as "protected cultural property". Their taking is rigorously forbidden. This protection may be extended to their habitats and breeding, summering and wintering places. The list includes a number of endemic and endangered species, such as the Japanese crested ibis (Nipponia nippon).

Listed species and their habitats, where also protected by that Act, come under the jurisdiction of the Ministry in charge of cultural property. However, listed mammals and birds are also prohibited game under the Wildlife Protection and Hunting Law which is enforced by the Environment Agency. In addition, most of the birds listed as cultural property are also covered by an Act which prohibits trade in and the transfer of rare or endangered birds, and species listed in the CITES Appendices are covered by CITES implementing legislation. These two laws are also enforced by the Environment Agency. Animal species other than mammals and birds and all plant species listed under cultural property law do not seem to be covered by any other legislation.

2. Territorial Jurisdiction

Jurisdiction in international conservation law has already been considered in the Introduction with regard to the Law of the Sea Convention, and in Chapter I above in respect of species *per se* and migratory species. The following section deals with the often confused nature of jurisdiction over species in national legislation.

Complex questions of territorial jurisdiction generally arise in federal or regionalised States. Nevertheless, in certain unitary States, jurisdiction over the protection of species may be delegated (as opposed to transferred) to lower levels of Government. In Sweden, for example, counties may adopt their own regulations for the protection of wild plants, but central Government retains the power to designate protected plant species at national level.

In federal or regionalised States, the distribution of powers between federal or central Government and the federated entities or regions is determined by the Constitution of the State in question. Where the Constitution is silent on the subject of wildlife, as is often the case particularly for older Constitutions, jurisdiction over wildlife is generally considered to belong to the federated entities.

Where powers for the protection of species are expressly provided for under Constitutions, a considerable variety of jurisdictional situations may result.

a. Primary Jurisdiction of the Federal Government

At one end of the scale, the federal Government has powers over wildlife and hunting under the Constitution. However, the federal entities in such countries are usually empowered to supplement federal legislation in matters of detail. Brazil, Mexico and Venezuela are amongst the countries which have adopted this jurisdictional structure.

However, the position has recently changed in Brazil. Pursuant to article 225 of the new national Constitution of 1988, the Constitutions of individual States which are adopted after that date may grant increased powers to States in the field of wildlife protection.

Changes are also apparent in Mexico where individual States, such as Chiapas, have started to enact their own environmental protection legislation.[13] However, very few provisions for the protection of species have so far been enacted at State level.

Germany has a federal Hunting Act and a federal Nature Conservation Act which the Länder must observe. Species covered by this federal conservation legislation are protected throughout the country, and the Länder cannot derogate from these provisions. However, the Länder may decide on open and closed seasons in respect of game species listed under the federal legislation.

In Switzerland, there is a list of species protected under the federal legislation. The Cantons may protect additional species, but they are not empowered to exempt any species from protection which are protected by the federal Government. Spain has a similar system.

b. Primary Jurisdiction of the Federated Entities or Regions

At the other end of the scale, the federal Government has no jurisdiction whatsoever over species, except with regard to imports and exports. In Austria, for instance, jurisdiction over species conservation, hunting and fishing is entirely vested in the Länder. Each Land therefore has its own legislation on these matters and draws up its own lists of protected species and game species.

A similar situation exists in Australia, except that the federal Government has the power to protect species in areas under federal jurisdiction. These areas include the outlying territories and marine areas outside the territorial jurisdiction of the States. The federal Government is empowered to regulate imports and exports. However, it may not regulate inter-State commerce, as the Constitution prohibits any restriction on such trade. As a result, species protected in one State may be freely traded in another.

c. Concurrent Jurisdiction at Central and Regional Level

Between these two extremes, there are a number of intermediate situations in which the federal or central Government and the federated or regional entities are vested with concurrent jurisdiction on wildlife.

An interesting example of concurrent jurisdiction is that of Italy. The Constitution empowers the central Government to enact framework legislation in the field of nature conservation, which the Regions must observe. On the basis of that legislation, the Regions may then pass their own legislation to deal with all matters of detail not provided for by the national enactments.

Where national framework legislation has not been enacted, the Regions are free to adopt their own species protection legislation as they wish. This freedom is in principle subject to the provisions of any international instrument to which Italy is a Party. However, this requirement seems to be particularly difficult to enforce.

Apart from marine fisheries legislation which comes under central Government in all but three Regions, the only framework Act so far enacted is the Act of 11 February 1992 which replaces the Hunting Act of 27 December 1977. The new Act deals not only with hunting but also with the protection of mammals and birds. It does not cover any other taxa. The Act lists

---

[13] Ley de Equilibro Ecologico y proteccion al ambiente del estado de Chiapas (*Periodico Oficial* no. 150 tercera seccion del 31 de julio de 1991).

protected and game species and lays down general rules for their protection and for the regulation of hunting activities.

Jurisdiction over all species other than mammals and birds is therefore entirely vested in the Regions, pending the enactment of further national framework legislation which remains no more than a hypothesis at present. In practice, there are considerable differences between the Regions as to which species are protected in law. Many Regions have no species conservation legislation at all. Moreover, no control of inter-regional trade is possible under this system of Regional jurisdiction.

d. Returning Jurisdiction to Central Government

In certain federal States, the federal Government has tried to recapture jurisdiction over species protection, which has in practice been devolved to the States or Provinces. This return of jurisdiction to central Government is usually made possible by means of the adoption of national legislation to which federated entities may accede if they so agree.

This approach has worked well in India, where the Constitution expressly provides for such an eventuality. For example, the national Wildlife Act 1972 lays down general rules and lists protected species. Almost all Indian States have now accepted this Act, which means that nation-wide protection rules have effectively been adopted.

The same approach was attempted in Argentina but without success. The Act of 5 March 1981, which deals in particular with the protection of endangered species, was specifically enacted to enable the Provinces to accede to its provisions. However, very few Provinces have in fact chosen to do so.

Another means to give jurisdiction to the federal Government is for that Government to sign a treaty. Constitutions almost universally state that the implementation of treaties is a federal matter.

An early example is the Treaty on Migratory Birds of 1916 between the United States of America and the United Kingdom (in the name of Canada). Until the conclusion of that Treaty, jurisdiction over all wildlife belonged to the States of the USA and to the Canadian Provinces. This made the uniform regulation of hunting of migratory species in the two countries impossible. Following the signing of the Treaty, federal laws were enacted in both countries, listing protected and game species of migratory birds and empowering the federal Government in each country to make regulations to control the hunting of these species throughout the national territory.

It is only to be expected, however, that federated entities may oppose the signing of such treaties as this entails the loss of certain of their powers to the benefit of the federal Government. It has been stated, for instance, that one of the reasons why the United States has not acceded to the Bonn Convention on Migratory Species of 1979 is that there would be strong opposition from certain American States to the transfer to the federal Government of jurisdiction over other taxa than birds.

Although Australia is a Party to several treaties on migratory birds, including to the Bonn Convention, it has so far refrained from adopting federal legislation to implement these treaties. It prefers to encourage States to adopt uniform legislation rather than to impose such legislation through its constitutional treaty powers.

e. Other Examples of Jurisdictional Separations

In the United States, jurisdiction over endangered species, marine mammals and migratory birds is federal, whilst all other terrestrial and freshwater species come under State jurisdiction. In respect of marine species, jurisdiction is split between the federal Government and the States, as discussed below.

In 1973, the Congress adopted the Endangered Species Act which empowers the federal Government to list endangered species. The consequence of such listing is to impose prohibitions on both the taking of and trade in such protected species. The constitutional basis for this assumption of jurisdiction by the federal Government is that of inter-State commerce. The provisions of the Act are based on the need to preserve the possibilities of inter-State commerce in endangered species, as well as the inter-State movements of persons who wish to observe and study these species. The Constitutional legitimacy of these measures has been acknowledged by the federal courts.[14]

In certain other countries, jurisdiction over species is also split between federal Government and federated entities. There is often no apparent logic in the resulting division of powers, which means that the reasons behind such a jurisdictional split are presumably political.

One such example is peninsular Malaysia,[15] in which jurisdiction is federal for most terrestrial species but not for freshwater species or turtles. This jurisdictional rule seems to be applied very strictly, as the Protection of Wild Life Act 1976 does not list any freshwater species except for otters, which are specifically mentioned in the Fishing Act as coming under federal jurisdiction, and crocodiles. Moreover, not a single species of turtle or tortoise is listed under the Protection of Wild Life Act 1976, even where such a species is entirely terrestrial and even for non-indigenous CITES specimens. This is the result of a constitutional provision which expressly places turtles (the term is not defined) under State jurisdiction. The reason for this is probably the importance of sea turtles, in particular their eggs, for local subsistence and the economy of certain of the Malaysian States. Turtle protection rules are therefore made exclusively by the States.

This rigid separation of powers also applies to the import and export of CITES species, as the federal Government has no jurisdiction to control foreign trade in any species which do not come under federal jurisdiction.

The Republic of South Africa provides a further example of jurisdictional separation. Although this country is not in law a federal State, most powers in the field of species conservation are devolved to the four Provinces. There is no national wildlife legislation, not even for the implementation of CITES, as controls over imports and exports of wildlife come under provincial jurisdiction. Under the national forest legislation, on the other hand, the central Government may designate protected tree species.

---

[14] *Palila v. Hawaii Department of Land and Natural Resources* (471 F.Supp.985 (D.Ha.1979), 639 F 2d 495 (9th Cir.1981)). For more information on this subject, see M. Bean, *The Evolution of National Wildlife Law*; New York, 1983, pp.26–28.

[15] The States of Sabah and Sarawak in East Malaysia have jurisdiction over all their species.

f. Complexities of Jurisdiction over Marine Species

Further complications occur in federal or regionalised States in respect of jurisdiction over marine species. Federated entities often have jurisdiction over a certain breadth of the coastal sea, whilst the powers of central Government over such species extend beyond that limit to the outer boundary of the Exclusive Economic Zone.

Once again, the jurisdictional situation varies considerably from one country to another. In the United States and Australia, State jurisdiction extends to three nautical miles offshore, which was the limit of the breadth of territorial waters before their almost universal extension to twelve nautical miles under the new Law of the Sea Convention.

In other countries such as Canada, central Government has jurisdiction over all marine fisheries, although it may delegate these powers to the federated entities. In Spain, the Autonomous Communities only have jurisdiction in internal waters, which are defined as those waters which are located landwards of the baseline from which the breadth of the territorial sea is measured. However, there is one exception to this rule, as Catalonia does have jurisdiction in the territorial sea.

In Italy, marine fisheries also come under national jurisdiction. However, there are three Regions (Sardinia, Sicily and Friuli-Venezia Giulia) which have jurisdiction over fisheries in their territorial seas.

Finally, jurisdiction over marine species under international law is governed by the Law of the Sea Convention of 1982.[16] A coastal State has jurisdiction over marine resources not only in its internal waters and territorial sea, but also within its Exclusive Economic Zone (EEZ) which extends up to a maximum of 200 nautical miles from the baseline. The State accordingly has sovereign rights for the purpose of exploring, exploiting, conserving and managing the natural resources found therein. In contrast, no State has jurisdiction over the marine resources found in the high seas beyond the outer boundary of the EEZ.

Coastal States also have sovereign rights over the continental shelf as far as the outer limit of the shelf, even when it extends beyond the outer limit of the EEZ. These rights apply in particular in respect of all sedentary species, namely

"all living organisms which, at the harvestable stage, are either immobile on or under the sea bed or are unable to move except in constant physical contact with the sea bed or the subsoil".

The above definition, which was taken verbatim from the Geneva Convention on the Continental Shelf of 1958, is far from clear, however, as the concept of "constant physical contact with the sea bed" is open to different interpretations. As a result, disputes have arisen in the past especially with regard to certain crustaceans such as lobsters and crabs.

These rights over sedentary species as thus defined are absolute and are not affected by the jurisdictional status of the superjacent waters. Where the continental shelf extends beyond the 200-mile outer limit of the EEZ, the coastal State has no jurisdiction over any free-swimming species in the superjacent waters, but does retain sovereign rights over all sedentary organisms on the shelf.

---

[16] The general jurisdictional provisions of the Law of the Sea Convention are summarised in the Introduction to this paper.

This somewhat lengthy review of jurisdictional problems would seem to be essential to illustrate the complexity of the existing situation in many countries. Jurisdictional separations, whether functional or territorial but particularly the latter, obey no ecological or biological logic. They are mostly political or administrative. The effect of dividing populations and habitats by artificial jurisdictional boundaries is often to make the rational conservation and management of wild species very difficult.

CHAPTER III
THE LISTING PROCESS

The method used universally to provide legal protection for wild species consists of laying down prohibitions or restrictions in an Act of Parliament or equivalent instrument. Species to which these rules apply are generally listed in a Schedule to the Act or else separately in statutory instruments made under the Act.

For a long time, and this remains the case in many countries, measures for the protection of species were incorporated into hunting and fishing legislation, through the establishment of close seasons and other restrictions (e.g. bag limits and size limits for fish). Where a game species was thought to be depleted or endangered, protection was provided by establishing year-round close seasons.

However, this type of legislation generally applies only to those game species or aquatic species which are subject to exploitation by conventional hunting or fishing methods. Other threats to species, including the deliberate destruction of or disturbance to their habitats, are usually ignored. Similarly, non-game species are often ignored, even when collected for trade or other purposes.

With the advent and gradual spread of nature conservation legislation, or legislation specifically geared to species protection, these attitudes are changing. More recent legislation aims to protect endangered species as well as species whose taking is no longer socially acceptable (e.g. small birds), rather than traditional game species.

Lists of protected species form the cornerstone of these laws, which makes it important to analyse the means by which such lists are drawn up.

A. The Content of Lists

1. Legislative Criteria for Selection

Legislation very rarely provides any criteria for listing species to be protected. Their selection is instead carried out through an internal administrative process. This is usually based upon consultation with the relevant scientific bodies as well as, in some cases, upon scientific research and lists of endangered species, whether national or international.[17]

Nevertheless, the power of the department responsible for the listing of species is almost universally discretionary. In consequence, there is no obligation to protect a species which is known to be endangered, nor is there any restriction against listing a species which is not endangered.

At national level, the criteria for listing are either non-existent, as mentioned above, or else so broad as to make almost any species eligible for listing. In Greece, for example, the

---

[17] Such as the IUCN Red Data Books.

Environmental Protection Act of 10 October 1986 allows species to be protected where they are endangered, vulnerable or are of particular ecological, scientific, genetic, traditional or economic value.

In France, the Nature Protection Act of 10 July 1976 authorises the protection of those species which are of particular scientific interest or whose conservation is necessary for the preservation of the national biological heritage.

Certain countries provide notable exceptions to the above rule. In the United States, the Endangered Species Act of 1973 is limited to species in danger of extinction and species likely to become endangered within the foreseeable future. The Act also authorises the listing of those species which are so similar in appearance to listed species that the effective protection of the latter requires the former to be listed. It is not clear what consequences flow from the listing by the Secretary of the Interior of a species which does not meet the above criteria. There has not been any court ruling on this question, although it could presumably be demonstrated to be an abuse of discretion.

A further interesting aspect of the Endangered Species Act is that listing must be based on the best scientific evidence and that the Administration has no discretion to refuse to list a species if the requisite conditions are met. Congress has made it clear that no extraneous considerations, such as the economic impact of listing, should be taken into consideration.

In the Australian State of Victoria, the Flora and Fauna Guarantee Act of 1988 specifies the criteria which make species eligible for listing. The taxon must either be in a demonstrable state of decline which is likely to result in extinction or be significantly prone to future threats which are likely to result in extinction. Regulations made in 1990 for the implementation of the Act have developed these two criteria in more detail.

The position in international law is broadly similar to that of national law, as Parties approve lists of species on an entirely discretionary basis.

However, the Convention on International Trade in Endangered Species of Wild Flora and Fauna of 1973 (CITES) has developed listing criteria. This was done in the early days of the life of the Convention at the First Meeting of the Conference of the Parties, held in Berne in 1976. Such criteria were deemed to be of particular importance to avoid overloading the Convention's appendices with species which were either not traded in practice or were not endangered by trade.[18]

Appendix I includes all species currently threatened with extinction which are or may be affected by international trade for any purpose, whether scientific or otherwise. Appendix II lists those species which, though not necessarily threatened with extinction, might become so unless trade in such specimens is made subject to strict regulation to avoid utilisation incompatible with their survival.

These criteria are in fact very broad and are now considered to be unsatisfactory. The Eighth Conference of the Parties, held in Kyoto in 1992, accordingly decided to revise the criteria and a proposal to this effect is to be submitted to the next meeting of the Conference of the Parties.

It must be stressed that all the preceding examples of the setting of criteria are exceptional cases. With regard to CITES, the inclusion of too many species or appendices would have made the Convention unmanageable. Moreover, unnecessary restrictions on trade should always be

---

[18] Conference Resolution 1.1.

avoided. The Acts of the United States and the State of Victoria are specifically designed for endangered species and, as a result, contain fairly far-reaching restrictions aimed at the conservation of the critical habitats of the species concerned. It was therefore necessary to impose strict criteria for the listing of species, as otherwise there would have been even stronger opposition than was aroused in any event.

There is usually less resistance where restrictions are imposed on taking and trade, which means that it is not necessary to bind the listing agency to list species only if certain criteria are met. Such an approach would reduce the number of species which may be listed at present and would be in breach of the precautionary principle.

Independently of the criteria for selection, a further question arises as to the taxonomic level of taxa, and the categories of taxa or groups of animals and plants which may be included for protection on lists under national and international law.

2. Taxa or Categories of Species which may be Listed

a. Taxonomic Levels

Although legislation is usually silent on this point, the power to list species is generally interpreted as the power to list higher and lower taxa. Many laws list whole genera or families, and sometimes higher taxa as well. They may also list sub-species. It is rare for separate populations to be mentioned, although this may be very convenient.

CITES is, once again, exceptional in that it does provide a definition of species, which includes

"any species, sub-species, or geographically separate population thereof".

In the United States, the Endangered Species Act and the Marine Mammals Protection Act both allow the listing of separate populations. This permits great flexibility, as it is possible to list populations at the margin of the species' range or isolated populations with particular genetic characteristics which may be in need of protection, even though the species may continue to be abundant elsewhere.

b. Categories of Species or Taxa

The listing of species is also determined by jurisdictional arrangements in a given country. As mentioned above, game, freshwater and marine species commonly do not fall within the jurisdiction of the listing authority.

Legislation may deliberately or inadvertently omit certain taxonomic groups. This is frequently the case for invertebrates. Plants are, of course, never covered where the governing legislation is the Hunting Act of the country in question.

Even where species in other groups may in law be listed, there is a strong trend for legislation to deal comprehensively with large mammals, birds, and increasingly amphibians and reptiles, but only to list a relatively small number of species in other groups. For example, few invertebrates are listed under the United States Endangered Species Act of 1973, although there is nothing in the Act to prevent their inclusion: moreover, in practice, the American Fish and Wildlife Service clearly gives priority to vertebrates. Priority amongst plants is similarly given to vascular plants, whilst mosses, lichens and fungi are for the most part ignored.

Turning to international conventions, the Berne Convention does contain some listed invertebrates, which are mainly dragonflies and butterflies. Only one spider is listed under the Convention!

3. Positive and Negative Lists

Lists are usually positive, in that they contain lists of the names of protected species or other taxa. However, this approach has some disadvantages:

- Lists tend to become very long. Many species may only be recognised by a handful of specialists, whereas the public cannot easily tell what is protected from what is not.

- Accidental species, namely species which occur only occasionally, such as vagrant migratory birds, are usually not covered.

- Species which are not known in the country in question, but which are subsequently discovered there, will not be protected unless the list is amended accordingly.

- The same problem also affects newly-discovered species.

As a result of these shortcomings, the trend is now moving towards negative lists, which list whole groups of species together, where necessary, with exempted species to which the protective measures do not apply. This method was originally adopted primarily for birds, because many countries now protect large numbers of bird species and negative lists were therefore much more convenient.

The negative listing system has now been extended to other groups of species. In Germany, for example, the Nature Conservation Act of 1976 protects all mammals, birds, amphibians and reptiles with a small number of exceptions. Nevertheless, some species which fall under the hunting legislation are still excluded from the scope of that Act.

A further example of negative listing is provided by the National Parks and Wildlife Act of 1974 in the Australian State of New South Wales. Under this Act, all mammals, birds and reptiles are protected except for those species specifically listed as unprotected. In contrast, amphibians are covered by a positive listing system, which means that no species of amphibian is protected other than those species which are explicitly listed.

Despite the above, negative listing also has its disadvantages, of which the main one is that no distinction is made between endangered and common species. Both categories are protected by the same rules. The same problem of course applies to positive lists, except that these do permit species to be placed in different categories of threat to which different rules apply. In a negative listing system, the only solution is to include positive lists of endangered species which benefit from stricter rules than the other species on the list. However, the simplicity of the negative listing approach is then lost to a great extent as the system may become very complicated.

By way of example, the National Parks and Wildlife Act in New South Wales, contains a list of unprotected fauna (as an exception to the general rule of protection) and a long list of endangered species, broken down into various categories which benefit from special treatment. All other mammals, birds and reptiles are protected, but not listed. In contrast, protected amphibians and plants are included on a positive list, which means that all those which are not listed are unprotected.

4. Categories of Protected Species

Certain laws have introduced different categories of protected species, with the purpose of fine-tuning protective measures in accordance with the needs of different species. This approach also makes it possible to set penalties commensurate with the risk of extinction affecting each species.

For example, the above-mentioned Act of New South Wales has three categories of endangered fauna. These are respectively those which are "in imminent danger of extinction", "threatened" and "of special concern". Nevertheless, there are no criteria in the Act for attributing species to any one of these categories.

The Endangered Species Act in the United States provides for two categories, those of "endangered" and "threatened" species. As mentioned above, "endangered species" must actually be in danger of extinction, whereas "threatened species" are those which are likely to become endangered in the foreseeable future.

Several American States have their own endangered species legislation to which they have added additional categories, such as rare species and species of special concern. Indiana, for instance, identifies "species of special concern" as do Minnesota and Florida. Connecticut has a list of "rare species".

Spain is another country which makes use of different categories in its legislation. The Act on the Conservation of Natural Areas and of Wild Flora and Fauna of 27 November 1989 requires lists to be drawn up of endangered species, vulnerable species and species of special interest.

In Peru, the Decree of 31 March 1977 makes use of the IUCN categories of threatened species, which are respectively endangered, rare, vulnerable and indeterminate.

In Mexico, the Order of 13 May 1991, which implements the relevant provisions of the Act of 1988 on the Ecological Balance and the Protection of the Environment, lists species as endangered, threatened, rare and subject to special protection.

Other countries have categories for protected and specially protected species. These include Finland, under its Nature Protection Act of 1923 as amended in 1987, and Italy, under the new Act of 12 February 1992 on the Protection of Warm-Blooded Wild Animals and Hunting.

A further category which has emerged recently is that of "species vulnerable to the destruction of habitats". The EC Birds Directive lists species for which Special Protection Areas must be established. Similarly, Annex II of the EC Habitats Directive lists species in respect of which Special Areas of Conservation must be established to ensure the protection of their habitats.

To implement the above Directives, the Spanish Act of 1989 on the Conservation of Natural Areas and of Wild Flora and Fauna requires the compilation of lists of species which are "sensitive to the deterioration of their habitats". However, these lists have not yet been drawn up.

5. The Inclusion of Non-Indigenous Species

Although most countries only list protected indigenous species, the increasing need to control international trade in endangered species, especially for the implementation of CITES, has made apparent the importance of listing foreign species as well.

The listing of foreign species is usually done separately in another instrument, although some countries do use the same instrument. In the latter case, all the prohibitions applicable to indigenous species will also apply to the non-indigenous species concerned. These will include restrictions relating to domestic trade, transport and possession. Taking will also be prohibited, of course, although such a provision is meaningless in respect of non-indigenous species.

The list contained in the Endangered Species Act in the United States includes many non-indigenous species.

Malaysia has recently amended the list of protected species under its Protection of Wildlife Act 1972 to include almost all CITES species.

6. Taxonomic Problems

Almost all lists of protected species use scientific names as well as local common names. However, there are sometimes considerable doubts as to which species are actually referred to. Four particular problems should be mentioned in greater detail.

Firstly, the use of synonyms instead of the present valid or accepted names often happens when the list of species is relatively old and has not been revised to bring it into line with present taxonomy. Sometimes the synonym is so old or little-used that the species is almost impossible to identify.

Secondly, identification may again be problematic where subspecies, or even varieties in the case of plants, are listed as full species.

Thirdly, genera or even families which have recently been split up by taxonomists will remain on the list until the necessary modifications are carried out. Where the new genus is not added to the list, the legal effect of such a split is that all the species in that particular genus will lose their protected status as a result of the taxonomic change.

Fourthly, a similar loss of protection occurs where a subspecies or a variety is elevated to the rank of full species, having belonged before the change to its nomenclature to a species which is listed as a protected species.

By way of example, the elevation to full species (*Hyla sarda*) in 1983 of the Tyrrhenian subspecies of the tree frog (*Hyla arborea sarda*) resulted in the removal of protection from this frog in France, as the arrêté of 24 April 1979 only lists *Hyla arborea* as a protected species and has not been amended since *Hyla sarda* was described as a new species.

A further example relates to the listing of all 105 species of European orchids in Annex C–1 of the European Community Regulation of 3 December 1987 implementing CITES, for the purpose of prohibiting trade in all those species. At that time, the 105 species were the only orchids mentioned in *Flora Europaea*, the comprehensive European Flora. Since then, many new European orchid species have been described, mostly as the result of the elevation of subspecies or varieties to the rank of full species. Unless and until Annex C–1 to the EC

Regulation is amended, however, there would seem to be nothing in law to prohibit trade in these new species, many of which are known to be rare or endangered.

Nevertheless, it may not always be easy to amend legislation merely for the sake of incorporating changes in nomenclature.

One possible solution would be the wider use of negative lists. Another is to insert a provision in the legislation stating that subsequent changes in nomenclature resulting from the splitting of taxa shall be deemed to be automatically covered. In other words, any new species resulting from the splitting of a listed species would be deemed to have the same status as that of the species already listed. However, no legislation has yet included this type of enabling provision.

It would of course be ideal if countries agreed to use a standard list of scientific names and to revise their lists at appropriate intervals in the light of changes made in the standard list. The Conference of the Parties to CITES has already agreed to the use of standard lists in respect of certain groups of mammals, birds, amphibians and certain plants. Unfortunately for many groups, such as reptiles, fish and invertebrates, these standard lists do not exist.

B.   Listing Procedures

In most countries, the law does not specify any particular procedure for the listing of species as this is a purely discretionary matter. This does not mean that no consultations take place: the opposite is usually true, at least with regard to scientific bodies or non-governmental organisations (NGOs). However, there are rarely any mandatory procedural requirements which, if not observed, would result in a listing being unlawful or ultra vires.

In some countries, certain consultations are required by law. In France, listing may not occur without prior consultation with the National Council for Nature Protection.

Other countries have developed more elaborate listing procedures, which will now be examined in greater detail.

1.   The Right of Initiative

An interesting feature of certain laws is the right of initiative given to the public in respect of the listing of species.

The Endangered Species Act in the United States provides that listing, delisting or changes of category (from endangered to threatened, or *vice versa*) may be initiated not only by the Secretary of the Interior, or the Secretary of Commerce in respect of marine species, but also pursuant to a petition from any interested person.

In the latter case, the Secretary of the Interior must determine within 90 days whether the petition contains sufficient information to enable him or her to start the procedure. If this is so, he must carry out a review of the status of the species concerned and determine within 12 months whether to propose a listing. He may only decline to do so if he finds that the listing is unwarranted or if he lacks the resources to proceed immediately. Such a finding is subject to judicial review.

Under its Endangered Species Act of 1984, California has also conferred the right of initiative upon any interested person.

The Flora and Fauna Guarantee Act of 1988 of the Australian State of Victoria also authorises any person to propose the listing or delisting of species.

Pursuant to the Spanish Act on the Conservation of Natural Areas and of Wild Flora and Fauna of 27 March 1988, the Royal Decree of 30 March 1990 provides that the initiative for listing, delisting or change of categories belongs primarily to the national nature conservation authority (namely the National Institute for Nature Conservation or ICONA).

However, such an initiative may also come from any of the regions, known as Autonomous Communities, or the National Commission for the Protection of Nature. In such cases, ICONA must still set the procedure in motion. In addition, the initiative may be taken by conservation NGOs, in the form of a petition accompanied by scientific justification. ICONA may freely decide to accept or reject the petition.

2. Public Consultation

Where special procedures do exist, their purpose is generally to inform the public of the proposed listing and to hear possible objections.

Under the Endangered Species Act in the United States, the Secretary of the Interior must publish a notice of the proposed listing in the Federal Register, notify both the State and local Governments concerned, publish a summary of the proposal in the local newspaper, notify scientific organisations and, if so requested, hold a public hearing. The final determination as to listing must be made within a year.

In the State of Victoria in Australia, the Flora and Fauna Guarantee Act of 1988 provides for a preliminary examination by the Scientific Advisory Committee and the publication of the proposal in the Official Gazette and in State and local newspapers, and permits observations to be made by members of the public. The recommendation of the Minister must also be published in the Official Gazette and two newspapers. The decision is then taken by the Minister and adopted by means of statutory instrument signed by the Governor.

Under the Wildlife and Countryside Act of 1981 in the United Kingdom, local authorities and other persons affected by the proposed listing of a species for protection must be given an opportunity to submit objections or representations. Provision is made for a public inquiry to be held, where necessary.

The Spanish legislation does not provide for any public consultation. However, the listing proposal must be submitted for comment to the National Institute for Nature Conservation and must be based upon detailed scientific information on the range of the species, its habitat, the threats to its conservation status and the conservation measures required. Listing is carried out by Order of the Minister of Agriculture.

The disadvantage of complex consultation procedures is that they are both time-consuming and costly. Indeed, in the United States, there is a long list of candidate species which the Secretary of the Interior has not even had the time to consider. This is a very grave situation as some of the species concerned could well become extinct before they have been listed.

As a result, an emergency listing procedure in the United States has been introduced under certain Acts[19] for species faced with a significant threat of extinction. This is only a temporary measure: if emergency listing is not replaced by a permanent listing in compliance with the regular procedure, it will lapse after 240 days.

Other provisions to improve the effectiveness of listing procedures in the United States include conferring some legal status upon candidate species. The Fish and Wildlife Service must closely monitor the status of these species so that, if necessary, an emergency listing can be carried out. All other federal agencies must work with the Fish and Wildlife Service to ensure that their programmes do not affect candidate species.

Candidate species are those for which the Fish and Game Commission has agreed to consider the petition. If the petition is rejected, a species ceases to be a candidate species. Candidate species benefit from the same protection as listed species, except that the provisions relating to assessments of projects which may affect species are not applicable. However, unofficial consultations are possible in respect of any project that may affect a candidate species.

Proposals to list a species or to amend an appendix in any other way (such as a deletion or the transfer of a species from one appendix to another) can only be made by the Parties. A proposal should generally be accompanied by reasons based on the best available scientific evidence, as required under the Bonn Convention.

The proposal should be communicated to the Secretariat a certain time in advance of the meeting at which it will be examined: both the Bonn Convention and CITES require at least 150 days. The Secretariat must properly communicate the proposal to all Parties. This allows time for national consultations to take place. CITES also provides for the communication of the proposed amendments to interested bodies, which enables organisations such as IUCN and the WWF to comment on the amendment proposals.

Comments from the Parties and other bodies must also be sent to the Secretariat not less than a certain number of days before the meeting of the Parties: these intervals are 30 days under CITES and 60 days under the Bonn Convention. After the last day for the submission of comments, the Secretariat immediately communicates all comments received to all the Parties.

The result of these procedures is that each Party is informed in sufficient time not only of the proposed amendments and their reasons, but also of the comments by other Parties or interested bodies, and may make its decision to approve or veto the amendment accordingly.

Although it is only Parties which are entitled to propose formal amendments, discussions for the revision of appendices may very well be carried out by the Conference or subsidiary bodies, as is done by the Scientific Council to the Bonn Convention. The Berne Convention has a Group on Reptiles and Amphibians, Invertebrates and Plants which reviews species for possible listing.

As described in the chapter on international conventions, the amendment of appendices is done by simplified procedure not requiring ratification and binds all Parties except those which have entered a reservation or objection.

---

[19] Including the federal Endangered Species Act of 1973 and the Californian Endangered Species Act of 1984.

3. Delisting Procedures

The procedure for delisting is always the same as for listing. However, the logic of the precautionary principle would suggest that a stricter procedure should be followed for delisting than for listing. For example, listing could be effected by regulations whereas delisting would require an Act of Parliament.

Several countries do have stricter rules for de-establishing protected areas than for creating them. However, a survey of national legislation does not reveal any provisions of this type with regard to the delisting of species.

Delisting is always more difficult under treaties than listing because of the qualified majority requirement. A proposal to amend the list may be blocked if more than one third of the Parties oppose the amendment. In the case of a proposal for delisting, however, it is necessary to secure a two-thirds majority of the Parties present and voting before the delisting can be approved.

C. The Evolution of Listing

As the number of endangered species increases and awareness of the importance of their conservation grows, a trend has developed towards longer and longer lists.

Lists originally featured large mammals and birds, and gradually incorporated small mammals, reptiles and amphibians. In contrast, invertebrates are only now starting to be included. An increasing number of countries have now also drawn up long lists of plants.

Freshwater fish and other animals have also begun to appear on lists in certain countries, although this is not always possible for jurisdictional reasons. Jurisdictional divisions also explain why marine species are almost universally excluded from the scope of nature conservation legislation, apart from sea turtles and marine mammals in certain cases. Fishing legislation is generally not concerned with the conservation of species of no economic importance, which means that endangered marine species are for the most part completely ignored.

As long lists of species, many of which can only be identified by specialists, came to be seen as counter-productive, the use of negative lists has become more widespread. In one jurisdiction, the Swiss Canton of Vaud, it was found simpler to enact a law to protect the whole of its fauna, including all invertebrates except for a few pest species. This example takes the use of the negative list to its limit. In practice, however, it has not been established whether this methodology makes much difference in terms of results.

The listing of plants poses a separate set of problems. The number of jurisdictions which have lists of protected plants is still relatively small: these are mostly in Europe, Australia and some of the American States. Many countries only list spectacular plants which, though not necessarily endangered, are attractive to the public and are therefore more likely to be picked on a large scale.

A few countries do have long lists of endangered plant species, often small range endemics, which again are frequently difficult to identify. For example, Greece lists more than 650 protected plant species. It is evident that the effective enforcement of that sort of legislation is well nigh impossible.

The Swiss Canton of Vaud has implemented an original system of great interest. A short list of attractive and spectacular plants is widely publicised for the public, whilst a long list of 180 very rare and mostly unspectacular species is simply deposited with the local natural history museum, where it may be consulted by botanists and any other interested person.

In addition, there is a growing trend to impose a general prohibition on the taking of animals, and sometimes of plants, without good cause. This blanket prohibition covers otherwise unprotected species.

For example, the Luxembourg Act of 20 August 1982 on the Protection of Nature and Natural Resources prohibits any exploitation, utilisation, injury to or destruction of unprotected animals or plants without reasonable justification.

Under the German Nature Conservation Act of 20 December 1976, as amended in 1986, it is prohibited to kill, injure, capture or disturb intentionally any wild animals, or to remove, destroy or reduce populations of wild animals, without good cause. The Spanish Act of 27 March 1989 on the Conservation of Natural Areas and of Wild Flora and Fauna also prohibits the killing, injuring or disturbance of wild animals, except for game species.

CHAPTER IV
TAKING

Traditional societies, in which resources were usually owned collectively, have controlled the taking of wildlife since time immemorial. Rules were designed to avoid over-exploitation and to ensure the perennity of the resource.

Some species were fully protected for religious reasons. In Madagascar, for instance, many species, including all lemurs, were considered to be sacred and could not be killed. For the most part, however, restrictions were usually only applicable to those species which were subject to exploitation, whilst other species were ignored.

The traditional split between game and non-game species was retained as hunting and fishing evolved from a subsistence activity into exploitation carried out on a commercial basis, and remained in place even when hunting and fishing later came to be pursued for recreational and sporting purposes as well. Most laws dealt exclusively with game species and ignored non-game species.

In consequence, the coverage of non-game species that has developed is usually provided by nature conservation legislation which is fairly recent. Moreover, such legislation is still limited to a relatively small number of countries.

For a long time, the only legal protection afforded to plants was provided by forestry legislation, which many countries used to protect certain tree species. The collection of other plants could only be controlled in public forests. The protection of endangered species of plants became more widely possible after the enactment of nature conservation legislation, but controls over the exploitation of wild plants of commercial importance remain relatively undeveloped. In certain countries, it was therefore found necessary to pass special legislation to establish exploitation controls in respect of such plant species.

A. Taking Prohibitions or Full Protection

1. Animals

Taking prohibitions always include prohibitions on hunting,, killing, injuring, capturing or collecting protected species. However, attempts at any of the above are more rarely prohibited, which may lead to greater difficulties in the enforcement of such provisions.

Some laws supply a broader definition of taking. For example, under the German Act of 20 December 1976 on the Protection of Nature, it is prohibited to disturb animals belonging to endangered species or their nests or breeding places, such as by photographing or filming them.

The Endangered Species Act of 1973 in the United States provides a much fuller definition of taking, which also prohibits the harassment and pursuit of or harm to such species. "Harm" is defined by regulations made by the Fish and Wildlife Service as an act which actually kills or injures wildlife, and includes "significant habitat modification or degradation where it actually kills or injures wildlife by significantly impairing essential behavioural patterns, including breeding, feeding or sheltering."[20]

This very extensive definition of taking is nevertheless exceptional. It covers a very wide range of activities, including pollution, and is also directly related to habitat protection.

Most nature conservation legislation also applies to all life forms or developmental stages of protected species, such as their eggs, young, larvae and pupae. In addition, it may often cover structures built or used by animals for breeding, sheltering or other purposes, such as birds' nests, burrows, dens, beaver dams and houses, and ant hills. The scope of this protection may sometimes be extended in a general way to all breeding, wintering and nesting sites.

An example is provided by the Wildlife and Countryside Act of 1981 in the United Kingdom. Under the Act, it is prohibited intentionally to damage, destroy or obstruct access to any structure or place which a protected animal uses for shelter or protection. An exception is also made for animals in dwelling houses.

The Act also contains interesting provisions with regard to bats. Nothing may be done in a building which may adversely affect bats, other than in the living areas of dwelling houses, unless English Nature[21] is first notified of the proposed activity or operation and has advised the proponent whether it should be carried out and, if so, the method that should be used.

In respect of birds, the prohibition applies to the taking, damage to or destruction of the nests of wild birds, other than a few unprotected species, while these nests are in use or are being built. No exemption is provided for nests which are situated on or in dwelling houses.

a. The General Exclusion of Unintentional Taking

The protection afforded by legislation is almost invariably against intentional or deliberate taking. In contrast, inadvertent or incidental taking in the course of a legitimate activity is usually neither punished nor punishable.

In Germany, under the Nature Protection Act of 20 December 1976, the prohibition on taking or destroying protected species does not apply to activities undertaken either in the course of the proper use of land for agriculture, forestry or fishing, or as a result of an authorised "intervention in nature". The latter term refers to any activity or project for which a permit has been issued.

In the United Kingdom, the Wildlife and Countryside Act of 1981 specifies that only persons who intentionally kill, injure or take protected wild animals or birds shall be guilty of an offence. In addition, intentional taking is lawful if the act was the incidental result of a lawful operation and could not reasonably have been avoided.

---

[20] Code of Federal Regulations, n° 50, Part 17 § 3.

[21] Following the Environmental Protection Act of 1990, the Nature Conservancy Council was divided into English Nature, Scottish Natural Heritage and the Countryside Council for Wales under the auspices of a Joint Nature Conservation Committee (JNCC). All references to English Nature in this paper should be taken to denote the appropriate territorially competent body.

It follows from the above that protected species are not protected from human activities that may affect them, other than deliberate taking and wanton vandalism. The merit of the United Kingdom is that it does at least state this clearly, whereas most laws are silent on this point. In any event, it would be unimaginable in almost every country to prosecute for the killing of a protected animal, say, during farming operations, the felling of trees or the construction of a road.

Nevertheless, the freedom to destroy protected animals in the United Kingdom has been somewhat reduced by the two conditions imposed under the Wildlife and Countryside Act of 1981, namely that an operation must be "lawful" and that the act could not reasonably have been avoided.

Certain countries go a little further and punish the destruction of protected species by negligence. For instance, the Swiss federal Act on Hunting and Protection of Wild Mammals and Birds of 20 June 1986 punishes negligence which results in the killing of protected species. Similar provisions are set out in the Hungarian Law-Decree n 4 of 1982.

The Endangered Species Act of the United States goes even further. In its amendment of 1978, the term "wilfully" was replaced by "knowingly", in line with the broader interpretation of "taking".

Incidental taking is defined by US Federal Regulations as any taking which would otherwise be prohibited, if such taking is incidental to and for the purpose of the carrying out of an otherwise lawful activity. A permit is required for the incidental taking of endangered species. Permits may only be issued if the Director of the Fish and Wildlife Service is satisfied that, *inter alia*, the applicant will, to the maximum extent practicable, minimise and mitigate the impacts of the taking and implement a conservation plan, and that such a taking will not appreciably reduce the likelihood of the survival or recovery of the species in the wild.

b. Categories of Species and Taking Prohibitions

Where the law contains different categories of protected species, such as endangered species, threatened species and species of special concern, taking prohibitions almost always apply to all of them. The differences relate to the penalties imposed for violations of the prohibition for each category of species, to whether there is a requirement to make recovery plans, and to various other factors.

Under the Endangered Species Act in the United States, all prohibitions laid down by the Act apply fully to all species categorised as endangered, whereas none of the prohibitions apply automatically to threatened species. In respect of the latter, the Secretary of the Interior, of or Commerce as the case may be, may issue such regulations as he or she deems necessary and advisable to deal with the conservation of threatened species. In this way, it is possible to tailor protection measures to the specific requirements of species.

In practice, however, there is very little difference between the two categories, as all the prohibitions under the Act have been made applicable to most threatened species. Special rules have been enacted for some of these species to relax certain of the prohibitions.

c. Specific Exemptions to Taking Prohibitions

In addition to the usual exemption for unintentional taking, legislation universally provides for more specific exemptions. One of these is self-defence, but it would normally be necessary to adduce proof to support such a claim. Other exemptions generally require permits.

Exemptions may be classified into several groups:

- **Public health and safety and the protection of property**

This exemption usually relates to the protection not only of the public directly but also of crops, forests, fish and game. Rogue animals, man-eaters, animals preying on domestic animals, animals in game farms, fish in fish farms and aquaculture, and disease carriers may thus be controlled. For example, a very large number of animals were slaughtered in Africa for the purposes of controlling the tse-tse fly by elimination of the parasite from reservoirs. The exemption would similarly be applicable to flocks of birds at airports or to animals destroying crops.

- **Scientific or educational purposes**

This exemption covers the collection of specimens for museums, zoos and research, and extends to ringing and marking such specimens.

- **Conservation**

This exemption covers the collection of specimens for captive breeding, reintroduction and restocking. It may also encompass the culling of over-abundant populations of a species which were threatening other species or destroying their habitat.

It must be emphasised that all these exemptions are very specific and usually require permits.

Legislation may sometimes allow for broad exemptions at the discretion of the permit-issuing authority, through the use of such catch-all phrases as "for overriding reasons of public interest". However, this type of blanket exemption seems to be on the way out. For example, such exemptions formerly featured in both Belgian and Spanish legislation but have now disappeared from more recent legislation which was introduced in compliance with the EC Birds Directive of 1979 which only permits specific exemptions.

Nevertheless, the Australian State of New South Wales still uses a blanket exemption in its National Parks and Wildlife Act of 1974, which authorise the grant of permits to take protected fauna "for any other specified" purpose.

In similar vein, the Berne Convention on the Conservation of European Wildlife and Natural Habitats of 1979 authorises exemptions to taking prohibitions for "other overriding public interests". Furthermore, exemptions are allowed

"to permit, under strictly supervised conditions, on a selective basis and to a limited extent, the taking, keeping or other judicious exploitation of certain wild animals and plants in small numbers".

Parties must report every two years on the exemptions that have been granted under these provisions.

The phrasing of such exemptions is generally designed to accommodate some traditional methods of bird trapping in certain Parties. The EC Birds Directive accordingly contains similar provisions, although the general exemption for other reasons of overriding public interest has not been retained in the Directive. However, the more recent EC Habitats Directive of 21 May 1992 has reinstated this form of exemption, but in a more explicit form. The exemption thus refers to

"imperative reasons of overriding public interest, including those of a social or economic interest... [or with] beneficial consequences of primary importance for the environment".

Nevertheless, all the above texts do set limits to these exemptions. For example, a derogation for whatever purpose may only be granted under the Habitats Directive if there is no satisfactory alternative and if it is not detrimental to the maintenance of the populations of the species concerned at a favourable conservation status in their natural range. In other words, even overriding public interests must in principle give way if the conservation status of the species concerned might become unfavourable if such an exemption were to be granted. This approach is highly innovative.

d. Exceptions in Favour of Subsistence Hunters and Fishermen

The exemption in favour of traditional bird trapping in Europe has survived from the time when such trapping constituted a subsistence activity. However, in many countries where subsistence hunting and fishing remains important to local populations, it has been necessary to incorporate exemptions for such activities into the legislation.

In many African countries, subsistence hunting coexists with hunting for sport. Customary hunting rights have been widely recognised by legislation and special rules have been established to protect them: for instance, hunting licences are not usually required. However, customary hunters are not allowed to take protected animals.

The situation is different in certain industrialised countries in which indigenous peoples continue to live on the basis of a subsistence economy. Special rules, including the right to take protected species, may apply to these populations.

In Canada, the law authorises Indians and Inuks to take protected species without a permit in certain areas and under certain circumstances. These provisions apply to migratory birds, seals, belugas, narwhal and walrus. The United States legislation also confers exemptions upon Indian tribes, Eskimos and Aleuts under the Marine Mammals Protection Act of 1972, for instance, whilst the Endangered Species Act of 1973 grants exemptions to permanent residents of Alaskan indigenous villages. However, the competent Secretary may make regulations to control such taking if this has negative effects upon the species concerned.

Other countries which have enacted legislative exemptions in respect of indigenous peoples are Australia for the Aborigines, the Scandinavian countries for the Lapps and Greenland for the Eskimos.

India provides an example of a developing country which has legislated for a similar exemption. The hunting rights of the tribes on the Nicobar Islands in the Bay of Bengal have not been affected by the Wildlife (Protection) Act of 1972.

Many international treaties also recognise this category of exemption. Although the International Convention for the Regulation of Whaling of 2 December 1946 does not itself make such provision, the Commission has agreed to make certain exemptions. For example, the Bowhead Whale has been protected from the inception of the Convention, but an exemption was subsequently allowed in favour of Alaskan coastal villages. Similar concessions have been agreed for the Grey Whale in favour of Siberian communities and for the Humpback Whale in Greenland.

Under the Oslo Agreement of 15 November 1973, the taking of polar bears by local populations is authorised through the use of traditional methods and in the exercise of their traditional rights.

2. Plants

Much of the preceding discussion about animals is also applicable to plants. However, it is important to emphasise that the protection of wild plants is still often considered to be less important than that of wild animals. Plants therefore remain the 'poor relations' of conservation legislation.

Taking prohibitions for plants generally include a ban upon the picking, uprooting, digging up, cutting, destruction and removal of specimens. However, two particular problems arise in respect of the protection of plants.

Firstly, legislation often prohibits the destruction of protected plants. However, this type of provision is ambiguous. Other than in cases of deliberate and purposeless destruction which are admittedly rare, the destruction of wild plants usually occurs as a result of a large range of legitimate activities, such as agriculture, forestry or public works. The destruction of protected plants is thus incidental to the destruction of habitats. The loss of habitats inevitably entails the destruction of plants, whereas animals may sometimes be able to protect themselves by moving to another suitable habitat, should one exist and sufficient time be available to them.[22]

Suffice it to say here that legislation either provides for broad exceptions or lists specific exemptions within the law. An example of the first approach is the Wildlife and Countryside Act of 1981 of the United Kingdom, which once again provides that the destruction of protected plants is lawful where this is the incidental result of a lawful operation and it could not reasonable have been averted.

The second approach is illustrated by the legislation of certain American States. Under the Native Plants Protection Act in California, no public agency may regulate agricultural operations, including the clearing of land, for the purposes of protecting native plants. Similarly, timber operations, fire control measures, the removal of plants from ditches and rights of way, and works carried out by public agencies or publicly or privately-owned utilities may not be restricted because of the presence of protected plants. However, landowners have an obligation to notify the Fish and Game Department at least ten days in advance of proposed land-use changes to allow for the plants to be salvaged.

In France, the only exception to the taking prohibition is that made for ordinary farming activities on habitually-cultivated land. This should mean that activities such as the clearing of new land for cultivation or construction are not exempted. Where a permit is required for a given activity, it may be easier to refuse it if this activity will affect protected plants. Where permits are not required, however, as is the case for most agricultural activities, how will farmers know that protected plants grow on the land they propose to clear, and who would prosecute a violation of the taking prohibition? No prosecutions have been brought in France to date.

More generally, legislation is silent on the question of exemptions and simply prohibits destruction. This does not of course mean that legitimate operations are thereby rendered impossible.

---

[22] This question is discussed in greater detail in Part I, Chapter VII, which deals with the Integrated Protection of Species and their Habitats.

The second aspect peculiar to the protection of plants concerns the problem of ownership. Unlike animals, which are legally characterised either as *res nullius* or as public property, plants are held to be the property of the landowner, whether public or private.

Ownership by the landowner has several consequences.

Firstly, the owner may prohibit the collection of plants on his or her land by third parties. In certain countries such as Sweden and Switzerland, however, the law reflects the custom and specifically allows the picking of wild plants on any land, including private property, except protected plants.

Even where picking is not specifically allowed, it is commonly practised and it is therefore difficult for landowners to oppose it. It is only recently in France, for instance, that the courts have convicted mushroom pickers for collection on private land, which has been characterised as theft. However, the legal argument was chiefly centred on the fact that fungi have a commercial value.

As a general rule, landowners have neither the will nor the means to enforce their rights and thereby to ensure the protection of wild plants.

Secondly, it is difficult to enforce taking prohibitions against the owners of the land and therefore the plants.

Although the legislation protecting plants in most European countries is binding upon landowners, enforcement is clearly problematic from both the political and practical points of view. In certain other countries, particularly common law countries, legislators have generally chosen not to impose plant taking prohibitions on landowners, presumably because this was felt to be an undue and even unconstitutional restriction of property rights. In consequence, the taking prohibition is usually only applicable to taking on public lands.

Under the Endangered Species Act in the United States, the collection or wanton destruction of protected plant species is only prohibited on federal lands. This provision was somewhat extended in 1988. It is now a federal offence to take these species on any land where they are protected by State legislation or have been collected on private land in violation of a State trespass law.

The situation is slowly altering as recent legislative changes have made taking prohibitions for protected species applicable to landowners. Countries that have adopted this approach include the South African Provinces, the United Kingdom, certain American States and Western Australia.

B.   Taking Restrictions or Partial Protection

1.   Introduction

Man has made use of wild species for food, clothing, shelter and medicine for as long as he has lived on the planet. Traditional societies intuitively developed their own rules to limit the taking of such species to maintain the perennity of the resource.

The disruption of traditional societies has led to the increasingly generalised application of the legal model inherited from Roman law, namely that of Western societies, based for animals on the *res nullius* concept. As a result, there is not only no incentive to conserve the resource but, on the contrary, there is an incentive to take as much of it as possible in case other members of the society get there first. Based on this analysis, resources held in common are inevitably doomed to overexploitation, since each resource-user places self-interest above community interest as part of the process described as the 'tragedy of the commons'.[23]

Furthermore, resources have become increasingly scarce not just through overexploitation but also through other factors, particularly the destruction of natural habitats. The enormous increase in demand for natural resources has been stimulated by rapid population growth, opening of new markets, especially in affluent societies, and new technologies for transport and storage, particularly refrigeration. The response to this increased demand has been to try to boost supply by means of new technologies and investments.

As a result, the need for sustainable exploitation has gradually become apparent and has now been incorporated as a principle embodied in many international instruments, from the Stockholm Declaration to the Convention on Biological Diversity. It is also one of the bases of the Law of the Sea Convention of 1982, and is set out as a fundamental principle of the World Conservation Strategy.

The market is clearly unable to prevent overexploitation, as is the present system of ownership of resources. This begs the question as to whether there is an alternative system that would succeed any better. It would appear that the only means to control overexploitation is by regulation, but this presents major difficulties.

The ideal would be to prevent the taking of more than the natural increase in a wild population in successive seasons. However, this presupposes that this quantity is actually known, so that regulation can be designed to ensure that these limits are not exceeded and effective enforcement measures may be carried out.

In order to determine harvestable quantities, scientists have developed the concept of Maximum Sustainable Yield (MSY). This is defined as

"the greatest harvest that can be taken from a self-regenerating stock of animals year after year while still maintaining a constant average size of the stock."[24]

The calculation of this quantity is not easy. It relies on the study of the population dynamics of the species concerned, which requires good knowledge of the total biomass of a stock, its natural mortality and the environmental factors that may affect recruitment.

The exploitation of a species, even at MSY level, may have important effects on other species, such as predators and prey, and on the ecosystem as a whole. Lower levels of exploitation may therefore be necessary to minimise these effects. This requirement has led to the development of a new concept, that of Optimum Sustainable Yield, which is embodied in the Law of the Sea Convention of 1982. Even if the Optimum Sustainable Yield can be accurately calculated, harvesting restrictions will also depend upon detailed knowledge of mortality induced by mankind. Accurate statistical returns on harvesting figures are necessary for this purpose, but are not always easy to obtain.

---

[23] L. Hardin, *The Tragedy of the Commons* (1968), *Science*, 162, p.1243–8.

[24] S. Holt and L. Talbot eds., *The Conservation of Wild Living Resources*, Report of Workshops held at Airlie House, Virginia, February and April 1975.

The situation is particularly difficult in respect of migratory species, as the rates of breeding or levels of harvesting in one country will obviously affect the situation in other countries.

Finally, it is necessary for the appropriate regulatory measures to be accepted, implemented and enforced.

The example of the International Convention for the Regulation of Whaling of 1946 demonstrates just how difficult such regulation is in practice. The Preamble to the Convention states that

"it is essential to protect all species of whales from further over-fishing" and that

"it is in the common interest to achieve the optimum level of whale stocks as rapidly as possible without causing widespread economic and nutritional distress".

Despite the above provisions, the International Whaling Convention has always fixed its quotas at too low a level, thereby favouring short-term interests and the avoidance of economic distress. The result has been that whaling has been all but stopped.

In view of the difficulties in setting levels for MSY or Optimum Sustainable Yields, it is not surprising that such methods have only been used for important commercial species, almost exclusively in the sea. As far as all other harvested species are concerned, harvesting controls are seldom based on scientific findings and are therefore largely empirical.

Moving on from these general considerations, it is of course true that the exploitation of wild species can take many forms. There has undoubtedly been a widespread evolution from subsistence harvesting to commercial and recreational taking. In industrialised societies, subsistence harvesting has indeed become negligible. Damage has in practice occurred when former subsistence activities were put onto a commercial footing as access to distant markets improved, resulting inexorably in overexploitation.

Apart from fisheries, especially marine fisheries, commercial exploitation may itself tend to decrease in industrialised societies as a result of the depletion of natural resources and legislation prohibiting trade in wild species. On the other hand, recreational taking, particularly through hunting and fishing, has increased significantly. These sectors of activity now represent a very important economic activity, in terms of the supply of weapons, gear and equipment and other expenditure.

In developing countries, the situation is very different. Subsistence exploitation remains very important, but the disappearance of traditional rules means that it is little regulated. Recreational taking is relatively undeveloped, except in a few countries which encourage foreign tourists and hunters to come on safari as they constitute an important source of foreign earnings.

Commercial exploitation may also be important as a source of foreign earnings. However, this is increasingly regulated, because legislation implementing CITES would otherwise be largely ineffective. Commercial harvesting and trade controls are of course inseparable, and will be discussed in greater detail in Part I, Chapter V which deals with Trade.

This section of the paper will therefore concentrate on other forms of exploitation, particularly those carried out for sporting purposes.

2. Subsistence Exploitation

In many developing countries, subsistence exploitation remains an important source of food and other products, although it is generally ignored by the relevant legislation. This means that traditional subsistence hunting is subject in principle to the same rules and restrictions as hunting for sport.

This approach is confirmed by the laws of several African countries, such as that of Senegal. However, most African countries have legally recognised the right of villagers to hunt unprotected animals without a permit, provided that the game is used only for the subsistence of hunters and other members of the village community. Moreover, only traditional weapons or methods of hunting may be used, which consequently excludes the use of firearms.

Examples of such legislation include the laws of the Central African Republic of 27 July 1984 and Togo of 7 November 1990, and the Decree of Gabon of 4 March 1987 regulating the exercise of customary rights of use.

Elsewhere, pursuant to the Natural Resources Code of Colombia of 1975, no hunting licence is required for subsistence hunting provided that its only purpose is to provide food for the hunter and his family.

3. Hunting Rights

The system of hunting rights varies considerably from one country to another. In many countries, these are vested in the State which then concedes the right to individuals through the issue of hunting licences. As a result, at least in certain countries, permit holders are entitled to enter private property to hunt.

Italy provides one example of this approach. The only way for a landowner to stop hunting on his or her property is to erect a high fence. However, it should be added that since the enactment of the new Act on Hunting and the Protection of Mammals and Birds of 11 February 1992, this rule is no longer absolute. It will henceforth apply only to those areas where "programmed game management" is carried out, which is a clear incentive for the development of such plans.

Under an alternative system, hunting rights belong to the State or a local authority which then leases them to a single person or group in a particular area of a minimum size. In consequence, only one person or the members of the group will be entitled to hunt in the area in question. This approach means that hunting rights are completely dissociated from land ownership.

Conversely, in many countries hunting rights belong to landowners. In countries such as Austria, Germany and the Netherlands, hunting rights may not be exercised unless the land is of a minimum size, which is relatively large. Landowners lease their hunting rights to hunters, who will usually need to conclude leases with several landowners in order to reach the size threshold required for hunting to be authorised.

Elsewhere, however, the general rule is that the owner has hunting rights and that nobody may hunt on another person's property without the owner's consent.

Whatever the system, the State always retains the power to regulate hunting under its police powers.

4. Game Species

Legislation determines which species are classified as game, and usually limits this category to mammals and birds. Under hunting legislation, no species may be hunted unless an open season has been declared in respect of that species. In several countries, no open season has been provided for many listed game species, which are therefore effectively protected in the same way as non-game species.

There has been a gradual decrease in many countries in the number of species which may be lawfully hunted. Indeed, some countries have very short lists, as is now the case in Belgium and Luxembourg.

a. Total Bans

Some countries have now banned hunting totally. This may be because the territory is small and largely urbanised, as is true in Monaco, the Canton of Geneva in Switzerland, the Brussels Region in Belgium and Hong Kong. Alternatively, such a ban may be to try to prevent or slow down the depletion of populations of game animals, as has been done in certain African countries.

For example, the Wildlife (Conservation and Management) (Prohibition on Hunting of Game Animals) Regulations of 1977 in Kenya contain two short sentences:

"Hunting of game animals is totally prohibited. All current hunting licences are hereby cancelled with immediate effect."

Elsewhere, the Decree of 4 November 1975 in Paraguay, which is thought still to be in force, bans all hunting except for those animals declared to be harmful by annual regulations made by the Minister of Agriculture. In Guatemala, the Ministerial Order of 26 September 1988 suspends all hunting temporarily. In São Paolo State in Brazil, the Constitution of 1989 completely bans hunting for any purpose.

The effect of these total bans on subsistence hunting is unknown.

b. Time Limitations: Open and Close Seasons

Legislation always provides for close seasons. In the past, close seasons tended to be the same for all species, but these are now increasingly adapted to the individual requirements of each species. Whilst this approach is ecologically sounder, it is of course more difficult to enforce.

The open season may be very short for certain species. In the Canton of Vaud in Switzerland, for example, the chamois can only be hunted for six days per year in the Alps and four days in the Jura.

Migratory birds pose a particular problem, as close coordination between countries along the migration route is essential. However, for a long time, the arrival of migrants was considered to be a windfall as they could be freely taken.

Some bird treaties include restrictions upon open seasons. Under the Convention on the Protection of Migratory Birds, concluded between the United States and Canada in 1916, there must be a close season of not more than three and a half months between 10 March and 1 September of each year. The Paris Convention on the Protection of Birds of 18 October 1950 states that all birds must be protected during breeding seasons and, in addition, migrants must be protected during their return journey to their breeding sites, especially in March, April, May, June and July.

Under the EC Birds Directive, those migratory species for which hunting is authorised must not be hunted during their period of reproduction or during their return to their rearing grounds.

There still remain considerable discrepancies between countries as to the duration of the hunting season for migratory birds, although the trend is definitely towards shorter open seasons. The recent Italian Hunting Act of 11 February 1992 closes the hunting season for migratory birds on 31 January, as does the Spanish Decree of 8 September 1989. However, to quote divergent examples from the Mediterranean, Greece closes its season on 10 March, Turkey on 15 March, Morocco on 3 March, and Malta does not do so until 1 June!

In parallel, there has been a decrease in the number of days during open seasons on which hunting is allowed. Hunting is often forbidden on Sundays, perhaps to protect those relaxing at the weekend, and frequently on other days as well. For example, the Italian Act of 1992 provides that the number of hunting days in a week may not exceed three. Tuesdays and Fridays are designated as non-hunting days throughout the country. Individual hunters may therefore chose their three days from the other five days available.

c. Area Limitations: Closed Areas

Hunting legislation may impose a complete ban on hunting in certain zones, but this does not involve any habitat protection measures. Areas may be closed to hunting either on a permanent basis, as in game reserves, or merely on a temporary basis.

The Italian Act of 1992 provides that in each Region, 20% to 30% of the territory must be designated as non-shooting areas, in which additional positive management measures must be taken.

d. Limitations on the Number of Hunters

Limitations of this type imply some form of discrimination which may be contrary to the constitutional principle of the equality of citizens. Some countries only grant hunting rights to the residents of a certain area in respect of certain species: this is the case in certain American States and Canadian Provinces.

Limitations may also be placed on the number of licences granted for the shooting of certain rare animals. The maximum number of animals of the species concerned which may be hunted is generally established by regulation. Licences are usually granted on the basis of first come, first served, or may be allocated by drawing lots or organising a lottery.

Under the Hunting and Fauna Protection Act of 1918 of Japan, as amended, hunters must be registered and the total number of registered hunters may be limited. Hunters obtain a badge upon registration which they must wear when hunting.

In some American States and Canadian Provinces, limitations are imposed on the number of licences granted for the hunting of certain species. In Saskatchewan, the number of licences for big game species is fixed by annual regulations for each Wildlife Management Zone. Applicants are selected by drawing lots.

A similar system exists in the Canton of Vaud in Switzerland. In 1989–1990, the number of hunters authorised to shoot red deer was 40 and only 6 for the ibex. Selection is also made by drawing lots, and is restricted to residents. With regard to the ibex, only those who have not been selected in previous years can apply. Each hunter may shoot only one animal, and the animal concerned is also chosen by drawing lots.

e. Restrictions on Animals that may be Taken

i. Sex and Age

Young animals are almost universally fully protected. A rather large number of countries protect females, particularly in Africa in respect of big game shooting. However, it may be difficult to make a distinction in the field between the sexes of certain species. Protection is also usually extended to females which are accompanied by their young.

ii. Bag Limits

Bag limits may restrict the number of animals of a certain species or group of species that may be taken in a day's hunting or during a certain period of time. Many different systems are used to fix such limits. In the case of big game, the bag limit is often fixed for the whole season, whereas it is more usually fixed by the day for small game.

Bag limits are usually accompanied by possession limits, which may be higher than bag limits. In the United States and Canada, possession limits are usually twice the bag limit. However, possession is interpreted in a very broad sense, as it includes those specimens which are kept at home in the freezer. Once the possession limit is reached, the hunter may not exceed this by shooting additional specimens, even if the daily bag limit has not been exhausted.

In California, the bag and possession limits are the same. "Possession" covers the possession of specimens in any form, whether fresh, frozen, dried, salted, smoked, pickled and so on. The trade and transfer of such specimens is also prohibited, with the result that it is not permitted to hunt for specimens over and above the limits of what can be consumed.

The Gabon Decree on the Protection of Fauna of 4 March 1987 imposes a daily bag limit of not more than three mammals of the same species or four mammals of different species, or a weekly bag limit of not more than nine mammals in all, whatever the species.

The Yukon Territory in Canada only allows non-residents to shoot one grizzly bear in their lifetime, whereas residents are allowed to shoot one every four years. Different bag limits apply to other game mammals and birds.

Bag limits may be combined with limitations on the number of hunters and other restrictions. This is the case in Newfoundland in Canada for the moose, *Alces alces*, the shooting of which is only allowed in certain areas. The open season for moose lasts for two months per year, and the bag limit is one animal per hunter for the whole season. No more than 5 to 25 permits are granted per year, although this varies from one zone to another. Permits are restricted to residents and are allocated by drawing lots.

In some circumstances, a global ceiling may be imposed for certain species or for certain areas. The International Whaling Convention operates this type of quota. In Canada, annual quotas are fixed by region for belugas, narwhals and walrus. The season is closed as soon as the quota is reached.

Bag limits may be enforced by two means. Firstly, individual game or hunting registers or books may be kept, in which each hunter must keep a record of the animals caught. Hunters must usually carry these registers at all times and produce them on demand. This system applies in certain African countries, at least in respect of all big game animals shot, wounded or captured.

Examples include the Kenyan Wildlife (Conservation and Management) Act 1976[25] and the laws of Senegal, Togo, Benin and Gabon.

In Gabon, the Decree of 4 March 1987 requires holders of hunting licences for both small and big game to register all partially protected animals that they have shot, together with details of the date and place of shooting and sex of the animal. Within 15 days, the hunter must declare the shooting of the animals to the competent authority and have his game register stamped. After use, the register must be surrendered to the authorities, to enable accurate statistics to be collated. The registers will be used as the basis for issuing certificates of origin, which are deemed to be proof that the specimen was lawfully obtained and possessed. This in turn will serve as the basis for granting export permits for trophies.

Swiss Cantons have a similar system. Any animal shot must be immediately entered in the register, which must be surrendered at the end of the season.

The second method of enforcement covers various kinds of marks to be affixed on dead animals, such as tags, bracelets, rings, seals and so on. These vary according both to species and to country.

The system works along the following lines. The hunter is supplied with a number of tags corresponding to the number of animals he or she is authorised to kill within a defined period of time. It is forbidden to possess or transport an animal without a tag. Once affixed, tags cannot be removed from the animal and must therefore remain in place until the animal is consumed. Tags and rings are also issued for certain species of fish, especially salmon, in France and Canada.

f. Hunting Methods

Certain methods of hunting are prohibited, particularly those which may result in excessive hunting pressure and are thus considered too unfair or destructive. Once again, the scope of such prohibitions varies widely from one country to another, although some prohibitions do seem to be universally accepted. These include night hunting, the use of motor vehicles, the use of poison, explosives, snares and automatic weapons. The use of nets for mammals and birds is also generally banned, as are most non-selective methods of hunting.

Prohibitions on certain hunting methods, particularly on indiscriminate means of capture and killing, are also set out in some international conventions, such as the Berne Convention.

g. Hunting Licences

Hunting licences are almost universally required, at least for certain species. One of the rare exceptions concerns the Wildlife Protection Act of 1980 of Saint Lucia, under which hunting appears to be unrestricted during the open season and no licences are required. Further exceptions are made for traditional or subsistence hunting: no licence is required in most countries where such hunting still occurs.

It is beyond the scope of this paper to describe in detail all the various types of licence which exist in different countries. Separate licences are often granted for big and small game, especially in African countries, and a distinction is frequently made between residents and non-residents.

---

[25] At least before the total ban on hunting which was imposed under the Wildlife (Conservation and Management) (Prohibition on Hunting of Game Animals) Regulations of 1977 in Kenya.

Some countries may have a large number of different types of licence, such as Zaïre which has nine.

Special licences are provided for in exceptional circumstances. In Senegal, for example, the President issues licences in respect of lions and hippopotami. Licences for the hunting of other rare species are issued by the Minister in charge of Nature Conservation: these species include the giant eland, *Taurotragus derbianus*, the red-fronted gazelle, *Gazella rufifrons*, and Denham's bustard, *Neotis denhami*.

As mentioned above, special licences of this kind are designed to limit the number of hunters of certain species and thereby to limit the global taking of such animals.

By and large, however, the purpose of hunting licences is to prevent unsuitable persons from hunting, such as minors or others lacking the requisite capacity or those who have already been convicted of hunting offences. The issue of licences also makes it possible to collect a tax from hunters.

Provided that an individual meets the conditions laid down by the relevant legislation, a licence should be issued and the competent authority has no discretion to refuse its grant. However, some countries do confer discretionary powers for this purpose upon the competent authority, as is the case under the National Parks and Wildlife Act of 1974 in New South Wales. Where an application for a licence is refused, the applicant may appeal.

An increasing number of countries have now instituted a hunting test, which must be passed before an applicant may be awarded his or her licence. Tests help to ensure that applicants are properly qualified to hunt. Applicants must be able to recognise protected and game species and must be familiar with the hunting legislation.

Many European countries have now instituted such a test requirement, including Belgium, Luxembourg, the Netherlands, Denmark, Italy, France, Germany, Spain, Norway, Portugal, Greece and Switzerland. The requirement also exists in Israel, Japan, some Argentinean and Canadian Provinces, some American States such as California, and in Togo and Zaïre.

The requirement is usually only applied to new hunters. Hunters who were originally issued with licences before testing became compulsory are generally exempted from the test.

h. Game Management

The above measures are largely empirical: although the objective is to reduce hunting pressure, the basis for calculating the level of rational or sustainable exploitation is usually lacking. Harvesting ceilings and bag limits do represent a step in the right direction, but these should be based on scientific findings rather than on rule of thumb as is probably the case in most countries.

In addition, the preservation of game habitats is hardly ever taken into consideration. Indeed, it is hard to see how it could be otherwise, as the protection of habitats is in law a matter for the landowner rather than the holder of hunting rights. Wherever the landowner and hunter are unconnected, there is in principle no incentive for the landowner to protect the habitats of game species.

A more rational system would imply closing hunting grounds to outsiders and creating some form of incentives to landowners to preserve habitats. The central European system of leasing hunting rights satisfies both these conditions. It is in the lessee's interest to maintain game populations at a level which will ensure him or her good hunting year after year. It is in the lessor's interest to maintain habitats as this will result in higher income from leases.

With regard to habitat protection, much obviously depends on land-use. In agricultural areas, the continuation of hunting depends on the maintenance of natural and semi-natural landscape features which are essential to provide shelter and food to some species. In areas which are not suitable for intensive agriculture, hunting can be a form of land-use which provides the landowner with an income, provided that hunters do not alter the natural vegetation. In all cases, the areas concerned must be sufficiently large, otherwise the game population will move in and out of the area and no rational management will then be possible. Within such areas, shooting or game management plans can be instituted.

Amongst examples of existing systems, hunting rights in Spain belong to the State which concedes them to private individuals. Landowners may lease their land to holders of hunting rights who are generally grouped within hunting societies. Small landholders may also band together to form a larger hunting area. During the last year of the currency of the lease, hunters may not shoot more than the average of the preceding years, to avoid exterminating the game before the lease expires. These large areas are called "cotos de caza". Hunting societies may propose a shooting plan, which is provided by the appropriate Government department. Such a plan may be compulsory for big game.

Similar systems are to be found in some Latin American countries, as for instance in Argentina.

In Norway, the Hunting Act of 29 May 1981 provides that hunting rights belong to the landowner but may be leased out. The legislation encourages adjacent properties to be joined together to form a common wildlife area. A Wildlife Board is established in each municipality, upon which both landowners and hunters are represented. The Board determines which properties should be joined together in this way. If this cannot be secured by the voluntary agreement of the landowners concerned, it may be decided by a majority of landholders and this decision will be binding on the minority.

Minimum sizes of the hunting area, varying in accordance with local circumstances, are established in respect of all species of deer, including elk, *Alces alces*, and beaver. The minimum area is calculated according to the number of animals and the living conditions of the species in the area concerned, as well as the damage caused by the species. The minimum size requirement means that owners of smaller properties cannot shoot the species concerned unless their property has been joined together with other properties. As described above, the Wildlife Board will try to achieve voluntary agreement by majority decision of the landowners represented on the Board. However, if this proves impossible, the Board may in the last resort make a Recommendation to the Directorate of Wildlife that the latter should order the properties to be joined.

Once the common wildlife area has finally been established, the holders of hunting rights may by majority decision prescribe rules governing the practice of hunting, the implementation of wildlife management measures, bag limits and a system for sharing costs. All such decisions must be approved by the Wildlife Board, and appeals are possible to the Directorate of Wildlife. If the owners cannot agree on questions such as hunting quotas and bag limits, the decision will once again be taken by the Wildlife Board.

Hunting permits in Norway may specify the sex and age of the animals which may be shot.

Austria, Germany and Switzerland are all federal States in which regulations may vary from one Land or Canton to another. In general, the shooting plan is prepared by the lessee but must be approved by the appropriate authority. The plan specifies the minimum and maximum number of animals which may be shot. The minimum size of hunting area is laid down by legislation.

In the Austrian Land of Tyrol, for example, these minimum limits are 200 hectares for one person and 500 hectares for a hunting society. A shooting plan must be drawn up for each year, in respect of all ungulates except wild boar, and marmots and certain game birds.

Shooting plans may also be developed in countries where the lease system does not exist. Such plans are generally prepared and adopted after consultation with hunters by the competent authority. This is the case in France, for instance, and in the United States. In California, plans must be drawn up for deer herds. The Department of Fish and Game designates deer herd management units and makes plans for each unit to maintain or restore deer to a good condition of health. The plans contain provisions for the identification and preservation of critical habitats and for the improvement of habitat quality. Each year the Department recommends the number of deer which should be shot in each unit.

Other types of system that have been developed in Africa include the leasing of public land to a licensed professional hunter or safari guide, as in Cameroon. The professional hunter must manage the area he has leased on good ecological principles to maintain his wildlife capital at an optimum level. He must make an annual survey of the fauna and draw up a shooting plan which must be approved by the Government authority. The number of animals shot must not exceed that laid down in the plan. The local population in the area concerned will continue to enjoy customary hunting rights.

This kind of system has advantages for everyone. Outsiders may not come in, whilst the professional hunter is given quasi-ownership and has a strong incentive to prevent poaching. The Government receives not only rent but also taxes on the animals shot, and general benefits accrue to the economy from tourists and hard currency.

However, such a system is of little benefit to local populations. In contrast, the Parks and Wildlife Act of 1975 in Zimbabwe vests nominal ownership of hunting rights in the State, but declares private landholders to be the "custodians" of wildlife on their land and authorises them to manage and exploit the wildlife as they wish. The only exception concerns the small number of totally protected species. This system has encouraged cropping systems on many large estates which now have mixed economies, combining cattle and wildlife.

However, large areas of the country consist of communal land owned by the State. To ensure that the benefits from wildlife would accrue to local populations, the Act was amended in 1982 to confer the status of landholder in respect of wildlife upon District Councils. The Councils may delegate their authority to wards or villages. This scheme, known as CAMPFIRE,[26] has been extremely successful as revenues from game cropping, sport hunting or game viewing are directly paid to the villagers and therefore make a substantial contribution to local development. As a result, wildlife which was once considered a nuisance is now seen as an asset.

5. Fresh Water Fishing

Much of what has already been said about hunting is also valid for freshwater fishing of wild aquatic resources *mutatis mutandis.*[27] Some controls are obviously specific to fish, such as

---

[26] Communal Areas Management of Indigenous Resourses.

[27] Aquaculture is excluded from this discussion as it counts as an agricultural operation.

minimum sizes: there is generally an obligation to return undersized fish or protected species to the water. A fishing test, along the lines of the hunting test, has been introduced in Germany.

6. Plants

The subsistence collection of plants for food, such as mushrooms or berries, or for medicinal purposes remains almost universally practised. The danger to wild plants lies in their commercial exploitation for such purposes as well as for ornamental reasons. In developed countries, the risks posed by recreational activities have also grown considerably, with improved access to the countryside and the increased collection of wild flowers by the public.

With regard to species traditionally collected for household consumption, legislation has been developed to prohibit their taking in commercial quantities. The use of mass collection instruments, such as combs for berries or rakes for mushrooms, is also prohibited.

a. Mushrooms

In certain Western European countries, there is a 'bag limit' which limits the specimens that may be collected either by number or by weight (e.g. 2 kg per person per day). Legislation may also require that collection permits be granted, at least to non-residents, and may establish close seasons, close areas, protected species and a prohibition on the use of tools which may damage the humus layer or the mycelium.

b. Other Plants

Some Italian Regions regulate the collection of wild berries, mosses or lichens by means of the maximum weight which may be taken on any one day.

Many countries prohibit the collection of medicinal plants without a permit, and also regulate the taking of certain ornamentals, such as holly, Ilex aquifolium, which is at great demand at Christmas, and Christmas trees. Southern States in the United States strictly regulate the collection of desert plants.

c. The Collection of Wild Plants by the Public

A certain number of European countries have a system of partial protection for listed species of wild flowers, including Belgium, Austria, Italy and Switzerland. This is generally achieved by imposing prohibitions on collection for commercial purposes and by establishing a prohibition on digging up, removing or damaging the subterranean parts of the plants.

The legislation may also lay down a daily collection limit, say, three or five specimens for a given species. Alternatively, as in certain Austrian Länder, the number of flowering stalks that may be picked must not exceed the quantity that may be held in the hand, or between the thumb and forefinger.

A few countries grant partial protection to all wild plants. This generally includes a prohibition against uprooting any species and sometimes, as in certain Austrian Länder, Italian Regions and Swiss Cantons, against picking the aerial parts of plants except in small numbers. Precise numbers often appear in certain of these laws, ranging from five to twenty according to the jurisdiction. Other jurisdictions apply the 'small bunch' rule, namely the quantity that may be held in the hand.

In countries where it would be difficult to apply these kind of restrictions to private landowners, as in many common law countries, the legislation only applies to public land. However, it may also specify that on private land, the collection of wild plants by third parties is prohibited without the permission of the landowner. In the United Kingdom, for instance, the Wildlife and Countryside Act of 1981 imposes a general prohibition on uprooting wild plants which applies to all persons except for landowners on their own land, persons authorised by landowners and persons with a permit issued by the District Council.

In certain Australian and American States, the written permission of the landowner is required for any collection. However, as landowners may not be interested or able to enforce these rules, it has sometimes been necessary to establish a dual permit system. In the American State of Texas and the Australian State of Victoria, the collection of wild plants on private land requires both the written permission of the landowner and a permit from the State. In Washington State in the United States, permits are issued by landowners and must be validated by the local Sheriff for the purpose of protecting owners against theft of certain plants, particularly conifers for Christmas trees.

7. Marine Fisheries

Although marine fisheries are mainly commercial, there is still a need to regulate traditional or small-scale fisheries and, increasingly, sport fishing, particularly underwater fishing.

a. Traditional Fisheries

Older systems of regulating such fisheries are collapsing almost everywhere, and desirable as these may be, it is improbably that they can be reinstated in any more than a very limited way.

In developing countries, this collapse is due to the pressure of population growth and the need to produce greater quantities of fish. These factors have encouraged the modernisation of techniques, such as the use of motor craft, the opening up of formerly closed systems to outsiders, the loss of traditional knowledge and the disappearance of customary rules and institutions.

In developed countries, all the above factors have also played their part in the past. In present-day circumstances, small-scale fishing systems would in any event be unprofitable, faced as they are with competition from industrial fisheries, unless certain areas were to be closed to industrial fishing.

Where self-regulation by fishing communities is no longer possible, regulation by Government has to take over. Once this occurs, the rules applicable are little different from those governing other types of fisheries. However, some countries do try to preserve their small-scale fisheries by various means.

One example concerns the Australian State of New South Wales, which has enacted fisheries legislation incorporating the concept of Restricted Fishing which is applied to the commercial fishing of certain shellfish, such as abalone and sea urchins. This concept establishes a requirement for a special licence, and the total number of licences granted may be limited. Licences may be allocated by drawing lots or by competitive bidding. The maximum quantity of the organism concerned which may be taken may be fixed by regulations.

Some laws have also tried to put an end to competition from industrial fisheries, particularly trawlers. These are also destructive of coastal marine habitats, the conservation of which is essential for the maintenance of the ecosystem and coastal fisheries. This type of restriction is now found in an increasing number of countries.

By way of example, both Spain and Italy now forbid trawling on the sea bed at a depth of less than 50 metres. In France, trawling is prohibited within 3 nautical miles of the coastline in the Mediterranean. Cyprus has a similar prohibition in waters shallower than 30 fathoms. Both the United Kingdom and Australia have completely closed certain waters to trawling. In Cameroon, the prohibition applies within 2 nautical miles of the coastline, and no fishing boats over 250 tonnes are allowed within the territorial sea.

b. General Regulation of Commercial Fisheries

These regulations are usually applicable to all forms of commercial fishing. The aim is always to try to reconcile the prevention of overexploitation with the need to maintain and develop fisheries as an economic activity.

The distinguishing characteristic of fisheries is that of competition between fishermen, which encourages over-investment in an attempt to catch more fish than one's rivals. In the past, intense competition also existed between national and foreign fishermen, as the coastal State had no jurisdiction beyond the three-mile limit of territorial waters.

Following the establishment of the Exclusive Economic Zone (EEZ), however, coastal States almost everywhere[28] have now acquired a monopoly over their coastal resources. Moreover, foreign fishermen are excluded unless they are specifically authorised to fish in a State's EEZ, for which they must generally pay a fee. In consequence, more than 90% of fisheries resources have now been nationalised.

Pursuant to the Law of the Sea Convention of 1982, the coastal State has two major duties:

- to prevent overexploitation; and

- to make available any surplus over and above what is taken by national fishermen to the fishermen of other countries.

In order to calculate any surplus, it is necessary first to calculate the total allowable catch, which must be based on the Optimum Sustainable Yield. These findings must then be translated into regulations. However, the enforcement of such limits is notoriously difficult.

The main purpose of fisheries regulation is to limit fishing effort to a level such that the optimum sustainable yield will not be exceeded. For a long time, such limits were established on an empirical basis, using closed areas, close seasons and a prohibition on certain destructive methods. These restrictions continue to be applied almost everywhere: in addition, scientific data is increasingly used to determine the minimum size of fish that may be caught by regulating the size of the mesh in the nets. Mesh-size limitations are essential to allow the escape of certain fish needed for the recruitment of the next age class.

New fishing laws also use the negative listing system to define allowed fishing methods. Under the Algerian Fisheries Act of 23 October 1976, for example, only those means, methods and gear that are specifically authorised by regulations are allowed, whilst all others are prohibited. The same system exists under the Mauritius Fisheries Act of 23 May 1980, which

---

[28] With the notable exception of the Mediterranean.

prohibits the use and possession on board at sea of any fishing gear other than that authorised by the law.

As these first steps to limit catch often proved insufficient to prevent stock depletion, further measures became necessary. Limits were accordingly imposed on tonnage, power and sometimes the number of fishing vessels. Fishing vessels must be licensed in almost all countries of the world. A global catch limit was established by species in each EEZ.

Individual ship quotas continue to meet with considerable opposition, as this would mean the end of free competition between fishermen. Whether this is the necessary price to pay to ensure sustainable utilisation of fisheries remains an open question.

Another important problem in the sea is that of by-catch, namely the incidental taking of non-targeted species in a fishery. With regard to commercial species, by-catches complicate matters as allowance must be made for them in establishing the total allowable catch. Regulations implementing authorised by-catch rates may be very complex.

The incidental taking of endangered or protected species with drift nets is clearly an international matter which has now been more or less resolved by the compromise which restricts the maximum length of nets to 2.5 km.[29]

Incidental catches by discarded nets or floating plastic debris are more difficult to control, although the discharge of such objects into the sea is in principle prohibited by the MARPOL Convention[30] in an optional Annex. However, the difficulties of enforcing such prohibitions are enormous.

c. Sporting or Recreational Fisheries

Recreational fishing is of considerable economic importance to certain fisheries, particularly for tuna, swordfish, marlin and other big game fish, which means that sport fishing is increasingly regulated.

In the Bahamas, under the Fisheries Resources (Jurisdiction and Conservation) Regulations of 1986, a permit is required for foreign vessels to engage in sport fishing. A bag limit is in force and the maximum number of lines that may be in the water at any one time is limited to six. Sport fishing tournaments must be approved by the Minister.

Underwater fishing, including spearfishing, may also be very destructive of the rich coral reef fauna which are an important tourist asset to many coastal States. Such fishing is accordingly subject to increasing regulation, which may include prohibiting such fishing for commercial purposes as well as the use of aqualungs. Some areas may be closed to such fishing.

Some countries have now totally banned spear guns, such as Bermuda under its Fishing Regulations of 1972. In Mauritius, under the Fisheries Act of 1980, any underwater fishing is subject to the permit requirement. Permits may only be granted for scientific purposes or for the collection of aquarium fishes. The import, manufacture and possession of spearguns is only authorised under permit. Both the Seychelles and the Maldives prohibit the import of spearguns.

---

[29] A more detailed discussion of this topic is set out in Chapter I(D)(3) above.

[30] International Convention for the Prevention of Pollution by Ships, signed in London on 2 November 1973.

8. Species causing Damage

This is a complex problem which must be discussed here, albeit briefly, in view of its interesting evolution.

Animals considered to be harmful to crops, livestock or game were characterised as pests or vermin. These were formerly viewed as species for which active destruction measures were either encouraged, often by the payment of bounties, or were frequently mandatory. As a result, the destruction of such species, including their eggs and their young, was authorised at all times. Some countries still have lists of pest species, although the payment of bounties is definitely on the way out.

There has been a gradual evolution from this outlawing type of legislation in order to prevent excessive taking of such species. This evolution has involved the gradual recognition that no species should become extinct. For example, the EC Birds and Habitats Directives permit the taking or destruction of wild species, including protected species, to

"prevent serious damage, in particular to crops, livestock, forests, fisheries and water and other types of property"

as well as for reasons of public health and safety. However, derogations may only be granted if they are not detrimental to the maintenance of the populations of the species concerned at a favourable conservation status in their natural range.

As a further step in the evolution of thinking, harmful species may now be considered as harmful only in certain areas or circumstances. Exceptions may sometimes then be made to allow the killing even of certain protected species in these circumstances. Examples include the presence of grey herons in fish farms, moles in gardens and stone martens in dwellings, pursuant to regulations made under the Hungarian Law-Decree on Nature Conservation of 1982.

The evolution of thought has included the development of various measures to avoid overkilling. These include the prohibition of trade in pest or unprotected species or the imposition of strict trade controls. Permits may be required for the killing of species, or killing may be placed under the control of the competent authority. Alternatively, the authority may itself proceed to an organised killing of such species in cases of proven damage.

Where protected species are concerned, some countries now make no exceptions to the no-killing rule. However, where proven damage has taken place, compensation may be payable. Priority is increasingly given instead to scaring away animals.

The problem of compensation presents considerable difficulties. Several countries that have assumed ownership of wild animals by law have used the same piece of legislation to disclaim any liability for damage caused by such animals.[31] Under the Nature Conservation Act of 1992 in Queensland,

"the State is not legally liable for an act or omission merely because protected animals and plant plants are the property of the State."

Such damage must accordingly be considered as a natural risk.

---

[31] For example, the Wildlife Act of 23 July 1982 of British Columbia, mentioned in Chapter II(A)(1) above.

However, a distinction must be made in this context between protected species, game species and unprotected species.

If protected species cause damage, that damage could of course be avoided if it was permissible to shoot protected species. Some jurisdictions therefore provide for the right to compensation in these circumstances. For example, in certain Italian Regions, damage caused by bears, wolves and golden eagles will be fully compensated, provided that the animals concerned have not been shot or killed. In the latter case, there would be no compensation and criminal penalties would be applicable.

Where game species are concerned, hunters are often made liable for damage caused by such animals, particularly where hunters are individually or collectively the lessees of a hunting territory. Hunters are in effect considered to be the owners of game animals and thus responsible for any damage. Under this approach, the recreational pursuit is linked to the damage: if hunters did not hunt, there would either be no animals or they would all have been exterminated. If animals are to be kept for the pleasure of hunters, hunters should therefore pay the cost.

Insurance for damage caused by wildlife does not seem to be available anywhere. In France, a compensation fund has been established which is financed by hunters' contributions.

In respect of completely unprotected species, such as invertebrates, nobody is liable for damage caused unless fault can be proved. Where there is an obligation to destroy pest species, liability for damage caused may result from the non-performance of this obligation.

A distinction should be made, however, between native and introduced species. As the latter may often cause considerable damage to indigenous fauna and flora, their taking is generally authorised all year round, and precautions against risk of extinction do not of course apply. Certain jurisdictions, particularly in Australia where introductions have been particularly harmful, make the author of the introduction liable for any damage that may result. In international law, the State which has allowed a harmful introduction to spread to a neighbouring State could be held liable.

CHAPTER V
TRADE

Prohibitions or restrictions on trade in protected species are now in force almost everywhere. Their purpose is to complement protection measures which forbid or limit the taking of such species, and to eliminate financial incentives for the overexploitation of partially protected species. Where trade is permitted, restrictions and procedures are designed to try to ensure sustainability of exploitation.

The following discussion will consider the law relating to both domestic and international trade.

A. Domestic Trade

1. Prohibitions on Trade in Protected Species

Trade bans are imposed at domestic level to complement prohibitions on the taking of certain species. In addition to banning taking, the relevant legislation will also generally prohibit the transport, sale, exchange and purchase of specimens of protected species. In most cases, possession is also prohibited.

The legislation may sometimes contain a long list of prohibited activities with a view to being as comprehensive as possible. In such cases, the prohibition will extend to the shipping, sending, exhibition, offering for sale, including by advertisement, utilisation and processing, taxidermy, serving in restaurants and consumption as food of such specimens.

From a strictly legal point of view, the prohibition of possession is enough in principle, since all the other activities listed above are covered by that global term. Nevertheless, it is certainly better to have comprehensive lists of prohibited activities to avoid any ambiguity. For example, the prohibition on transport clearly applies to the transporter, whereas if only possession is forbidden, it could be argued that the transporter is not the 'possessor' of the specimens.

Exemptions from the prohibitions on possession, transport or other activities may of course be granted for specimens which have been lawfully taken, for example under a scientific research permit.

2. Restrictions or Prohibitions on Trade in Partially Protected Species

Legislation increasingly prohibits trade in game and other partially protected species, except where commercial harvesting is authorised. A complete ban on all trade is generally imposed during the close season. There is usually also a ban on trade in specimens obtained through recreational hunting and fishing, as it is now generally felt that such activities should not be exercised for profit. This avoids providing incentives for hunters or fishermen to take more than what is required for personal consumption.

There are certain exceptions to this rule. Some laws continue to allow trade in some common game species.

One example is the EC Directive on the Conservation of Wild Birds. In respect of all naturally occurring wild birds, Member States of the Community must prohibit the sale, transport for sale, keeping for sale and offering for sale of live and dead birds, and of any readily recognisable parts and derivatives of such birds. This prohibition applies whether a species is fully or partially protected.

Exceptions are permitted for a small number of species, provided they have been lawfully killed or otherwise legally acquired. There is a short list of only seven species in which trade is authorised throughout the Community, and an additional list of 19 species for which trade may be authorised by individual Member States after consultation with the EC Commission.

Trade prohibitions are also increasingly applied to unprotected vertebrate species and even to listed harmful species, in order to eliminate financial incentives for their mass collection.

The rules applicable to possession are not of course identical for fully protected species as for partially protected species and unprotected species. In respect of the latter two categories, it is common to prohibit or control trade but not to restrict possession except during close seasons or if specimens have been unlawfully taken, which raises the question of proof.

Many laws reverse the burden of proof by instituting a presumption of unlawful possession if certain conditions are not met. In many African countries, for instance, it is common practice to institute a system of certificates of origin or certificates of lawful possession. If the requisite certificate cannot be produced, possession is automatically deemed to be unlawful unless evidence to the contrary can be adduced by the individual in possession.

For example, under the Act on Nature Protection and Hunting of 11 February 1980 in Benin, no animal, whether live or dead, and no trophy can be transferred, possessed, transported or exported without a certificate of origin certifying possession and allowing for its identification with sufficient accuracy. The certificate must therefore specify features such as the species, size, sex and any distinguishing marks. The certificate of origin is issued after verification and provided that the specimen is duly included in the applicant's hunting register. The issue of the certificate must also be entered in the Register.

In Zambia, the National Parks and Wildlife Act of 6 September 1991 institutes a system of certificates of ownership. These are granted to the person in lawful possession of any trophy,[32] and also apply to live game, protected animals and the meat of such animals. Strict rules are applied to prescribed trophies, namely ivory and rhino horn and any other kind of trophy listed as a prescribed trophy by regulations. The possession of a prescribed trophy without a certificate of ownership is an offence. The transfer of a trophy to another person must be accompanied by the transfer of the certificate of ownership, duly endorsed with the date of the transaction and the name of the transferee.

A similar system also exists outside Africa and would appear to be spreading. In India, the Wild Life (Protection) Act 1972 establishes a requirement for a certificate of ownership issued in respect of the lawful possession of wild animals and animal products. In Bangladesh, under the Wildlife (Preservation) (Amendment) Act of 1974, any transfer of animals covered by the

---

[32] Trophy is very broadly defined to include any durable part of a game species or a protected animal.

Act requires a certificate of lawful possession. Acquisition is not lawfully possible if this certificate is not transferred to the person acquiring the specimen.

In the Spanish region of Andalucia in Europe, possessors of protected animals, whether live or mounted, or of dead animals to be mounted, may obtain certificates of legal possession provided that the animals were acquired before the entry into force of the nature conservation legislation.

In Colombia, the Decree of 2 October 1981 prohibits the transport of wild animals or their products without a transport permit. Any new movement of the same specimen requires a new permit. The permit is deemed to be proof that the specimen was lawfully obtained and must be transferred to the person acquiring the specimen, the processor or the taxidermist together with the specimen concerned.

3. Licensing of Certain Activities

There are many activities involving the utilisation of or trade in wild animals or plants. In order to control trade, specific rules are required for the regulation of these activities. Those carrying out these activities include live animal dealers, processors of wild animal and plant products, those dealing in such products, taxidermists and others dealing in trophies and raw skins, those trading in game meat, breeders of wild animals in captivity, those propagating plants artificially, and the owners or operators of zoos.

Most of the above activities are now subject to licences in a large number of countries. It is important to ensure that no unlawfully taken animals or plants are used in the course of these activities and to be able to institute the necessary controls to this end.

Special regulations are generally applicable to most of these activities in many countries, and it is not possible to give full details of such measures here. However, the rules normally applicable to these activities are based on the following elements:

- licensing;
- the keeping of books and records. All transactions in protected specimens must be registered, with the names and addresses of suppliers and clients. Certain exceptions may be made for retail sales;
- general prohibitions on trade in protected species; and
- special rules on the conditions in which live animals may be kept in captivity.

The fourth of these elements merits a more detailed discussion.

4. Wild Animals in Captivity

Over past decades, there seems to have been a considerable increase in the demand from the public for live wild animals, and a corresponding increase in the pet trade, including at international level, and in the number of private zoos.

There were for a long time no restrictions on the keeping of captive wild animals in virtually any country. However, such keeping is now increasingly regulated for four principal reasons:

- conservation, including the implementation of CITES controls;

- public safety: the keeping of dangerous animals, such as leopards or poisonous snakes, may lead to serious damage if an animal escapes;

- the prevention of the accidental introduction of alien species which may cause damage to native species and ecosystems; and

- animal welfare: it is important to regulate the conditions of captivity and transport wherever the keeping of wild animals is authorised.

In addition to legislation which is specifically designed to implement CITES and which is therefore only applicable to CITES specimens, certain other laws serve to illustrate the above objectives.

a. Public Health and Safety

With regard to the risk to public safety posed by dangerous animals, the Dangerous Wild Animals Act of 1976 in the United Kingdom prohibits the keeping in captivity of certain animals listed under the Act, unless a permit has been obtained. Such animals include kangaroos, many monkeys, large carnivores, seals, many ungulates, ostrich, cassowaries, emu, crocodiles, venomous snakes, lizards, spiders and scorpions.

In Italy, the Act of 7 February 1992 prohibits the keeping of and trade in mammals and reptiles which may present risks for public health and safety. The list published by Ministerial Order of 18 May 1992 includes all mammals except cetaceans, on the theory that even where these are not dangerous, they may still create problems for public health if they are disease or parasite carriers. The list also covers many reptiles, including crocodiles, large monitor lizards and many kinds of venomous snakes.

b. Animal Welfare

More general regulations dealing with wild animals in captivity include the Swiss federal Act on the Protection of Animals of 9 March 1978. This Act is chiefly concerned with animal welfare and applies to the keeping of wild animals by professionals as well as by private persons.

Regulations made under that Act on 27 May 1981 require permits for the keeping of all mammals, except insectivores and small rodents; certain birds, including all raptors; many reptiles, including crocodiles and venomous snakes; and fish which are longer than one metre. Permits for certain animals whose keeping is considered to be particularly difficult will only be issued if proof of adequate conditions of captivity is presented by an approved expert. Licences in all such cases are subject to conditions. Facilities for captivity must be adapted to the specific needs of the species concerned, and detailed minimum specifications are provided for facilities required for different species.

Perhaps the most radical system is that found in Norway, pursuant to Regulations dated 20 November 1976 implementing a provision of the Animal Welfare Act of 20 December 1974. These Regulations prohibit the import, trade in and possession of exotic mammals, reptiles and amphibians to be used as pets or placed in any other form of captivity. However, permits may be granted if the conditions of captivity are shown to be satisfactory.

In Israel, no keeping of live wild animals is permissible without a permit.

c. Captive Breeding and Artificial Propagation

Captive breeding of protected animals and artificial propagation of protected plants are generally carried out as commercial operations. It is therefore important that wild animals and plants which are caught or collected are not put on the market under false pretences.

This problem is addressed in several countries by a system of licensing breeders, horticulturists or nurserymen, and a requirement that books should be kept in which transactions are recorded. This system is in force, for example, in South Africa which grows many species of wild flowers and where horticulturists and florists who grow protected species must be licensed. All quantities bought and sold and the names of the buyers and sellers must be recorded. The buyer must retain the invoice as long as he is in possession of the plants.

In the United Kingdom, the Wildlife and Countryside Act of 1981 requires that where birds of certain listed species are born in captivity, they must be ringed at their birth with close rings which cannot be removed. Only ringed birds may be kept, transported, bought or sold.

5. Trade in Federal States

Problems may occur where jurisdiction over wild animals or plants is vested in a country's individual States or regions.

In Switzerland, for instance, many animals and plants are protected by federal legislation. Few problems arise in respect of these species, since trade controls are applied throughout the country. The situation is very different for those species which are protected by the Cantons under their own legislation. As soon as specimens have crossed the boundary of the Canton, they may be freely traded in other Cantons if the species concerned is not protected there. No legal action may be taken where a species is traded in a Canton in which it is not protected. If they so wish, the Cantons may of course protect the species concerned or, as is the practice in the Canton of Solothurn, may prohibit the sale of wild plants which are protected in other Cantons.

Nevertheless, it would still be simpler to have nationwide legislation where this is constitutionally possible. This is done in the United States, where the Lacey Act, called after its promoter, was adopted as long ago as 1900 to prohibit inter-State commerce in wild animals taken in violation of State law. The Act, as subsequently amended, applies to all wild animals, including animals bred in captivity, and to the taking, transportation, sale and possession of any such animals. With regard to plants, the Act only applies to indigenous endangered species, those listed under State legislation and those featuring on the Appendices to CITES.

In Canada, the new federal Wild Animal and Plant Protection Act of 17 December 1992 regulates inter-provincial trade in specimens of CITES species. In particular, the Act prohibits the transport from one province to another of any such specimen, where the animal or plant in question was taken or the specimen was possessed, distributed or transported in contravention of any provincial Act or regulation. No transport of a specimen from one province to another is allowed without a permit.

The federal Government is empowered to extend the application of the Act to non-CITES species which come under federal jurisdiction, or to species protected by a particular Province at the request of that Province.

In certain federal States such as Australia and Austria, as mentioned earlier, wildlife matters are entirely under State or provincial jurisdiction and the federal Government has no powers to regulate inter-State commerce in wildlife. In these countries, there is no means to regulate such trade except by voluntary cooperation between individual States. The best solution would seem to be that these States agree on a list of species for which they would all implement the same trade control measures, even if such species were not present on their territory.

B. International Trade

International trade in protected species is largely governed by CITES, the Convention on International Trade in Endangered Species of Wild Flora and Fauna, which was signed in Washington on 3 March 1973. However, export and import controls imposed at national level formed part of the current panoply of conservation instruments long before CITES was concluded. CITES itself principally provides for the international coordination of trade controls through the acceptance of obligations under international law.

1. Exports

The prohibition or control of exports is the logical consequence of domestic restrictions on taking and trade. The purpose of regulating exports is to ensure that unlawfully-obtained specimens of protected species are not exported.

A large number of countries have instituted export controls in their legislation by requiring export permits, which are usually granted where proof of lawful taking of the specimen in question can be produced. Such proof is usually provided by a certificate of origin or ownership. In most of these countries, however, these controls are only applicable to protected species. As a result, the export of all other species is wholly unregulated.

In contrast, a few countries have enacted more comprehensive legislation. Some require export permits for the export of all animals belonging to certain groups, generally vertebrates, or for all animals except for fish (Gambia), or for all animals (Cameroon).

Some countries, such as Brazil and Ecuador, also prohibit exports unless these are for scientific, educational or other non-commercial purposes. Under its foreign trade legislation, India has banned the export of any wild animal for any purpose, subject only to very minor exemptions.

Export controls are more rarely applied to plants other than protected plants. Congo requires an export permit for the export of any wild animal or plant specimen or product.

The country which has developed perhaps the most comprehensive system of export controls for all indigenous animals and plants is Australia. The federal Wildlife Protection (Regulation of Exports and Imports) Act of 1982 provides that no native wild animal or plant or part or product thereof can be exported without a permit.

A few exceptions are made to this general prohibition. These include fish and other marine animals used for food and, in respect of plants, seeds, spores, fruits, timber, wood chips, bark, articles derived from timber and eucalyptus oil. Permits can only be granted for specimens which have been lawfully obtained. Exports may only be authorised if this will not be detrimental to, or contribute to trade which is detrimental to the survival not only of the species concerned, but also of any species or subspecies of animal or plant.

Moreover, permits will be granted only if specimens have been taken or collected in accordance with an approved management programme. This is a unique feature of Australian conservation legislation, which is intended to ensure sustainable exploitation of the species concerned. Management programmes are approved by the appropriate federal Minister upon the production of evidence that the taking of specimens will not be detrimental to the survival of the species concerned; will be carried out at minimal risk to the continuing role of that species in the ecosystems in which it occurs; will be carried out so as to maintain the species in a manner that is not likely to cause irreversible changes to, or long-term deleterious effects on the species or its habitat; and that the management programme provides for adequate periodic monitoring and assessment of the effects of the taking of specimens on the species to which they belong, their habitat, and other species likely to be affected by that taking.

It may be that this requirement was eventually considered to be too burdensome, as the Act was amended in 1991 to allow for derogations in a rather large number of cases. These concern invertebrates, fish, plants and dead vertebrates. All species listed under CITES continue to require management programmes, as do live native mammals, birds, reptiles and amphibians.

2. Imports

Controls over imports of wild animals or plants may have several objectives.

Firstly, such controls prevent unlawfully-taken specimens which have been fraudulently exported from being legally imported at a later date into their country of origin and lawfully sold as 'imported goods'.

Second, they prevent the introduction of alien species which may cause harm to indigenous species and ecosystems.

Third, they restrict imports to specimens which have been lawfully exported from their country of origin. This double check system is fundamental to the functioning of the CITES Convention.

Fourth, such controls prohibit or restrict the import of species which are, in the opinion of the importing country, in need of particularly strict conservation measures.

Most countries which have instituted import controls limit these controls to specimens of their native species: unfortunately, this is also the case for many of the Parties to CITES! Some countries require an export permit and/or a certificate of origin from the exporting country before they will accept imports, and some have instituted a system of import permits. Some examples of the types of legislation in force are given below.

In Argentina, there is a prohibition on importing specimens of native species and their products. In Benin, no imports may be authorised of animal species without a certificate of origin and a health certificate. In Gabon, the import of any wild animal or product requires a certificate of origin and an export permit from the State of origin, on the basis of which an import permit will be issued. There is a total ban on importing species which are protected in Gabon.

In Switzerland, in addition to CITES requirements, an import permit is required for all live mammals, birds, reptiles and amphibians. Where the habitat of the species is restricted or its populations are small, an export permit may be required before an import permit can be granted.

Under the Endangered Species Act of 1973 in the United States, import controls only apply to listed species. However, the list contained in the Act does include many species which are non-indigenous to the United States. A permit is required for the import of any specimen of a listed species. In addition, the Lacey Act of 1900, as amended, prohibits the import into the United States of animals obtained or exported in violation of the conservation legislation of their country of origin.

The new Wild Bird Conservation Act of 1992 in the United States prohibits imports of live specimens of certain CITES-listed bird species. Other CITES species may continue to be imported for one year. After that period, imports will be prohibited unless the species features on an approved list to be drawn up by the Fish and Wildlife Service. Listing will be governed by the effectiveness of the conservation programme in the countries of origin. If the domestic legislation and enforcement measures in exporting countries are considered to be unsatisfactory, the import of the species concerned into the United States will be prohibited.

Under the Wildlife Protection (Regulation of Exports and Imports) Act of 1982 in Australia, import permits are required for specimens of all species listed under CITES,[33] all cetaceans, and for live animals of any origin with a small number of exceptions. Live plants are exempted, provided that their introduction into Australia is in accordance with the Quarantine Act. For all CITES species and cetaceans, import permits may not in principle be granted unless these species are under an approved management programme. However, in this context, it is the management programme of the foreign country that must be approved by Australia. Some derogations from this provision are now possible under the amendments of 1991 to the Act, as described above in the case of exports.

3. International Conventions

International trade controls on protected species are imposed under two international conventions on the conservation of wild fauna and flora. As early as 1940, the Western Hemisphere Convention required that its Parties control the import, export and transit of species protected under the Convention and they prohibit the import of specimens of fauna and flora which are protected in their country of origin unless they are accompanied by an export permit.

The African Convention of 1968 requires that the export of species protected by the Convention be subject to an authorisation and, in addition, makes their import and transit subject to that authorisation.

In contrast, more recent conventions such as the Berne and ASEAN Conventions have not provided for international trade controls. Since the signing of CITES in 1973, it has been increasingly accepted that all matters relating to international trade in endangered species should be left to this specialised treaty and that it would lead to unnecessary duplication if the same issues were dealt with in regional wildlife conventions. In practice, this is not always the case, as species protected at regional level will often not be listed on the CITES Appendices. For example, most of the species covered by the Berne Convention are not covered by CITES, which means that their international trade remains unregulated. As a result, certain endangered species

---

[33] Including those species for which CITES does not require such a permit.

which are fully protected in their country of origin continue to be lawfully imported and traded in countries to which they have been illegally exported.

Since the substantive obligations of CITES are well known, they will not be set out in detail here. In summary, Appendix I of the Convention lists species threatened with extinction, for which no commercial trade is allowed. The export of specimens of such species requires an export permit which may only be granted if the country of import has already granted an import permit. Both import and export permits may only be issued if Scientific Authorities[34] in the countries concerned have advised that the transaction will not be detrimental to the survival of the species. However, these provisions only apply to wild species. In respect of species listed in Appendix I, animals which have been bred in captivity and plants which have been artificially propagated for commercial purposes are treated as if they were on Appendix II. They are therefore subject to all the controls applicable to species listed in that Appendix.

Appendix II covers species which may become threatened with extinction unless their trade is strictly regulated to avoid utilisation incompatible with their survival. An export permit is required for the export of such species and may only be issued if the export will not be detrimental to the survival of the species. The import of the specimen is subject to the presentation of the export permit. Subject to these conditions, trade in Appendix II specimens is authorised.

Appendix III lists species listed by individual Parties which require the cooperation of other Parties to control the trade. The export of a specimen of a species on that Appendix from the State which has listed the species requires an export permit. The import of such a specimen from any other State requires the presentation of a certificate of origin, to prove that the specimen does not originate from the country which has listed the species concerned.

Export permits are not required for captive-bred animals and artificially propagated plant specimens belonging to species listed on Appendices II and III. The Convention provides that the State of export must instead issue a certificate stating that it has been satisfied that the specimens in question have been bred in captivity or artificially propagated. The same rule applies to specimens of species listed in Appendix I which have been bred in captivity or artificially propagated for non-commercial purposes.

The Convention applies to trade between Parties and non-Parties. Permits and certificates issued by non-Parties can only be accepted by Parties if they substantially conform to the requirements of the Convention.

Although the transit of specimens through the territory of a Party is specifically excluded from transactions to which the Convention applies, the need to control shipments in transit has become so necessary that the Conference of the Parties has now recommended that the existence of valid CITES documentation be checked. This also applies to transit between non-Parties, in which case the documentation should substantially conform to the CITES specifications.

Each Party must designate at least one Management Authority responsible for the issue of permits, as well as a Scientific Authority to advise the Management Authority on a number of matters, particularly on whether a proposed export would be detrimental to the survival of the species concerned. Should that be the case, the Management Authority is bound by that advice. This right of veto by the Scientific Authority when the survival of a species is at stake is an

---

[34] These are official bodies designated by each Party to the Convention which are described below.

essential element of the Convention. However, few Parties seem to be actually implementing it as their legislation establishes discretionary powers to their Management Authority to issue or deny permits irrespective of the advice of the Scientific Authority.

The Convention authorises Parties to adopt stricter measures for the import or export of listed species or to apply the same controls as those mandated under CITES to other species than those listed in the CITES Appendices. Pursuant to these provisions, as mentioned above, Australia requires import permits for all Appendix II species, as well as for many non-CITES species, and export permits for all native flora and fauna. The EC Regulation implementing CITES in the European Community obliges Member States to require import permits for all CITES listed species. In addition, some Appendix II species are to be treated as if they were listed on Appendix I.

The Conference of the Parties to CITES plays a particularly important role in the implementation of the Convention, by adopting resolutions which often go far beyond the treaty provisions. The legal basis for the exercise of these powers is the right of the Parties to take stricter measures than required under the Convention: there is nothing to prevent them from deciding that they will take such measures collectively. Resolutions are not binding in law, but there is obviously a strong moral obligation upon a State that has voted in favour of a resolution to apply it effectively. This quasi-legislative activity of the Conference of the Parties has permitted the development of the Convention far beyond what was initially envisaged by the negotiators.

One example of the scope of the Conference's activities relates to animals bred in captivity, which are largely exempted from the requirements of the Convention. There is therefore a considerable risk of fraud, as importing countries have no means of checking the veracity of certificates issued by the country of export relating to animals which are allegedly captive-bred. The Conference accordingly adopted a precise definition of the expression, "bred in captivity", and decided to establish a register of all operations breeding Appendix I species in captivity for commercial purposes. Parties are invited not to accept "bred in captivity" certificates for specimens of such species originating from unregistered operations.

Another interesting set of decisions of the Conference concerns ranched specimens, namely specimens taken from the wild and raised for commercial purposes in a controlled environment. It appeared that for certain species, this solution was preferable to breeding specimens in captivity, which would entail the establishment of self-reproducing captive populations. The latter approach does indeed tend to eliminate the need to maintain wild populations and may therefore be detrimental to the survival of the species. This is particularly true where the species is considered dangerous or harmful, as is the case with crocodiles.

Ranching, on the other hand, provides an incentive to maintain wild populations as a source of eggs or young for raising in captivity. The Conference took important decisions dealing with crocodiles, by allowing the export of crocodile skins obtained from ranching operations in a number of countries under annual quotas established by the Conference, as well as an uniform marking system for these skins to avoid fraud.

CHAPTER VI
ENFORCEMENT

A. Practical Difficulties of Enforcement

Legislation for the conservation of species is notoriously difficult to enforce for several reasons.

Long lists of species are of relatively little use if few people can recognise the species concerned. The negative listing approach may help considerably in this respect, as public and enforcement personnel have only to be able to recognise a relatively small number of authorised species, as has the 'small bunch' rule for plants.

An element of intention is often required in order for an offence to have been committed, yet this may be extremely difficult to prove and it is more usual for good faith to be presumed. In this context, the hunting proficiency test is particularly valuable. Hunters who have passed the test have learned about protected species and would therefore have greater difficulty in pleading ignorance of the specimens they may have taken or damaged unlawfully.

Reversing the burden of proof can contribute significantly to effective enforcement. The effect of such a reversal is that possession of a specimen of a given species is deemed to be unlawful, unless the possessor can prove otherwise. For example, under the Customs Code in France, the possessor of CITES specimens must be able to prove that these have been lawfully obtained. However, great care must be taken if an offence is deemed to be one of strict criminal liability, as there may be a serious risk of injustice incompatible with democratic societies.

The enforcement of trade restrictions is easier than that of taking or collection restrictions, provided that there are both sufficient funds and personnel available. It is necessary to provide for the licensing of breeders, growers, taxidermists, processors, sellers and so on, as well as for procedures and equipment for marking and the issuing of tags. The holding of registers in which all transactions must be recorded must also be organised. This approach can work if there are enough inspectors, but it is nevertheless costly and cumbersome. Few countries go that far in practice.

Enforcement personnel must be specially trained, which is already provided for in certain countries. The Fish and Wildlife Service in the United States has its own agents. Italy and Spain have special units under the *Carabinieri* and the *Guardia civil*, but this is exceptional. Forest guards and game wardens are more commonly used in most countries.

An important aspect of enforcement is to educate the public. It is vastly preferable to inform and explain rather than to embark upon criminal proceedings and impose fines. Teaching school children about protected species is now a legal requirement in a few countries. Public education is supported by the widespread use of posters and pamphlets.

In a rather large number of jurisdictions, the legislation provides for the possibility of appointing honorary or voluntary wardens on the basis of their competence and interest in the matter. Such appointments are often made from amongst the members of naturalist societies or conservation NGOs. By way of example, there are honorary game wardens in many African countries. In Europe, some Austrian and German Länder, Italian Regions, Swiss Cantons and other countries such as Bulgaria have nature conservation wardens. The Czech Republic has Nature Guards. Similar systems are used in some Australian and American States.

Wardens are officially appointed as auxiliary police officers and have the same powers as the police, except that they are not usually authorised to make arrests. Their role should accordingly consist for the most part of providing information and giving warnings. Excessive zeal may be counter-productive, especially for minor infractions committed in good faith.

B. Penalties for Offences under Conservation Legislation

Penalties vary considerably from one country to another, which may be considered as a reflection of the way in which different societies judge the seriousness of offences committed against species. Some countries rely entirely on the imposition of fines. Other countries also make use of prison sentences, generally for short periods of time, although in some countries they may be longer or even very long.

Under the Endangered Species Act in the United States, a sentence of one year may be imposed for the taking of endangered species. In Australia, a maximum prison sentence of five years was originally established for the import or export of species for which a permit is required. However, this maximum has been raised to ten years under amendments to the conservation legislation made in 1991! China allows for capital punishment for serious offences in respect of wildlife.

It should be added that a higher penalty may sometimes be imposed for offences concerning endangered species. For example, penalties under the Endangered Species Act in the United States in respect of endangered species are double those for offences related to threatened species.

In addition to fines or imprisonment, the legislation usually provides for the confiscation of specimens which have been taken, held or traded illegally. Confiscation may also often be ordered in respect of the vehicles, equipment and weapons used in the offence.

Attempting to commit the offence is usually also punishable. Some countries punish offences committed through negligence. Certain laws may provide that each violation constitutes a separate breach of the legislation. Where a large number of specimens are involved, this may amount to very substantial sums in fines.

Furthermore, legislation may authorise the withdrawal of licences and impose a prohibition against the obtaining of new licences for a certain period of time, usually several years. Some laws provide that upon a second conviction, the withdrawal of the licence may be permanent. These provisions may apply not only to hunting and collection licences but also to traders, for whom even the temporary withdrawal of a licence might have serious economic consequences. In some jurisdictions, legal persons can be fined for offences against species conservation legislation. In Australia, for instance, the maximum fine that may be imposed on legal persons is double that which may be imposed on natural persons.

A certain number of jurisdictions have instituted a system of damages to be paid to the State, according to a scale of values attributed to different endangered or protected species or groups of species. These values are calculated on the basis of the rarity or degree of endangerment of each species concerned. This system of payments is not the equivalent of a range of fines, but should rather be seen as compensatory damages for the destruction of a valuable natural resource.

Countries that have adopted this approach include Bulgaria, Hungary, Spain, Turkey and Zimbabwe. It also exists at federal level in Australia and in some Australian and American States, such as Idaho. In the Australian State of Queensland, the new Conservation Act of 1992 provides for payment to be made for the "conservation value" of a species. This system existed prior to the new Act, but such payments were formerly referred to as "royalties". Payment is made to the State for any taking of a protected species, including taking under a licence or permit. In cases of unlawful taking, the level of such payments may be doubled.

Finally, perhaps the greatest problem of enforcement is the failure to prosecute offences. If law enforcement personnel are not really interested in that type of offence, public prosecutors are probably even less so, unless the offence is very serious and involves, say, large numbers of valuable specimens.

Even where prosecutions are brought, courts tend to be lenient in their sentencing policy. After all, offences against nature do not affect anyone's financial interests. Moreover, in many countries where environmental consciousness is still low, harsh penalties would be perceived by the public as illegitimate and not commensurate with what are generally viewed as minor offences. At any rate, information on prosecutions and convictions is generally not available in most countries, except for important cases that have been widely publicised.

CHAPTER VII
INTEGRATED SPECIES PROTECTION

As the preceding chapters have shown, nearly all national legislation for species conservation is centred on the protection of individual members of species. It is increasingly necessary for legislation to evolve from the protection of the individual to the conservation of the species as a whole, but very few States have yet done so: the rare examples are given below. In parallel, there is a need for States to develop integrated species protection that takes into account all relevant factors, such as habitat protection, control of damaging processes and the promotion of positive measures, such as recovery plans.

A. Restrictive Measures for the Protection of Species' Habitats

Most laws are limited to prohibitions or restrictions on the taking of and trade in certain species. In consequence, such legislation ignores other threats, of which the most significant is the destruction or alteration of their habitats. Even where legislation does contain a broad definition of "taking" which includes destruction, it often exempts those activities which are likeliest to result in the destruction of natural habitats. As mentioned above, such exemptions chiefly relate to agriculture, forestry, public works and the construction of buildings.

In both national and international law, the protection of natural habitats is more usually effected through the establishment of protected areas. Under international instruments, the obligation for Parties to protect species' habitats generally requires them to establish some form of protected area. For instance, the EC Birds and Habitats Directives require the creation of Special Protection Areas and Special Areas of Conservation respectively. The Berne Convention does provide an exception to this rule, but the obligation it sets out to preserve the habitats of protected species is notoriously poorly implemented because it is far too general.

It should be noted that there is only very rarely an obligation under international law or national legislation to establish protected areas for the purpose of protecting endangered species. Exceptions to this rule are found in article 194.5 of the Law of the Sea Convention and in the EC Birds and Habitats Directives.

1. Automatic Habitat Protection

A small number of laws provide for the automatic protection of habitats of protected species. In other words, the legal consequence of the presence of a protected species in a given area is that the habitat in which it lives must be protected without the need to enact any additional regulations.

Under the Nature Conservation Act of 1976 in Ireland, for instance, the wilful alteration of, damage to, destruction of or interference with the habitat or environment of a protected species of flora is prohibited. However, the need for an element of intention, indicated by the term, "wilful", raises doubts as to whether this provision is applicable to farming or other legitimate activities.

Some Canadian Provinces, such as Ontario, New Brunswick (for plants only) and Alberta, protect the habitat of protected species. Similar provisions apply in Quebec under its Act on Endangered or Vulnerable Species of 21 June 1989. In habitats of endangered or vulnerable species of flora, for instance, the Quebec Act prohibits any activity which is likely to modify existing ecological processes or biological diversity, as well as the chemical or physical characteristics of the habitat. Exceptions may be made by regulations. In addition, the Government may authorise any activity even if it is likely to modify the habitat concerned, if it considers that the fact that the proposed activity would not otherwise be carried out would be more damaging to the community than the alteration of the habitat of the species concerned.

Certain Austrian Länder, such as Vienna, prohibit the destruction of or interference with the habitat of fully protected plant species. Lower Austria has similar provisions, but exempts agricultural and forestry activities, including land improvement projects and changes in agricultural land-use.

In the Czech Republic, the new Act on the Protection of Nature and the Landscape of 19 February 1992 protects the biotopes of all specially protected animals and plants. As far as plants are concerned, an exemption is provided for agricultural operations but only to a limited extent, as it merely applies to "usual cultivation". The Act specifies that this does not cover interventions which may cause changes to the hydrological conditions of the soil, the surface soil or the chemical properties of the environment, other than those interventions which are carried out in the course of normal forestry work in accordance with a valid forestry plan. In addition, the exemption does not apply to critically and severely endangered plants, for which the method of "usual cultivation" used requires the prior advice of the nature conservation authorities. The latter may impose compensatory protective measures, such as the transfer of the plants.

With regard to animals, the exemption under the Czech Act applies to current work on real property. However, any intervention requires prior consultation with nature conservation authorities which may impose compensatory protective measures. In no circumstances may this exemption be applied to critically or severely endangered species.

A different approach is used in Finland, pursuant to the Nature Conservation Act of 1923 as amended in 1991. The amendment establishes the obligations of landowners with respect to the habitats of listed endangered and specially protected species. Where a recovery plan has been developed for any such species, the County administration must inform landowners of the presence of that species on their land and of the activities liable to affect it or endanger its survival. If the landowner wishes to undertake one of these activities, he or she must give notice to the County at least one month in advance. This requirement allows for negotiations for the establishment of a reserve, subject to compensation being paid to the landowner. The County has the power to make an interim conservation order for a period not exceeding two years to safeguard the habitat, before the reserve is formally established. Compulsory purchase of such land is also possible.

The Finnish system has certain resemblances with the system of Sites of Special Scientific Interest (SSSIs) in the United Kingdom.[35] SSSIs are designated by English Nature on the basis of scientific criteria related to the flora, fauna or geological or physiographical features which

---

[35] Discussed in detail in Part II, Chapter IV(C) below.

make the site of special interest. There is nothing to prevent SSSIs being designated to protect species' habitats, although such protection is more usually incidental. Exceptionally, where a site's features are of "special interest" and an SSSI is in need of extra protection, a Nature Conservation Order[36] may be made by the Secretary of State for the Environment, *inter alia*, for "the purpose of securing the survival in Great Britain of any kind of animal or plant".

Just as the scientific interest of a site unconditionally triggers conservation procedures under the SSSI system, the identification of a particular area in the United States as essential to the survival of an endangered species automatically sets in motion habitat protection measures under the Endangered Species Act of 1973. The Secretary of the Interior is required to designate the critical habitats of each listed species, namely the areas which are essential to the conservation of the species concerned, and their boundaries must be precisely defined on maps published in the Code of Federal Regulations.

The scheme is therefore both species-specific and site-specific. It is also mandatory, although the Act does allow for certain exceptions where disclosure of the habitats' locations to the public might expose the species to vandalism, collection or other threats, or where insufficient information is available about the species at the time of listing to designate the appropriate habitats. Subject to those exceptions, the instrument is automatic in that any area which is of such importance for the survival of a species must be so designated.

Furthermore, the designation may specifically refer to the primary constituents of the critical habitat, thus making it clear which features are of particular importance for the protected species and that any alteration of these components of the habitat would be unlawful. By way of example, the constituent elements of the critical habitat of the Conasauga Longperch, *Percina jenkinsi*, include high quality water, pool areas with flowing water and silt free riffles with gravel and rubble substrate, and fast riffle areas and deeper chutes with gravel and small rubble.

Pursuant to section 7 of the Act, federal agencies must ensure that any action funded, authorised or carried out by them is not likely to jeopardise the continued existence of any listed endangered or threatened species or to result in the destruction or adverse modification of their critical habitats. Similar provisions have now been enacted in the legislation of certain States, such as California, as discussed at section (3) below.

In an interesting development, an extended definition of "taking" under the federal Act and subsequent regulations has been adopted as a result of its interpretation by the courts. In the *Palila* case of 1979, proceedings were instituted by conservation NGOs on behalf of an endangered Hawaiian bird, the Palila, *Loxioides bailleui* (*Psittirostra bailleui*), against the Hawaii Department of Land and Natural Resources. The habitat of the species was threatened by the grazing of feral sheep and goats. The court ruled that the survival of the species was jeopardised by the destruction of the native forest by such grazing and that this amounted to a prohibited taking within the terms of the Endangered Species Act.

Significantly, the federal Act as drafted only protects designated critical habitats against federal actions, as mentioned above. The *Palila* ruling therefore broadens the scope of the Act considerably as it establishes a general prohibition on the destruction of essential habitat by any person including, in the case in point, the State of Hawaii.

---

[36] Under section 29 of the Wildlife and Countryside Act of 1981.

A very similar case occurred in the State of New South Wales in Australia. The case was that of *Corkill v. Forestry Commission of New South Wales* in 1991, but is also known as the *Chaelundi* case after the name of the forest concerned.

Legal proceedings were instituted to prevent logging in a Government natural forest, on the grounds that the proposed operations would infringe the National Parks and Wildlife Act which prohibits the taking of endangered species. The Act defines taking as including "disturbing". The court ruled that

"taking cognisance of the object and purpose of the Statute..., it is reasonable to conclude that [the word] 'disturb' was intended to cater for indirect interference with an animal, such as would adversely impact on its breeding, feeding or nesting so as to disturb or destroy the habitat and lead to a reduction in species population."

It is immensely significant that courts are beginning to interpret species conservation legislation in terms of the survival of the species as a whole, rather than its individual members. This evolution is paralleled by the recent legislation enacted in a few States to protect species as such. The most notable examples are the Flora and Fauna Guarantee Act of 1988 of the Australian State of Victoria, and the Spanish Act on the Conservation of Natural Areas and of Wild Flora and Fauna of 27 March 1989, which are still in the earliest stages of implementation. The objective of both these Acts is to "guarantee" the survival of endangered species.

A new and interesting approach is that of the South Australian Native Vegetation Act of 1991. The Act prohibits the clearing of native vegetation without a permit.[37] When deciding whether to grant a permit, the permit-issuing authority must have regard to certain principles laid down by the Act and must not make a decision that is seriously at variance with those principles. One of the principles is that native vegetation "should not be cleared if it includes plants of a rare, vulnerable or endangered species" which is listed as such under the National Parks and Wildlife Act.

Finally, an Act of 22 January 1992 of the Italian Region of Liguria requires the Region to protect all main breeding, feeding, wintering and summering sites of all species of reptiles and amphibians. All such sites must be identified and listed, and the list must be updated every five years. For each site, the list should contain a description, indication of the main biological components, the degree of vulnerability and, where required, specific management criteria. The list must be approved by the Regional Executive. Once this has been done, sites are covered by the regional territorial coordination plan (the major land-use instrument in the Region), together with the specific rules applicable to each site. It is generally prohibited in all listed sites to modify the hydrological balance and the vegetation, to make earth movements, to modify significantly the physical and chemical characteristics of the waters, to drain the land, to use herbicides, insecticides and other chemicals. Additional protective measures may be taken if necessary.

Under the Ligurian Act, the protection of species is not quite automatic as a list of all main sites must first be drawn up and formally adopted. Nevertheless, the Regional Government has no discretion as it must identify and list all main sites.

In most other legislation, discretionary power remains the rule.

---

[37] For further details on this Act as an instrument for the protection of natural areas, see Part II, Chapter V below.

2. Discretionary Protection of Species' Habitats

In this category, the law empowers the appropriate authority to declare protected habitats by regulations. However, there is no obligation to make such a designation in respect of any or all habitats in which protected or endangered species are found.

Several Cantons in Switzerland protect the natural habitats of protected species by making the destruction or alteration of habitats subject to planning permission under land-use legislation. Such habitats must also be included in the zones of municipal land-use plans which are covered by maximum restrictions. The Canton of Zürich provides that where land-use restrictions are not sufficient to prevent developments which may affect species' habitats, specific regulations may be made to prohibit activities which are not covered by planning legislation and which may be detrimental to the animals or plants concerned. Compensation to landowners may be payable in such circumstances.

In France, the Nature Conservation Act of 10 July 1976 contains a general provision prohibiting the destruction, alteration or degradation of the habitats of protected species. This provision seems to have been included to provide automatic protection for such habitats. However, the implementing regulations merely empower the Préfet (the representative of central Government in the Département) to designate areas in which certain activities are prohibited or restricted to preserve the habitat of protected species.

Such orders are called "arrêtés de protection de biotope". They may only be made if a protected species is present in the area concerned and may only concern the protection of the habitat of that particular species and not, for instance, disturbance. As there are hundreds of such species, an "arrêté de biotope" may be made almost anywhere. However, not very many have been made so far, as the total does not exceed 220. In practice, many protected species which clearly need habitat protection measures are not covered by such "arrêtés".

In the Canadian Province of Quebec, the habitat of listed endangered or vulnerable species may be designated by ministerial order. Once this designation has been made, it is prohibited to exercise any activity which may modify existing ecological processes, biological diversity or the chemical or physical components of the habitat concerned, unless the activity in question is specifically authorised by regulations.

The Australian State of Victoria uses the critical habitat concept which it borrowed directly from the Endangered Species Act in the United States. However, under its Flora and Fauna Guarantee Act of 1988, the designation of critical habitats is not mandatory. Such designation may only be carried out in respect of listed endangered species, whether animals or plants. The legal effect of this designation is to impose a general prohibition on the taking of protected plants in their designated critical habitat, which also applies to landowners.

However, the main effect of the designation is to empower the conservation authority to make interim conservation orders if the critical habitat of an endangered species, or a species which it is proposed to list as endangered, is in need of preservation. An interim order will normally contain not only prohibitions or restrictions but also positive obligations to do certain things. It may prohibit any activity within the designated critical habitat as well as any outside activity that is liable to be harmful to that habitat. The order ceases to be valid after two years. In the meantime, the conservation authority is expected to negotiate an agreement for the safeguarding of the site.

3. Damaging Processes and Activities

Although habitat protection measures are an essential element of integrated species protection, they are by no means sufficient. Many threats result from damaging processes affecting species and their habitats, which often originate far from the habitats in question, and these processes must be controlled if the species are to survive.

The Flora and Fauna Guarantee Act of 1988 in Victoria is perhaps unique in that it provides for the listing of potentially threatening processes, defined as processes which may have the capacity to threaten the survival, abundance or evolutionary development of any taxon or community of flora or fauna. Under the listing criteria set out in the Regulations, a process is eligible for listing if it poses or has the potential to pose a significant threat to the survival or evolutionary development of two or more taxa or of a community of taxa. Examples of listed processes are alterations to the natural flow regimes of rivers and streams, predation of native wildlife by the introduced Red Fox, soil and vegetation disturbance caused by marble mining, and the use of lead shot in cartridges for the hunting of waterfowl.

However, the listing of a process has relatively few consequences. The conservation authority must prepare an action statement for any listed process, setting out what has to be done to manage the process and what is intended to be done. The authority may also make a management plan for any process. The conservation authority may enter into an agreement with any public authority for the management of any potentially threatening process.

None of the above provisions are binding. Nevertheless, the action statement does constitute a declaration of policy, which draws attention to the importance of the matter. Management agreements concluded with other public authorities may provide an informal way of getting things done.

It is rare for conservation legislation to include controls over activities that may affect protected species and their habitats.

The most notable law in this respect is, once again, the Endangered Species Act in the United States, which as mentioned above, prohibits federal agencies from authorising, funding or carrying out activities likely to jeopardise the continued existence of any endangered species or to destroy or modify the critical habitats of such species.

The strength of this provision has been tested on several occasions in the courts, particularly in the famous *Snail Darter* case (*TVA v. Hill*) in which the Supreme Court ruled that a dam could not be put into operation as it would threaten the survival of a newly-discovered small fish.

As a result of this judgment, Congress amended the Act in 1978 to establish a consultation process between the Secretary of the Interior and other federal agencies, intended to avoid major conflicts of this kind in the future.

An essential element of this consultation process is the making by the Secretary of the Interior (or the Secretary of Commerce for marine species) of a "biological assessment" whenever a listed species is present in an area likely to be affected by a proposed federal activity. If the assessment concludes that the species will be affected, the Secretary must issue a "biological opinion" showing how the proposed activity will affect the species or its critical habitat and proposing reasonable and prudent alternatives to avoid this result. These rules apply whether or not an environmental impact assessment is required.

The Secretary does not have a right of veto over the proposed activity. The federal agency may therefore decide to go forward with the proposed action in spite of the unfavourable biological opinion. However, if its decision is challenged in the courts, the agency must demonstrate that the biological opinion was erroneous or else it remains bound by its duty under the Act not to jeopardise the continued existence of the species concerned.

The weakness of the Endangered Species Act is, of course, that its provisions only apply to federal actions: it cannot be otherwise in a federal country. However, some individual States have enacted their own endangered species legislation, along the same lines but applicable this time to actions authorised, financed and carried out by the State in question. This of course brings a large number of projects and activities under the ambit of the Act.

In California, for example, the State Endangered Species Act of 1984 provides that the State Fish and Game Department must determine, in respect of any project for which an EIA may be required, whether it would jeopardise the continued existence of a State or federally listed endangered species or would result in the destruction or adverse modification of its critical habitat. Projects which would likely result in the extinction of such a species cannot be approved. For other projects, the Act requires the taking of reasonable mitigation measures to minimise their impact upon habitats essential to the continued existence of endangered species.

This requirement that public authorities should not allow activities which may adversely affect protected species also appears in the legislation of several Swiss Cantons. In the Canton of Zürich, land-use and planning legislation requires that rare and endangered animals and plants and the habitats necessary for their conservation should be preserved. In the exercise of their functions, all public authorities must pay due attention to the protection of such species and areas and safeguard them intact where this is in the public interest. This is a general obligation which does not depend on the identification of specific sites to be protected or on the making of specific regulations. It applies, in particular, to construction, the making and approval of land-use plans, the issue of permits and the grant of subsidies.

Finally, the new Convention on Biological Diversity requires Parties to identify, regulate and manage processes and categories of activities which have or are likely to have significant adverse impacts on the conservation and sustainable use of biological diversity (articles 7(c) and 8(1)).

B. Positive Measures for the Protection of Species

The preceding discussion has concentrated entirely on prohibitions. Very few laws in fact provide for active measures to promote the recovery and management of species.

Once again, the Endangered Species Act in the United States contains a general obligation for federal agencies to utilise their authority in furtherance of the purposes of the Act, by carrying out programmes for the conservation of endangered and threatened species.

More specifically, the Act requires the Secretary of the Interior (or the Secretary of Commerce in respect of marine species) to prepare and implement Recovery Plans for listed endangered and threatened species which occur in the United States. Such Plans must ascertain the conservation status of the species, identify all threats, establish recovery objectives and propose actions to meet these objectives within a certain time frame. They may also require captive breeding or artificial propagation programmes, as have already been implemented for several species such as the Californian Condor, *Gymnogyps californianus*, and the Black-Footed Ferret, *Mustela nigripes*.

Recovery plans usually require the cooperation of many different parties at federal and State level, including public agencies, universities and NGOs. Their content is of course not binding, but is of great value in setting objectives, determining what must be done and allocating the various tasks amongst the most appropriate bodies. The development of Recovery Plans requires research, money and considerable time. It is therefore not surprising that the number of completed Plans lags woefully behind the number of species in need of Plans. At the end of 1991, there were a total of 681 listed species occurring in the United States in respect of which only 382 Recovery Plans had been adopted.[38]

Some American States or territories, such as Puerto Rico, also require the preparation of Recovery Plans. Countries with similar provisions include Spain, under its Act of 27 March 1989 on the Conservation of Natural Areas and of Wild Flora and Fauna. Recovery Plans are necessary for species which are listed as endangered. These Plans must determine the measures necessary to eliminate threats to such species. Conservation Plans must be prepared in respect of species listed as vulnerable.

In Finland, the Nature Conservation Act of 1923 as amended in 1991 imposes a duty on the Ministers of the Environment and of Agriculture and Forestry to monitor the status of endangered species and to prepare Conservation Plans for specially protected species.

No recovery plans are required by law in the United Kingdom. However, English Nature launched a Species Recovery Programme in 1991 with a view to achieving the long-term self-sustained survival in the wild of the species of plants and animals currently under threat of extinction. This Programme was launched in partnership with a wide range of individuals and organisations and covers a first group of 20 species with effect from 1992.

The Convention on Biological Diversity requires Parties to adopt measures for the recovery and rehabilitation of threatened species and for their reintroduction into their natural habitats under appropriate conditions (article 9(c)). By way of a more general positive measure, article 11 of the Convention states that

"each Contracting Party shall, as far as possible and as appropriate, adopt economically and socially sound measures that act as incentives for the conservation and sustainable use of biological diversity."

---

[38] *Endangered Species Technical Bulletin*, published by the US Department of the Interior Fish and Wildlife Service, Vol. XVI, nos. 9–12, September–December 1991.

CHAPTER VIII
CONCLUSION

A. Endangered Species

Most legislation continues to deal almost exclusively with protection from taking of and trade in individual animals and plants, which are generally understood in a narrow sense.

It is nevertheless clear that the protection of a species means the maintenance or restoration of conditions, in particular of its habitat, without which it cannot survive. Very few laws address the problem from the point of view of the species as such, rather than of its individual members. The rare examples include the Endangered Species Act of 1973 of the United States, the Nature Conservation Act of 1923 of Finland, as amended in 1991, the Flora and Fauna Guarantee Act of 1988 of the Australian State of Victoria, and the Spanish Act on the Conservation of Natural Areas and of Wild Flora and Fauna of 27 March 1989.

The latter two Acts are very recent and little experience has yet been gained of their implementation, which is proceeding very slowly in any event. However, an interesting feature of both Acts is that they have as their stated objective to "guarantee" the survival of endangered species. They are probably the only laws in the world to have embodied such an ambitious statement to date.

It is clearly essential that legislation should address all threats to a species. The taking of and trade in a given species may be of little or no importance in many cases. It is therefore open to question whether the compilation of long lists of species whose taking is prohibited is in fact very useful. Indeed, in certain circumstances, these lists may be counter-productive because they are impossible to enforce.

However, taking should of course be prohibited where this may affect the survival of the species concerned or of other species, or may disrupt ecological relationships within an ecosystem. Prohibitions on the possession of and trade in endangered species are also necessary, even if no trade exists at a certain point in time, as a precautionary measure in case of the appearance of a sudden demand for the species concerned.

Certain texts have taken steps towards listing species whose habitats need particular protection, irrespective of whether those species are protected from taking. For example, the Spanish Act provides that lists of species sensitive to the alteration of their habitats should be drawn up, although none have yet been prepared. The EC Habitats Directive of 1992 lists species for which Special Areas of Conservation must be established to ensure the protection of their habitats.

Whereas the above laws are no more than curative in their purpose, what is actually important is the prevention of these threats. Ideally, the need for an Endangered Species Act should never arise. If societies continue to wait until a species becomes endangered, there will be thousands of species to be dealt with on an emergency basis. In the United States, there are already thousands of candidate species awaiting listing under the Endangered Species Act. Moreover, if the purpose of the Act is achieved and the populations of the species do recover, they should in theory be delisted immediately as they are no longer endangered. Once the species is no longer protected, the vicious circle of destruction-endangerment-regulatory intervention-recovery-destruction may begin all over again.

Since loss of natural habitat is usually the main problem, though by no means the only one, and drastic emergency measures are often required to safeguard or rehabilitate the habitat before it is too late, it may be necessary to embark on the acquisition of such habitats. However, this is not only expensive but is often impossible: owners may not want to sell their land, whilst compulsory purchase may not be possible either legally or politically.

In similar vein, land-use controls may often not be willingly accepted for the sake of an obscure plant, bat, rat, lizard, venomous snake or invertebrate. Controls are perceived in these circumstances as illegitimate, because they affect property rights, free enterprise, business opportunities and employment.

As a result, it is almost impossible to implement an extensive system of species-based land-use controls. This difficulty is exemplified by the 'spotted owl' controversy in the western part of the United States. What is at stake is not the survival of the spotted owl, as only one of its subspecies is actually affected, but the biological diversity of old growth forest which is home to hundreds of species. As there is no legal mechanism to protect endangered or declining habitat types, the presence of the spotted owl in the forests in question was used as the legal basis to invoke the provisions of the Endangered Species Act. Such tactics have backfired as the issue is now viewed in terms of the symbolic opposition of an owl versus the regional economy and employment of thousands of timbermen.[39]

Other factors which should be brought under legislative control include the regulation of potentially damaging processes as they may affect particular endangered species, as has been done under the Flora and Fauna Guarantee Act of 1988 in Victoria. These processes should be identified and assessed, so that legal and other means may be developed either to ban them completely or to make their effects on the species concerned less destructive.

Preventive measures do not of course solve the problem of species that have already become endangered. 'Gap analysis'[40] in the United States has shown that protected areas generally fail to preserve areas with the greatest concentration of endangered species. Although of course protected areas still have the potential to play a crucial part in the conservation of endangered species, protected area policy will nevertheless have to be re-orientated in many countries so that such areas can better serve the purpose of preserving endangered species and biological diversity in general.

Protected areas will not suffice on their own to save endangered species. However, the protection of species-specific habitats seems difficult to achieve unless fair compensation can be

---

[39] At the time of writing, is was hoped that an environmental conference convened by the newly-elected United States administration had made progress towards resolving the deadlock.

[40] A technique whereby a map showing the range of endangered species is superimposed on a map showing the location of protected areas, in order to determine whether or not the boundaries of these areas are correctly drawn.

provided to private landowners. Even where accompanied by compensation, prohibitions or restrictions on land-use continue to be widely unpopular. These negative methods will not be sufficient where positive management measures are necessary to secure the continued existence of the species in question.

A preferable approach could well be to foster the conclusion of management agreements which should, wherever possible, be made binding upon successors in title.[41] Such agreements may be supported by the possibility of compulsory purchase of the land in question, to be used as a last resort if the landowner refuses to sign an agreement or does not comply with the conservation obligations set out in the agreement. However, the conditions of the agreement must be sufficiently attractive because if the survival of species is to be assured, it is essential that the presence of an endangered species should be considered by the landowner concerned as an asset rather than a liability.

Species recovery or management plans should be used to identify critical habitats of species which should then be given priority in the conclusion of management agreements.

On publicly-owned land, management agreements should also be developed between conservation authorities and the Government agency managing the land. This kind of arrangement exists almost nowhere, although the Flora and Fauna Guarantee Act of Victoria opens up interesting possibilities. If local communities are economically affected by conservation measures taken as a result of such agreements, appropriate incentives should be provided.

Recovery plans are essential for endangered species and should be developed everywhere. At present, they only exist in a small number of countries. Recovery plans should determine the conservation status of the species concerned, identify the threats to its survival and set out the conservation measures required in order of priority and with an accompanying timetable.

As far as is possible, recovery plans should apply to a species as a whole or to whole discrete populations of a species within the same country, irrespective of jurisdictional boundaries, whether territorial or functional. At international level, international recovery plans should be promoted through existing regional conventions, or at least through the few such conventions which have the institutional structures to make the development of such plans feasible and which could follow up their implementation. Such conventions include the Berne Convention, the Kuala Lumpur Agreement between the ASEAN countries and the Kingston Protocol.

B. Exploited Species

If sustainable exploitation is ever to be more than a form of words, it is imperative to limit the catch or harvest of a given species on the basis of scientific rather than empirical data. Three essential conditions must be fulfilled to the greatest possible extent.

---

[41] This is discussed in greater detail in Part II, Chapter VIII of this paper which deals with Voluntary Conservation.

1. Unit Management

Species, particularly migratory species, or at least discrete populations of a species, must be managed as a single unit irrespective of jurisdictional boundaries. This problem is particularly acute at international level, and can only be solved by treaties and the establishment of international institutions setting up a framework for international cooperation.

This has sometimes been done in the case of fisheries, for example by the International Convention for the Regulation of Whaling of 1946. The International Whaling Commission regulates the taking of whales anywhere in the world.

However, such objectives are much more difficult to achieve for terrestrial migratory species. It is obviously essential that all Range States should be Parties to any agreement concerning such species. The internal cooperation mechanisms which have been put in place for migratory bird species in North America, between Canada, Mexico and the United States, seem to be working well, but nothing analogous is to be found elsewhere. The Bonn Convention of 1979 was largely concluded for that very purpose, but its implementation has been extremely slow. It is nevertheless hoped that the proposed Western Palaearctic Waterfowl Agreement will be concluded in the not too distant future, providing for unit management of these bird species along the full length of their migration route.

Similar problems often arise in federal States, which may be resolved by federal entities. In the United States, for example, the Connecticut River Atlantic Salmon Compact has been concluded between the States of New Hampshire, Massachusetts, Connecticut, Vermont, and the federal Fish and Wildlife and National Marine Fisheries Services.

Still in the United States, eight Regional Fisheries Management Councils have been established. These are composed of all the coastal States of a certain region, such as the North Atlantic, the South Atlantic and the Gulf, and of federal Government representatives. The Council has authority over all fisheries in the marine area concerned, whether these are under federal or State jurisdiction or, in other words, whether they encompass the exclusive fishing conservation zone or territorial sea. Each Council must propose a fishery management plan for each fishery in its area of authority.

2. Rational Management

Rational management requires that the permitted taking level for a given species should be adjusted to the capacity of the exploited population at any point in time. For this to be possible, detailed research is necessary together with reliable statistics determining the maximum or optimum sustainable yield and procedures for adequate enforcement.

Shooting plans for big game seem to be working increasingly well in this respect. In Europe, for instance, deer populations have probably not been as abundant for a long period of time.

For small game, however, the position seems to be much more difficult. In the case of migratory birds, rational management is impossible in the absence of unit management, as the allowable level of taking will depend on nesting success in other countries. A joint body is necessary for the collection of information and statistics and for the development of management plans.

With regard to commercial fisheries, relatively reliable information is generally available, but regulatory measures are difficult both to establish and to enforce. There are significant problems involved in carrying out rational management in practice, because of short-term economic reasons and political considerations.

There are very few known instances of the rational management of plants.

The control of trade plays an essential role in rational management because trade is an powerful incentive to overexploitation. This begs the question of whether one should wait until a species is 'threatened' by trade before prohibiting or restricting its commerce. It would seem to be more appropriate to proceed on the basis of the principle that trade in wild species should be banned, except where taking is regulated under an approved management programme that guarantees rational exploitation.

This can be made applicable not only at national but also at international level. Australia allows imports of certain species only when it has approved the management programme in the country of origin.

In respect of international trade, it is essential to continue to improve the operation of CITES, especially by the enactment of domestic implementation legislation which is still all too often sadly lacking, and also by strengthening institutional means to enforce the legislation. There is in particular a pressing need in most Parties to establish independent Scientific Authorities, as required under the Convention, to advise Management Authorities on the issue of permits.

Further development of export quota systems through resolutions of the Conference of the Parties may go a long way towards developing rational management of CITES Appendix II species.

In addition, as a means to seek to ensure that no specimens illegally obtained in the country of origin enter international trade, the Lacey Act system in the United States should be adopted by all wildlife-importing countries. By way of reminder, that Act prohibits the import of and inter-State commerce in specimens of species which are protected in their country of origin. For the system to work, there should be an international data base of legislation for the conservation of wild fauna and flora in every country, including lists of protected species, such as the one developed by the IUCN Environmental Law Centre, which could be consulted by importing countries whenever they have reason to believe that a specimen has been illegally exported.

3. Ecological Management

As with protected species, the ecological conditions that are necessary to the life and development of exploited species must be preserved, yet legislation is almost universally silent on this matter.

Ecological conditions may admittedly be protected through other methods. These include the National Wildlife Refuges in the United States which safeguard large areas of game bird habitats. However, most hunting and fishing legislation is completely dissociated from the need to meet the ecological requirements of the species concerned.

This is often particularly true for marine fisheries, as essential spawning or nursery habitats, especially when located in estuaries, mangroves and all coastal areas, are almost universally ignored by fisheries legislation. This prompts the question as to what use it is to calculate the optimum sustainable yield of a fish stock if recruitment in the fishery is jeopardised by the destruction of critical habitats.

Similarly, harmful processes such as pollution are seldom taken into consideration with regard to their specific effects on exploited species.

Perhaps greater responsibility should be placed upon the users in trying to ensure that adequate ecological conditions are met. Resource harvesters generally count on Government to do the necessary research and expect Government to do what is necessary to maintain the resource on a sustainable basis, but object to the imposition of restrictions which may affect them.

The United States system of duck stamps and special taxes on weapons and gear for the acquisition of waterfowl habitat is an important step in that direction.

Finally, in addition to these three conditions, a last and perhaps even more difficult objective is to provide incentives for rational and ecological exploitation, by avoiding wasteful competition between harvesters and ensuring that benefits from the resource will go to those responsible for its conservation. This is necessary to avoid the tragedy of the commons, and can be done by assigning ownership of the resource to a single user or group of users. Alternatively, and this amounts to the same thing but may be easier from the legal point of view, a right of exclusive exploitation may be assigned through a concession contract or a lease, as with the leases over hunting areas in central Europe.

Such arrangements are of course subject to special rules that the user must respect. The CAMPFIRE system in Zimbabwe and the innovative system of wildlife ownership in the new Wildlife Conservation and National Parks Act of Botswana of 1992 are examples of what can be done.

PART II

AREA-BASED CONSERVATION AND THE LAW

CHAPTER I
THE INTERNATIONAL LAW OF PROTECTED AREAS

Most conservation treaties contain obligations to conserve natural habitats through the establishment of protected areas or otherwise. Some provide precise definitions of the types of protected areas that their Parties are required to create, together with specific rules which must be observed in respect of these areas. Others merely lay down a general obligation to preserve natural habitats, without specifying the means which must be used to achieve this result. A few conventions provide for a list of specific sites that Parties commit themselves to protect.

A. Treaties

1. Treaties laying down Obligations to establish Protected Areas

a. African Convention

The first treaty containing such an obligation was the London Convention of 1933 on the Preservation of Fauna and Flora in their Natural State. Contracting Parties were to explore forthwith the possibility of establishing national parks and strict nature reserves. Precise definitions were given of these two categories of protected areas and their legal regime was clearly specified. It was required in particular that such areas must be under public control and that the boundaries of national parks could not be altered except by the competent legislative authority. Contracting governments were to give consideration to the establishment of buffer zones. All parks and reserves created under the Convention were to be notified to the depository. Many of the greatest African national parks were subsequently established pursuant to these provisions.

The London Convention was replaced in 1968 by the African Convention on the Conservation of Nature and Natural Resources signed at Algiers. Most of the earlier provisions on protected areas were, however, taken up by the new Convention with little change.

b. Western Hemisphere Convention

The second convention in chronological terms to provide for the establishment of protected areas was the Convention on Nature Protection and Wildlife Preservation in the Western Hemisphere of 1940. Contracting governments were asked to explore at once the possibility of establishing protected areas of which four categories were defined: national parks, national reserves, nature monuments and strict wilderness reserves. The legal regime for such areas is largely similar to that laid down by the London Convention of 1933.

c. South Pacific Convention

The Convention on the Conservation of Nature in the South Pacific, signed in Apia on 12 June 1976 and in force since 26 June 1990, contains provisions relating to national parks and national reserves, the text of which is remarkably similar to the corresponding articles in the African Convention.

d. ASEAN Agreement

The ASEAN Agreement on the Conservation of Nature and Natural Resources of 1985 also defines national parks and reserves and sets forth specific rules for the conservation of these areas once they have been established. However, the text of these provisions is much stronger than in the earlier conventions in that the obligation this time is no longer merely to explore the possibility of creating these areas but actually to establish them.

The purposes of protected areas are also clearly stated. They are designed to safeguard *inter alia* the ecological and biological processes essential to the functioning of the ecosystems, satisfactory population levels for the largest possible number of species of fauna and flora belonging to these ecosystems, and areas of particular importance because of their scientific, educational, aesthetic or cultural interest.

Parties must take all possible measures in their power to preserve those areas which are of an exceptional character and are peculiar to their country or the region. Protected areas must be regulated and managed in such a way as to further the objectives for the purpose of which they have been created. In those areas, Parties must prohibit activities which are inconsistent with such objectives. There are also specific provisions on land planning and land-use and environmental impact studies. Parties are to cooperate in conserving and managing shared resources and contiguous protected areas.

The Agreement also requires Parties to prepare management plans for protected areas; to manage these areas on the basis of these plans; to establish buffer zones; and to endeavour to prohibit the introduction of exotic species into protected areas, the use or release of toxic substances or pollutants which could cause disturbance or damage to protected ecosystems as well as activities exercised outside protected areas where they are likely to cause damage to these areas. However, the Agreement contains few prohibitions relating to activities that may be detrimental to national parks and reserves.

e. The Protected Areas Protocols to Regional Seas Conventions

Protocols relating to the conservation of natural areas have already been concluded in respect of four regional seas: the Mediterranean (Geneva, 1982), East Africa (Nairobi, 1985), the South-East Pacific (Paipa, Colombia, 1989) and the Caribbean region (Kingston, Jamaica, 1990).

All four Protocols require their Parties to establish marine or coastal protected areas. However, in the first of these Protocols, that covering the Mediterranean, this is not a binding obligation: it is merely provided that Parties shall establish such areas only to the extent possible and shall only endeavour to undertake the action necessary to protect these areas. In contrast, this obligation is binding in all three subsequent Protocols. All four Protocols provide for the establishment of buffer zones, but this obligation is only binding in the South-East Pacific Protocol.

No Protocol defines any particular categories of protected areas, which accordingly leaves Parties free to use any category of their choice to meet their obligations. Nevertheless, all four Protocols contain non-exhaustive lists of detailed measures that may be taken to preserve protected areas established thereunder. There is no binding obligation to implement these specific measures, although in all but the Mediterranean Protocol, these lists are linked to a general obligation to achieve the conservation objectives for protected areas.

Environmental impact assessments in respect of activities liable to have adverse effects on protected areas are required by the South-East Pacific and Caribbean protocols but not by the two others.

In contrast to the Protocols, many other treaties—and this seems to be indicative of a new trend—now only establish performance standards, and do not include any provisions specifying the means to be used to achieve them.

2. Instruments laying down Performance Obligations

a. The Global Conventions

All the treaties which contain specific provisions relating to the definition, establishment and preservation of protected areas are regional treaties. They are presumably founded on the idea that it is both possible and desirable to achieve a fairly great degree of harmonisation at regional level of protected area concepts and protection rules.

In the early 1970s, when the first global conservation treaties were concluded, it seems to have been realised that it was no longer possible to lay down universal hard and fast rules applicable to all protected areas. It was therefore thought to be preferable to limit binding obligations to general principles or performance standards, leaving it to each individual Party to enact and apply its own legislation, according to its particular conditions.

This method is used by the two most important global treaties dealing with the conservation of ecosystems: the Convention on Wetlands of International Importance (the Ramsar Convention) of 1971 and the World Heritage Convention of 1972.

Neither of these two conventions defines any category of protected areas or lays down any particular rules relating to the kinds of protection measures that Parties should take to meet their obligations. Thus, the Ramsar Convention merely requires that Contracting Parties

"formulate and implement their planning so as to promote the conservation of wetlands" and
"promote the conservation of wetlands and waterfowl by establishing nature reserves ... and provide adequately for their wardening".

The World Heritage Convention provisions are even more general. Parties recognise that they have a duty to identify, protect, conserve and transmit to future generations the natural heritage situated on their territory and do all they can to this end to the utmost of their resources. In particular, they must take the appropriate legal, scientific, technical, administrative and financial measures that are necessary to achieve this result. There is no mention of protected areas or of any kind of specific protection measures that should be taken.

Another global treaty which lays down general rules for the conservation of natural areas is the United Nations Convention on the Law of the Sea of 1982, which is still not in force. As mentioned in the Introduction, article 194.5 of that treaty sets out a general obligation to protect fragile ecosystems as well as the habitat of depleted, threatened or endangered species and other forms of marine life. The Regional Seas Protocols were developed on the basis of this provision to deal specially with protected marine and coastal areas.

There are also a certain number of treaties relating to the protection of particular species which require their Parties to establish protected areas for the preservation of those species and their habitats. Provisions of this kind appear in several of the bilateral Conventions that have been concluded for the protection of migratory birds and in the Convention for the Conservation of Vicuna of 1969.

Other treaties, such as the Agreement on the Conservation of Polar Bears and the global Bonn Convention on the Conservation of Migratory Species, merely provide for the protection of the habitats of the species concerned, without specifying whether this should be achieved by the creation of protected areas or by other means.[1]

b. The European Convention

The Convention on the Conservation of European Wildlife and Natural Habitats of 1979 (the Berne Convention) goes even further in that it does not even mention protected areas. It simply requires its Parties to take appropriate and necessary measures to ensure the conservation of the habitats of wild flora and fauna species, especially those which are listed as fully protected in Appendices I and II to the Convention. Parties are also required to take measures for the conservation of endangered European natural habitats, such as peatlands, grasslands and salt marshes, irrespective of the endangered species which these may contain (article 4). In consequence, Parties to the Convention are completely free to use any legal instrument of their choice to fulfil these obligations.

The new Convention on the Protection of the Alps of 1991 only lays down general obligations, including with relation to the conservation of nature and maintenance of landscape. The draft Protocol on nature conservation, still to be adopted, adopts a similar approach to the Berne Convention by leaving the Parties free to enact measures of their choice for the protection of species and natural habitats.

c. The European Community Directives

The European Community Directive on the Conservation of Wild Birds of 1979 (n 79/409) sets out a list of species in Annex I whose habitats require special conservation measures. Member States are required to establish Special Protection Areas for the conservation of these habitats. For that purpose, Member States must identify and protect the most suitable territories in number and size. This obligation also applies to the breeding, moulting and wintering areas and staging posts of regularly occurring migratory species not listed in Annex I. Parties must also pay particular attention to the protection of wetlands, especially to wetlands of international importance.

The new Directive on the Conservation of Natural Habitats and of Wild Fauna and Flora (n 92/43) of 21 May 1992 requires that EC Member States apply special conservation measures, firstly to natural and semi-natural habitats which are threatened in the Community, independently of the species which they may contain, and secondly, to the habitats of species whose habitats are similarly threatened. These habitat types and species are listed in Annexes I and II to the Directive respectively.

---

[1] Conventions dealing with migratory species are dealt with in greater detail in Part I, Chapter I(A)(3) above.

The protection of such habitats is to be achieved by the establishment of a pan-European network of Special Areas of Conservation on the basis of scientific criteria listed in Annex III. The objective of this network, discussed further in section (d) below, is to

"enable the natural habitat types and the species' habitats concerned to be maintained or, where appropriate, restored at a favourable conservation status in their natural range." (article 3)

In keeping with this broad goal, there are once again no rules specifying either the legal status or regime that should characterise these areas or the prohibitions or restrictions which should be applicable therein.

d. The Nature and Implementation of the Performance Obligation

Whether they require the establishment of protected areas or merely the taking of habitat protection measures, most of these instruments leave each Contracting Party the freedom to decide which areas it will protect. More specifically and in contrast to the procedure usually adopted for the protection of species, habitat conservation conventions never contain annexes drawn up by the original negotiators listing specific areas which the Parties on whose territory they are situated are, *a priori*, bound to preserve.

In the few cases where a list of areas to be protected does exist, such as under the Ramsar or World Heritage Conventions, this list is always established *a posteriori* as a result of the listings made by individual parties and never forms an integral part of the conventions concerned. Even in these cases, therefore, Parties remain free in principle to designate sites of their choice and only become committed to preserve such sites once these have been formally listed.

There are of course usually guidelines and criteria on the selection and use of the most suitable areas for protection pursuant to the objectives of the convention concerned. Such guidelines may either be included in the convention itself or may be developed and adopted subsequently by the Parties. Nevertheless, these guidelines and criteria are most often of such a general nature that Parties continue to enjoy a considerable degree of latitude in their choice of areas to be protected. This is the case, for instance, where Parties are merely required to preserve representative samples of natural ecosystems, the habitats of endangered species, or sites of particular scientific importance.

There are, however, a few cases where the criteria contained in a given convention, or adopted at a later time by the Parties, are sufficiently precise to create an obligation binding the Parties to preserve all the sites which fulfil these particular conditions. The subsequent inclusion of such sites on a list of areas to be protected under the convention, where it provides for such a list, then simply constitutes the materialisation of a pre-existing obligation to preserve these sites. There will, therefore, be a violation of the convention in question not only if a Party fails to protect a listed site adequately, but also if the Party has failed to identify, preserve and list all the sites that meet the criteria in question.

The first example of this kind of situation is provided by the Kingston Protocol for the Caribbean Region. Under that treaty, Parties have the obligation to establish protected areas for the purpose, *inter alia*, of preserving the critical habitats of endangered, threatened or endemic species of fauna or flora as well as areas whose ecological or biological processes are essential to the functioning of the wider Caribbean ecosystems. It may be argued, however, that the obligation here is to create protected areas for these specific purposes, but not necessarily to do so in respect of all critical habitats of all endangered, threatened or endemic specie, or of all areas with essential ecological or biological processes.

Both the EC Birds and Habitats Directives require that EC Member States establish protected areas for the preservation of the habitats of certain species and, in the case of the Habitats Directive, of certain endangered habitat types.

Under the Birds Directive, Member States are simply required to designate as Special Protection Areas the most suitable territories in number and size for the conservation of listed bird species. Although this criterion is admittedly not very precise, it is clear nonetheless that the failure to designate areas widely recognised by ornithologists as being of significant importance for the bird species concerned would constitute a violation of Community obligations.

The new Habitats Directive is more innovative in that it provides for a List of Sites of Community Importance to be established by the European Commission, albeit with the agreement of the Member States concerned. A European network of Special Areas of Conservation (SACs), known as Natura 2000, will be established pursuant to this Community List.[2] A three-step procedure for the identification and designation of these Sites is set out in the Directive.

Firstly, each Member State must propose a list of sites, based on scientific criteria set out in Annex III (Stage 1), indicating which natural habitat types in Annex I and species in Annex II native to its territory the sites host. This information must be transmitted to the Commission within three years of the notification of the Directive, together with information on the site which includes a map and specified data.

Secondly, the Commission selects Sites of Community Importance from these national lists according to criteria set out at Annex III (Stage 2) and with the agreement of Member States. Sites may be of importance either for the entire Community or for one of the five biogeographical regions contained in the Community, namely Alpine, Atlantic, Continental, Macaronesian and Mediterranean. The Community List must be established within six years of the notification of the Directive and adopted by the Commission in accordance with the procedure laid down in article 21.

The Directive introduces the concept of "priority natural habitat types" and "priority species",[3] for which the Community has particular responsibility in view of the danger of the disappearance of such habitat types or the proportion of the natural range of such species falling within the European territory of Member States. The Community List will be based on the number of priority natural habitat types or species found on the sites in question. All sites containing priority habitat types and/or species are automatically classified as Sites of Community Importance.

Thirdly, Member States have up to six years formally to designate the listed sites on their territory as Special Areas of Conservation and to enact appropriate conservation measures to maintain and/or restore such habitats and wild fauna and flora at favourable conservation status.

---

[2] This network is discussed in Section B(1)(c) below.

[3] These priority habitats and species are indicated by an asterisk in Annexes I and II respectively.

The establishment of SACs is a performance obligation, as with any Directive. Member States are therefore free to choose any suitable method to avoid the deterioration of natural habitats and species' habitats as well as the disturbance of the species for which the areas have been designated. Such measures should include, where necessary,

"appropriate management plans specifically designed for the sites or integrated into other development plans, and appropriate statutory, administrative or contractual measures which correspond to the ecological requirements of the natural habitat types in Annex I and the species in Annex II present on the sites."

Article 5 sets out a procedure to be followed in the event of any disagreement over listing between the Commission and Member States. It establishes a bilateral consultation procedure for cases where a Member State omits from its national list a site which the Commission considers essential for the maintenance of a priority habitat type or the survival of a priority species. If no agreement can be reached, the final decision on whether or not to list a site is taken by the Council, acting unanimously.

The detailed procedures described above shows that Member States have relatively little discretion in the choices they may make. Once a Site is recognised as being of Community Importance, it must be protected. Disputes may of course arise with regard to the characterisation of a site as "of Community Importance", but not over the obligation to preserve the site if its importance is undisputed.

A very similar obligation is at the heart of the World Heritage Convention. The main duty of each Party to that Convention is to ensure the identification, protection, conservation, presentation and transmission to future generations of the cultural and natural heritage as defined in articles 1 and 2 of that treaty. "Natural heritage" includes natural sites, natural features, formations or sites, which are of outstanding universal value from the aesthetic, conservation or scientific point of view or which constitute the habitat of threatened species of outstanding universal value from the point of view of science or conservation. Whenever a site meets those requirements, the Party on whose territory it is situated is consequently under a duty to preserve it, irrespective of whether or not it has been included in the World Heritage List.

In the words of the High Court of Australia,[4] it is for a Party to identify for itself the cultural and natural heritage on its territory. The obligation to conserve such heritage does not flow from inclusion in the World Heritage List but from the identification made by that Party pursuant to its obligation. Even if the World Heritage Committee subsequently refuses to enter a site on the List, article 12 of the Convention makes it clear that this refusal cannot be construed as meaning that the site ceases to have an outstanding universal value and therefore to be part of the World Heritage. The obligations imposed by the Convention in respect of such sites therefore remain in force.

It follows that whilst the inclusion of a property in the World Heritage List may, in a practical sense, confirm the appropriateness of the identification of that site as part of the World Heritage, its only consequence under the Convention is to make that site eligible for international assistance.

---

[4] *Queensland v. Commonwealth*, 1989, referred to as the *Daintree* case.

A site can only be entered on the List if it satisfies the criteria laid down by the World Heritage Committee in its Operational Guidelines. As a result, practical problems arise if a site is not included in the World Heritage List, because there is no other standard by which a State Party's identification of a project as belonging to the World Heritage can be judged. If the Party concerned has erred in its identification, would it still be considered as bound by the Convention to preserve that site?

This problem would obviously be a matter for the courts to decide, should the question arise. In the *Daintree* case mentioned above, the High Court judged that the decision as to whether a property forms part of the World Heritage is an executive decision and not a matter for the courts to determine.[5]

In conclusion, it is essential to note that the legal duty to preserve World Heritage Sites derives from the intrinsic qualities of the area concerned, rather than from a decision to designate an area as such.

B. International Networks of Protected Areas

If providing for the establishment of protected areas is good, organising these areas into an international network may be even better. Indeed, the creation of such a network may have many advantages. Gaps become readily apparent, which may constitute an incentive to establish new protected areas. A network provides better opportunities for cooperation as well as exchanges of information and even of staff between areas within the network. It is easier to develop and apply harmonised criteria and standards, which give an international dimension to the areas which are included in the network.

Networks may be established by international instruments or simply by the action of international organisations.

1. International Instruments

a. Networks that are not based on Lists having Legal Effects

The Mediterranean, East African and Wider Caribbean Protocols provide for the establishment of networks of protected areas in their respective regions.

The Mediterranean Protocol, for instance, requires its Parties to establish, to the maximum extent possible, a co-operation programme to co-ordinate the establishment, planning, management and conservation of protected areas, with a view to creating a network of protected areas in the Mediterranean region. It further provides that there shall be regular exchanges of information concerning the characteristics of the protected areas, the experience acquired and the problems encountered. Parties must also exchange scientific and technical information,

---

[5] B.M. Tsamenyi and J.M. Bedding: *Implementing International Law in Australia*, Journal of Environmental Law, VI.2 N°1, 1990, pp. 108–123.

co-ordinate their research and endeavour jointly to define or standardise the scientific methods to be applied to the selection, management and monitoring of protected areas.

Guidelines for the selection, establishment, management and notification of coastal protected areas were adopted in 1987.

Meetings of the Parties must monitor the development of the network and adopt guidelines to facilitate its establishment and management and to increase cooperation between the Parties.

The Nairobi Protocol to the Eastern African Regional Sea Convention likewise provides that Contracting Parties must establish a regional programme to coordinate the selection, establishment and management of protected areas and the protection of wild fauna and flora, with a view to creating a representative network of protected areas in the Eastern African Region. It also requires that there be a regular exchange of information between the Parties.

The Kingston Protocol also calls for the establishment of a cooperation programme and the creation of a network of protected areas. However, since this network is based on a list of protected areas which is provided with certain legal effects, it will be considered in section (B) below.

The fourth Protected Areas Protocol on the South-East Pacific does not specifically refer to a network. Nevertheless, it requires its Parties to cooperate in the management and conservation of protected areas, and to exchange information on the programmes and research carried out in those areas and on the experiences gained in each area, particularly in the scientific, legal and managerial fields.

b. Networks based on Lists having Legal Effects

Treaties providing for the establishment of protected areas do not usually require that particular sites be specifically protected. Parties therefore remain free to remove protected status at their discretion from the areas they have designated. This is clearly an unsatisfactory situation, especially where the areas concerned are considered to be of international importance because of their outstanding scientific or biological value.

For this reason, certain treaties have established mechanisms providing international recognition for certain specific areas included on a list. The Parties to the treaty of course remain free to propose any site of their choice for inclusion on the list. Once an area has been listed, however, a Party has certain obligations to preserve it and it cannot delist the site without having at least to follow a certain procedure.

Among the three treaties providing for an international list of protected sites, the Ramsar Convention is the least demanding.[6] It simply requires Parties to formulate and implement their planning so as to promote the conservation of the wetlands included in the List. Parties may, in principle, designate any site of their choice. Some criteria have been developed and approved by the Conference of the Parties, but as the Conference does not have to approve listings, these criteria are little more than guidelines which Parties are free to disregard. As to delisting, Parties

---

[6] By way of reminder, the Ramsar Convention also requires Contracting Parties to take conservation measures, including the establishment of reserves, to preserve wetlands not included in the List which accordingly do not form part of the network.

are free to delete a site for reasons of "urgent national interest". They must, however, inform the Convention Bureau of such changes and "should as far as possible compensate for any loss of wetland resources" resulting from such a delisting.

The World Heritage Convention provides for a much stricter procedure. The inclusion of a site on the World Heritage List requires the approval of the World Heritage Committee. Criteria for inclusion are defined by the Committee. Sites which do not meet the criteria will therefore be refused. The Convention is silent on the question of delisting, but it is doubtful whether an individual Party may unilaterally delist a listed site and this has never yet happened.

In a famous law case in 1983 in Australia, known as the *Tasmanian Dams* case,[7] the High Court of the Commonwealth judged that the Convention imposed a legal duty on each Party to protect World Heritage Sites on its territory and that there was, as a result, an international obligation not to abolish the protection status of such sites. However, the World Heritage Committee may delist any site which in its judgement has lost its World Heritage values, pursuant to the Operational Guidelines it has adopted for the implementation of the Convention.

The Kingston Protocol on Specially Protected Areas and Wildlife in the Wider Caribbean Region provides for a third type of procedure. Each Party may nominate a protected area for inclusion in the List of protected areas. Such nomination must be made in accordance with the guidelines and criteria adopted by Meetings of the Parties. Each nomination must be evaluated by a Scientific and Technical Advisory Committee. If the guidelines and criteria are met, the area will be included in the List by the Meeting of the Parties. Delisting must follow the same procedure as listing.

In all three cases, of course, listed sites must be considered as belonging to a network and cooperation among the Parties for the development and strengthening of the network is required.

On the basis of the experience acquired from the operation of the Ramsar and World Heritage Conventions (the Kingston Protocol is not yet in force), it appears that a certain number of conditions must be met if an international listing system is to work effectively.

There is, first, a need to ensure that sites proposed for listing correspond to the objectives of the convention concerned and meet specified criteria with regard to their international importance and conservation status. Whether the listing of a site should or should not be effected would then be a decision taken by the Parties on the advice of scientific experts. This procedure is provided for under the World Heritage Convention and the Kingston Protocol but not under the Ramsar Convention.

The delisting of a site by an individual Party should, as far as possible, be subject to a similar procedure. At the very least, advance notice of the intention to delist, with the reasons for the proposal, should be submitted to the Convention Secretariat, before a site is delisted, and the matter should be put for discussion on the agenda of the next meeting of the Parties.

Listed sites should be monitored for any serious degradation of their conservation status and the Parties concerned invited to take remedial action. The World Heritage Convention has a procedure whereby a site can be included in the List of World Heritage in Danger, and the World Heritage Committee has adopted criteria for the inclusion of sites on that list. The Parties to the Ramsar Convention have established a monitoring procedure which, with the agreement of the

---

[7] *Commonwealth of Australia v. The State of Tasmania*, 46 Australia Law Reports, 624.

Party concerned, involves an on-the-spot appraisal of the status of listed wetlands that are known to be threatened. The Kingston Protocol requires its Scientific and Technical Advisory Committee to carry out periodic reviews of the status of listed areas. All three treaties also require their Parties to submit reports on the status of listed sites. In the Ramsar Convention, this requirement is limited to those cases where the ecological character of a listed wetland has changed, is changing or is likely to change as a result of human interference.

Nevertheless, there is always a risk that a Party, without requesting the delisting of a site or changing its legal status, will merely allow its destruction or degradation to a point where it no longer meets listing criteria. There is therefore the need for a procedure allowing Conferences of the Parties to delist a site, once its maintenance on the list can no longer be justified. The Operational Guidelines to the World Heritage Convention provide for this possibility as they establish a procedure for the deletion of properties from the World Heritage List. Deletion may be effected where a site has

"deteriorated to such an extent that it has lost those characteristics which determined its inclusion in the World Heritage List and where the intrinsic qualities of a World Heritage site were already threatened at the time of its nomination by action of man and when the necessary corrective measures, as outlined by the State Party at that time, have not been taken within the time proposed".

Provided these basic requirements are met, an international site-specific conservation system has many advantages compared to a simple obligation to establish protected areas. In particular, it enables international attention, as well as the efforts of the Parties concerned, to be focussed on the need to preserve particularly valuable ecosystems as a matter of international and national priority.

Not least among these advantages is the possibility of setting up special financial mechanisms to assist Parties to discharge their obligations in respect of listed sites. The World Heritage Fund, established by the World Heritage Convention has been extremely successful. The Parties to the Ramsar Convention have now set up a Wetland Fund along on similar lines to contribute to the preservation of Ramsar sites. The Kingston Protocol authorises the Meetings of the Parties to seek voluntary contributions from any source to be used for purposes connected with its implementation.

c. The European Community Networks

In the EC, the inclusion of an area in the network of Special Protected Areas under the Birds Directive gives rise to an obligation on the part of Member States to avoid pollution or deterioration of habitats, insofar as these would be significant having regard to the habitat conservation objectives of the Directive.

In addition, once an area has been designated as a Special Protection Area, Member States must send all relevant information to the EC Commission, so that it may promote appropriate initiatives to coordinate these areas into a coherent whole, which meets the conservation requirements of the species concerned. The network of Special Protection Areas should not merely juxtapose areas individually protected by Member States, but should also provide for a well-balanced distribution of protected areas, so that the habitats of all species in need of habitat protection are covered all along their migration routes.

The new EC Habitats Directive goes further in that it provides for an Community-wide network of Special Areas of Conservation, which explicitly includes all the Special Protection Areas designated under the Birds Directive. This "coherent European ecological network", to be called Natura 2000, must be established within 12 years of the notification of the Directive to Member States, namely by the year 2004.

The detailed procedures for the designation of areas for inclusion in the network have been discussed in section (A) above. It should, however, be noted that where the sites hosting one or more priority habitat types and species represent more than 5% of a Member State's national territory, that State may request that the criteria for selecting Sites of Community Interest for inclusion in the Natura 2000 network be applied more flexibly.

Following their designation of sites as Special Areas of Conservation, Member States must take measures to avoid the deterioration of the protected habitats or disturbance to the protected species, and must make provision for environmental impact assessments of any project liable to affect a site (article 6(3)). Subject to very limited exceptions in cases of overriding public interest where there is no alternative solution, the competent national authorities should only authorise such a project if it will not adversely affect the conservation objectives of the site. Where such a project is carried out, the Member State is obliged to undertake compensatory measures to protect the overall coherence of Natura 2000 and the Commission must be informed of these steps.

In addition, land-use planning and development policies are required to be modified where necessary to encourage the management of landscape features (either linear, such as rivers and their banks or stepping stones, like small woods or ponds) that are of major importance for wild flora and fauna (article 10).

The Directive only provides for a Special Area of Conservation to be considered for declassification where this is warranted by natural developments (article 9).

The Commission must develop programmes for the development, monitoring and strengthening of the Natura 2000 network. Most innovatively, financial provision for this purpose may be made from the new Community Environmental Fund ("LIFE"), which was established by Regulation 1973/92 of 21 May 1992.

Member States may apply for Community co-financing when they notify the Commission of their national list of sites. The Commission assesses the financial requirements of the proposed conservation measures, in relation to the concentration of priority habitats or species on a Member State's territory and the relative financial burden imposed on the Member State in question. It then draws up a "prioritized action framework", granting co-financing to certain projects.

The action framework incorporates a two-yearly review, and Member States which have not received funding for conservation projects may postpone such measures pending the review, provided that they refrain in the meantime from any actions likely to result in deterioration of those areas.

2. International Organisations

a. Biosphere Reserves

A small number of international organisations have undertaken specific programmes for the establishment and conservation of protected areas. The only one of these which has a worldwide coverage is the Biosphere Reserve network which was developed under UNESCO's Man and Biosphere (MAB) programme. As there is no treaty or any legally binding obligations governing the network, designations of biosphere reserves are made on a purely voluntary basis. Proposed designations by individual States must, however, be approved by the MAB Coordination Council. Unsuitable areas may therefore be refused. However, there are no criteria that require some degree of legal protection to be accorded to an area before it can be listed, not is there any legal obligation to protect an area once it has been listed. Notwithstanding, there is a strong moral obligation to do so, in view of the prestige of the network and its major scientific importance. In 1992, there were 313 biosphere reserves located in 76 different countries.

Unlike sites on the World Heritage List which must each be of exceptional value, biosphere reserves are selected because they are representative of different types of ecosystems. Their purpose is the conservation of these ecosystems and the species they contain, scientific research, monitoring, education and training. In addition, biosphere reserves must be integrated into their social, economic and cultural environment and for that purpose local populations should be involved as much as possible in their conservation and management.

Biosphere reserves must be zoned, with a core area devoted to strict protection and a buffer zone in which human activities may only be authorised to the extent that they are compatible with the conservation objectives of the area. Around the buffer zone, there must be a transition zone where specific cooperative links with the local populations should be developed and maintained. It follows that biosphere reserves are usually inhabited areas except for the core zone.

This concept has evolved over the years. The first biosphere reserves to be designated were mostly protected areas of the conventional type, such as national parks and nature reserves. Few had buffer zones and the transition zone requirement did not even exist at that time. As the concept developed, new types of areas were included into the network. Some of these do not in fact have any specific legal status under the national legislation of the countries that have designated them. Apart from the fact that there is no obligation for a country to give legal protection to its biosphere reserves, the reason for this is perhaps that only a small number of countries have yet adopted legislation applicable to what is in fact a new category of protected area, in which human occupation and activities compatible with the purpose of the reserve remain authorised but may be strictly regulated. Furthermore, the integration of a reserve into its socio-economic environment may also require new regulatory or management tools which are still lacking.

The result is that there are considerable differences between the legal status of the reserves included in the network. These range from the strict protection usually given by national park legislation to no protection at all, where a designated area does not fall into any of the categories defined by the law. Paradoxically, the biosphere reserves which enjoy the highest level of legal protection are almost always those which in their design are the furthest away from the new concept which was developed over time.

It would seem, therefore, that some improvements are necessary to provide better legal protection for those areas which require it and to ensure a greater degree of harmonization between the legislation of the countries concerned. This was recognised by the MAB Coordination Council which in 1990 recommended that the legal basis of the biosphere reserve network be strengthened.

Although there has been so far little discussion on how this could be achieved, some preliminary thoughts on this matter may be of some use.

A requirement could be established to the effect that only those areas which have a legal status consistent with the objectives of the network, as presently defined, would be eligible for listing. It should also be made possible to delist areas which do not benefit from adequate legal protection or, as is the case for World Heritage Sites, which do not have or which have lost the characteristics required for inclusion in the network.

In parallel, the States concerned could be encouraged to develop national legislation, where this does not exist, designed to provide the necessary legal basis for the new biosphere reserve concept. New laws of this kind should not necessarily establish a "biosphere reserve" category of protected areas at national level: indeed, it would be preferable not to do so because the system should remain as flexible as possible. In other words, the national legal category or form of protected area assigned to a biosphere reserve does not really matter, provided the objectives and criteria for inclusion in the network are met.

National authorities should therefore be able to use the type of national instrument which appears to be the best adapted to the particular situation of each individual site. In addition, if specific biosphere reserve legislation were to be enacted, there would be no way in which national legislation could be harmonised. The definition of biosphere reserves, as well as the legal regime applicable to them, would then vary from one country to another.

On the other hand, what is essential yet generally lacking is legislation enabling the competent authority to establish protected areas which correspond to the objectives and standards of biosphere reserves as they are now defined, together with their core, buffer and transition zones.

There are at least two ways in which the designation of such zones can be achieved. The first way is through the extension of conventional protected areas, by setting up buffer and transition zones around their boundaries in concentric circles. At the same time, legal regimes appropriate to these zones must be established to permit the control of all human activities in the buffer zones and the provision of special benefits to local communities in the transition zones. Management institutions having jurisdiction over all three zones may also have to be developed.

The second way is the nature park concept which is now developing rapidly in certain countries, such as Italy and Spain. Nature Parks are inhabited areas where many human activities are allowed to continue, whilst their zoning system includes reserve areas where all activities are prohibited or severely restricted. Nature Parks are discussed in greater detail later in Chapter VI below.

In the first case, a biosphere reserve accordingly constitutes a natural area extending to surrounding inhabited zones, whereas in the second, it is an inhabited area comprising natural, strictly protected zones.

In both cases, there will always be a need to adapt existing legislation to these new requirements in accordance with national or even local conditions. Pilot projects may be necessary to determine the conditions under which each of these two approaches is likely to work better, or indeed whether another model would have better chances of succeeding.

b. Council of Europe

The Council of Europe, an organisation of European States with 26 Members as of March 1993, has for a long time devoted a part of its efforts to the conservation of natural habitats through a specialised Committee for the Conservation and Management of the Environment and Natural Habitats (CDPE).

Two programmes relating to protected areas were initiated by that Committee: the European Network of Biogenetic Reserves and the European Diploma.

The European Network of Biogenetic Reserves was set up in 1976 with the objective of conserving representative samples of natural areas and critical habitats of endangered species, promoting scientific research and helping to raise public interest in conservation. It is designed to complement the world biosphere reserve network at European level.

To be eligible for inclusion in the network, a site must be of European interest for nature conservation and have an effective protection status. Sites are proposed for inclusion by the Member States concerned. Proposals are screened by the Committee and accepted if they meet the criteria. Sites may be withdrawn from the network at any time. They may also be removed by the Committee if irreversible changes to their biological values have occurred. The Committee may also recommend that the Government concerned take such steps as may be necessary to bring a site in conformity with the objectives of the network. By December 1992, 286 sites in 17 Member States had been included in the network, covering almost 4,400,000 hectares.

The European Diploma was instituted as long ago as 1965. It may be awarded, at the request of the Member State concerned, for natural areas of European interest which are adequately protected.

The Diploma may only be awarded after an on-the-spot appraisal has been carried out by an independent expert, accompanied by a member of the Council of Europe Secretariat. It is awarded by a decision of the Committee of Ministers of the Council, subject to any conditions or recommendations that may be considered necessary for the adequate safeguarding of the site. After the Diploma has been awarded, the government concerned must submit annual reports to the Committee on the status of the area. The Diploma is only awarded for a period of five years. It may, however, be renewed after a new on-the-spot appraisal. New conditions and recommendations may be attached at that time. If the protection status of a area covered by the Diploma is no longer adequate, or if the conditions attached to the award have not been fulfilled, the Diploma may be withdrawn or may not be renewed.

The policy of the Committee since the inception of the institution has been to be very selective in the granting of the Diploma. Only 38 Diplomas (in 14 different countries) have been awarded in the 28 years of the existence of the institution. Many applications have been denied. By and large, the Governments concerned have complied with the conditions and recommendations laid down by the Committee and the Diploma has been a subject of pride for both the Governments and the managers of the diplomed areas. Only one Diploma has been withdrawn. There are periodic meetings of managers of Areas covered by the Diploma which are organised by the Council of Europe.

C. Transfrontier Protected Areas

Many protected areas have been established in frontier regions. The reason for this is that these regions are often sparsely populated, access is difficult and they are remote from major population centres, especially where the international border follows a mountain range. They remain relatively pristine and may have major biological and scenic values. It may therefore be easier to establish protected areas in these regions, as less local opposition and fewer political, economic and social problems may be expected.

These protected areas usually stop at the border and there is no corresponding park or reserve at the other side of the international frontier in the neighbouring country. However, it appears that in a surprisingly large number of cases, protected areas have actually been created on both sides of a border. A study undertaken by IUCN in 1988[8] shows that there are at least 70 examples of such a situation and that no less than 65 countries are involved. Some of the protected areas concerned are among the most famous in the world and are included in the World Heritage List. The Argentinian Iguazu and Brazilian Iguaçu national parks are the only examples of World Heritage sites established on the two sides of a border.

On the other hand, there are several cases of World Heritage sites on one side of the frontier with a still-unlisted adjoining protected area on the other side. Examples include the Manas Wildlife Sanctuaries in Bhutan and India, the Indian area being the only one listed; the Amistad national park in Costa Rica and Panama, the Costa Rica park being the only one listed; and the Darien national park in Panama which is listed, whilst the adjoining protected area in Colombia is not listed.

From a legal point of view, there is of course no difference between a protected area situated along the border and any other protected area in the same country. Where there are adjoining protected areas situated in two (or more) neighbouring countries, the international border constitutes at the same time a jurisdictional boundary between the management authorities of the protected areas concerned and the line at which the laws of the countries in question cease to be applicable. It follows that although the areas on either side of the frontier may constitute a single ecological unit, they will necessarily be administered and managed under different laws and regulations and by different administrations or managing bodies.

This is clearly an unsatisfactory situation and it is now widely recognised that there would be many advantages if protected areas that are located on either side of an international border could at least be managed under a harmonised set of rules and a joint management plan. This may be difficult to achieve for many reasons, however, and in any event always requires the conclusion of specific agreements between the countries concerned.

Failing joint management measures, there is a minimum requirement for the provision of consultation and cooperation mechanisms between the States or management bodies in question.

Although it is always possible for States to consult with each other and to cooperate on any matter outside the framework of a formal agreement, it is clear that the existence of an international obligation to do so may facilitate matters to a considerable degree. As to cooperation between the administrative authorities responsible for the management of the protected areas concerned, this will often require some legal basis or at least a political decision at Government level. Normally such authorities would not even have the right to meet informally without first going through their respective Ministries of Foreign Affairs!

Here again, a cooperation requirement established pursuant to a treaty could provide a much-needed incentive to the development of closer links between the management bodies concerned.

Provisions along those lines can be found in several conservation treaties. They usually lay down obligations for the Parties concerned to consult with each other where one Party intends to establish a protected area contiguous to the frontier of another Party, and to cooperate after

---

[8] J. Thorsell (ed.): *Parks on the Borderline: Experience in Transfrontier Conservation*; IUCN Protected Area Programme Series n°1, Gland, 1990.

the creation of the park or reserve, or in case where a protected area was already established before the treaty came into force.

The first treaty to contain a provision of this kind was the London Convention of 1933. The Algiers Convention of 1968 did not specifically take up this requirement and merely imposes upon its Contracting Parties an obligation to cooperate whenever necessary to give effect to the provisions of the Convention and whenever any national measure is likely to affect the natural resources of any other State. Proposed amendments to the Convention, which are now open for discussion by the Parties, go further in providing that whenever a natural resource is of common interest to two or more contracting Parties, these States undertake to cooperate in the conservation, development and management of such resources. Although there is no mention of border parks, it is clear that these could in many cases be considered to be of common interest.

Article 5 of the Ramsar Convention requires Contracting Parties to consult with each other about implementing obligations arising out of the Convention, especially in the case of a wetland extending over the territories of more than one Party. This would clearly apply to listed Ramsar sites.

Under the Berne Convention, the Parties

"undertake to co-ordinate as appropriate their efforts for the protection of natural habitats when they are situated in frontier areas".

The ASEAN Agreement provides that Contracting Parties

"shall especially cooperate together with a view to the conservation and management of border or contiguous protected areas".

The most elaborate provisions relating to contiguous parks or reserves appear in three of the Regional Seas Protocols. The Mediterranean Protocol, for instance, provides that

"if a Party intends to establish a protected area contiguous to the frontier or to the limits of the zone of national jurisdiction of another Party, the competent authorities of the two Parties shall endeavour to consult each other with a view to reaching agreement on the measures to be taken and shall, among other things, examine the possibility of the establishment by the other Party of a corresponding protected area or the adoption by it of any other appropriate measure".

Similar articles are included in the Nairobi and Kingston Protocols. The latter refers, in addition, to the possibility of developing cooperative management programmes between the Parties concerned.

Useful as they may be, however, these provisions remain very limited in their scope as they require no more than consultations and a certain degree of cooperation between Parties. Only the Kingston Protocol requires the taking of joint management measures. Moreover, in most of the treaties concerned, these provisions are not binding. Parties are merely required to endeavour or to examine the possibility for consultation and cooperation.

In spite of these various provisions in international conventions inviting Parties to cooperate, very little seems to have been done so far to achieve it in practice.

Ideally, the whole of the area under protection should be treated as a single unit. This means that the international border would be merely symbolic, with immigration and customs controls moved back to the park boundaries. The same regulations would apply throughout the park and there would be a joint administrative body, possibly under an international commission, and a single management plan.

Although not theoretically impossible, this objective would nevertheless be very difficult to achieve, primarily for legal reasons. A transboundary park cannot be a legal no-man's land. The laws of each of the countries concerned, such as the civil law, the criminal law, the legislation relating to the powers of the law enforcement officers and so on, must continue to apply to the parts of the protected area which are under that country's sovereignty. This may well make it impossible to develop a unified set of rules.

There are also psychological obstacles, as the concept of an international park may well appear to be an unacceptable relinquishment of sovereign rights over a part of the national territory as well as an invitation to foreign interference in national affairs.

In any event, whether the goal is to institute some form of joint management or merely a certain degree of cooperation in the management of transfrontier protected areas, a treaty will generally be required if only to set out the respective rights and obligations of the States concerned in legal terms. Very few treaties of that kind have been concluded so far, however, and these almost invariably confine themselves to the establishment of cooperation mechanisms. Joint management remains apparently unattainable.

Despite the above, there is, as far as can be ascertained, one case of joint management which shows that this ideal is feasible, given the political will to put it into practice. This is the Roosevelt Campobello International Park, which was established by a treaty between Canada and the United States. The treaty creates an international commission which lays down the management policy for the park. It is financed by equal contributions from the two States. However, this is a very unusual case as the purpose of the park is to preserve a number of historic buildings of major importance to the United States as the island was for many years the summer home of President Roosevelt. Moreover, the park is established on a small Canadian island and cannot really be considered as a true transboundary park.[9]

All other existing treaties relating to border parks merely provide for cooperation obligations and occasionally some mechanisms for this purpose. In fact, the "first international park" was not designated by treaty but rather by two separate Acts, adopted by the Canadian Parliament and the U.S. Congress in 1932. This designation concerned the Waterton Glacier International Peace Park. The Park has two components: the Canadian Waterton National Park and the American Glacier National Park. The Acts in no way affect the legal status of the two parks concerned or the applicable rules. They were merely designed to establish a basis for informal cooperation between park authorities. Agreements were concluded in 1987 between the two park services to facilitate such co-operation in a number of specific fields.[10]

In Central America, an agreement was signed on November 1987 between El Salvador, Guatemala and Honduras declaring the Montecristo mountains an international biosphere reserve. The competent authorities of each of the three countries will continue to be in charge of the management of the areas under their jurisdiction but on the basis of a management plan which must be jointly formulated in a "homogenous way" by the signatories. It is unknown whether this agreement is now in force or if it is still awaiting ratification.

---

[9] *Ibid.* p. 32.

[10] *Ibid.* pp. 41–49.

In Europe, three transboundary parks have been set up along the Polish-Czechoslovak border pursuant to an agreement concluded in 1924, whose purpose was to bring to a peaceful end litigation between the two countries over the frontier line in the Tatras mountain range. Consultations and cooperation between the Park authorities of Poland and Czechoslovakia concern scientific research, visits by tourists, forest and wildlife management and the harmonization of park regulations. Joint councils have been established for each park as a framework for such co-operation.

Another example of transboundary co-operation in Europe is the joint nature park which was established by treaty in 1964 between Luxembourg and the German Land of Rheinland-Pfalz.[11] The treaty lays down a few basic obligations. It requires in particular that the total area of forest in the park must not be diminished. It establishes a Joint Commission to which the two Governments must submit their park management plans for information. The Commission may make recommendations to the Governments on future management programmes and for the harmonisation of national regulations and other measures.

The large area of mudflats called the Waddensee, which extends along the shores of Denmark, Germany and the Netherlands in the North Sea, provides another example of inter-State cooperation in the management of a shared ecosystem. Although calls for a treaty between the three countries for the coordinated management of the area have so far been unsuccessful, the Governments concerned have agreed on a Joint Declaration stating general conservation and management objectives. In addition, there are relatively frequent meetings of representatives of the three States. Finally, an Agreement for the Conservation of the Waddensee seals has been signed under the auspices of the Bonn Convention.

Given the many cases in Europe where increased cooperation in the management of protected border areas could be highly beneficial to their conservation, the Council of Europe has attempted to persuade the governments of its Member States to accept the idea of transboundary parks. In 1988, a Council of Europe Committee developed and adopted a draft model agreement for the establishment and management of transfrontier nature parks.

Specific recommendations were also adopted by the Committee of Ministers of that organisation in respect of areas holding the European Diploma. For example, the Committee recommended that in the case of the Swiss National Park, "contact be established with the

---

[11] Other European agreements instituting some form of cooperation for the establishment and management of transboundary nature parks include an Agreement between the Netherlands and the German Land of Nordrhein-Westfalen for the establishment and management of the Maas-Schwalm-Nette Nature Park of 1977, and an Agreement between Belgium and the Länder of Nordrhein-Westfalen and Rheinland-Pfalz on the creation and management of a nature park in the Hautes Fagnes-Eifel region. These two agreements are fairly similar to the one concluded between Luxembourg and Rheinland-Pfalz. They both provide for the establishment of a Joint Commission.

More unusual is a Protocol of Agreement between the Pfalz Nature Park in Germany and the Parc Naturel Régional des Vosages du Nord in France, a rare example of an agreement concluded between park authorities. It provides for the exchange of information and for joint programmes for the study and protection of certain natural habitats.

management of the adjacent Stelvio National Park in Italy, for the purpose of establishing an agreed protection policy", although apparently to no avail.

In the case of the Spanish Park of Ordesa and Monte Perdido and the French Western Pyrenees National Park, the Committee's recommendation went markedly further. It invited

"Spain and France to open consultations, under Council of Europe auspices, with a view to determining the legal basis and practical arrangements for cooperation between the two parks, and the form which a joint body to manage the whole of the protected area could take".

For the time being, however, only a Cooperation Charter has been signed between the Nature Conservation Administrations of the two countries, although the idea of joint management of the area in the future has not been abandoned.

The example of the Council of Europe does show that there is a need for an institution to encourage consultations and cooperation between countries in the field of border parks. It is to be hoped that whenever a conservation treaty lays down an obligation for the Parties to consult each other and to cooperate in the establishment and management of such parks, any body created to review the implementation of the that treaty could be instrumental in encouraging the necessary contacts between the Parties, and even between Parties and non-Parties.

The Regional Seas Protocols on Protected Areas would seem to be particularly valuable instruments in this respect. Likewise, article 5 of the Ramsar Convention could be used to encourage cooperation between Parties in respect of Ramsar border sites, of which there are quite a few. The matter was briefly considered at the fourth meeting of the Ramsar Conference of the Parties in 1990 and will constitute an important agenda item at its next meeting in 1993.

D.   Areas beyond National Jurisdiction

1.   The High Seas

The outer boundary of the Exclusive Economic Zone (EEZ) or Fishery Zone, which under the Law of the Sea Convention of 1982 may extend seaward as far as 200 miles from the baseline, constitutes the limit of national jurisdiction. Beyond that line, the old rule of the freedom of the seas continues to apply. Navigation, fishing and other activities cannot therefore be restricted, although a coastal State does have sovereignty over the sedentary resources of its continental shelf.

It is always possible, of course, for a State to regulate the activities of its own nationals or ships in the high seas. Thus, Italy has adopted a decree empowering the Minister in charge of Fisheries to designate areas in the high seas where fishing by Italian fishing vessels is prohibited. A closed area in the vicinity of the island of Lampedusa has, as a result, been so designated. Fishermen of other countries obviously cannot be affected by that order.

To be meaningful, fishing prohibitions or restrictions in a particular area of the high seas must therefore be accepted by all nations fishing in that area. This can only be achieved by the conclusion of a treaty. But, as treaties are only binding upon their Parties, it is impossible to prevent non-Parties from fishing in the area which is closed to the Parties.

Most fishery conventions provide for the possibility of closing areas to fishing by their Parties, in order to protect spawning areas or nurseries or to allow overfished stocks to recover.[12] This is usually achieved by means of a legal mechanism which empowers the Fishery Commissions established under the respective conventions to adopt regulations binding upon all Parties, except those which may have lodged an objection. One example of such a closed area is the Indian Ocean Sanctuary which was established by the International Whaling Commission.

Fishery conventions cannot regulate activities other than fishing. However, certain other activities in some areas of the high seas are dealt with by other treaties. The International Convention for the Prevention of Pollution from Ships of 1973 (MARPOL) establishes a certain number of special areas where oil discharges are completely prohibited. These areas are semi-enclosed seas, such as the Mediterranean, the Baltic, the Black Sea, the Red Sea and the Gulf.

The Law of the Sea Convention addresses the problem of deep-sea mining in the high seas. It may be recalled that the sea-bed, ocean floor and sub-soil beyond the limits of national jurisdiction are considered under the Convention to belong to the common heritage of mankind, and that an International Sea-Bed Authority is to be established to manage the mineral resources of these areas.

As the exploration and exploitation of these resources may have harmful effects on marine ecosystems, the Convention provides for requirements to minimise the effects, such as habitat destruction or pollution, of these activities. The International Sea-Bed Authority must adopt

"appropriate rules, regulations and procedures for the protection and conservation of the natural resources of the Area (i.e. the sea-bed and sub-soil) and the prevention of damage to the flora and fauna or the marine environment".

The Authority may withhold approval for exploitation from particular areas, in cases where substantial evidence indicates the risk of serious harm to a unique environment. In practice, this would be almost the same as establishing a protected area.

2. The Antarctic

The legally-binding "Agreed Measures for the Conservation of Antarctic Fauna and Flora", adopted in 1964 by the Parties to the Antarctic Treaty of 1959, specifically protect a certain number of areas of particular ecological value. Other protected areas were established subsequently by recommendations of the Parties. Over the years, about twenty Specially Protected Areas have been designated in this way, together with a certain number of Areas of Outstanding Scientific Interest. The latter are principally intended to preserve on-going scientific research in certain sites against outside human interference.

Under the Madrid Protocol to the Antarctic Treaty, adopted in 1991, activities in the Treaty area must be planned and conducted so as to limit adverse impacts on the environment and, in

---

[12] There is no such provision under the Law of the Sea Convention.

particular, to avoid detrimental changes to or degradation of areas of biological, scientific, historic, aesthetic or wilderness significance. An Annex to the Protocol, dealing with the conservation of fauna and flora, makes no mention of protected areas. As a result, existing areas will presumably continue to be governed by the Agreed Measures and subsequent recommendations.

E. The Convention on Biological Diversity

As explained in the Introduction, threats to biological diversity have increased almost everywhere in the world during recent decades, mainly as a result of the destruction of natural habitats. Requirements for the conservation of biodiversity have therefore developed far beyond what was envisaged when the first conservation conventions were concluded.

As part of a global approach to conservation, the Convention on Biological Diversity accordingly places far greater emphasis upon the conservation of ecosystems than upon the protection of species as such. Under article 6 dealing with *in-situ* Conservation, Parties are required, as far as possible and as appropriate, to:

- establish a system of protected areas or areas where special measures need to be taken to conserve biological diversity;

- develop, where necessary, guidelines for the selection, establishment and management of protected areas and areas where special measures need to be taken to conserve biological diversity;

- promote the protection of ecosystems, natural habitats and the maintenance of viable populations of species in their natural surroundings;

- promote environmentally sound and sustainable development in areas adjacent to protected areas with a view to furthering protection of these areas; and

- rehabilitate and restore degraded ecosystems and promote the recovery of threatened species, *inter alia*, through the development and implementation of plans or other management strategies.

The Convention contains no obligation for Parties to protect the areas which are most important for the conservation of biological diversity. Even attempts to provide for the listing of all globally important areas, purely as a means to focus attention at national and international level on the need to conserve them, failed during the negotiations.

Instead, Annex I to the Convention only provides guidance for area selection in the form of an indicative list of components of biological diversity important for its conservation and sustainable use. However, the list is so broad that it may be difficult to use it in practice to establish priorities.

The Parties are therefore free to protect whatever areas they choose.

There are no provisions either for the establishment of a global network of areas protected because of their importance for biological diversity, or even for the compilation of a list of areas protected in this way. However, these are matters which may be covered subsequently through the adoption of a Protocol or by decisions of the Conference of the Parties.

The conservation of ecosystems is also promoted through general obligations for the identification and monitoring of important components of biological diversity (article 7). Parties are required to identify processes and categories of activities which may have significant adverse impacts on the conservation and sustainable use of biological diversity. Environmental impact assessment obligations are set out in article 14, which is discussed in Chapter VII below.

CHAPTER II
BASIC LEGAL INSTRUMENTS FOR AREA-BASED CONSERVATION

A. Introduction

From a legal point of view, an area set aside for conservation is an area where specific land-use restrictions have been established to preserve certain or all natural features present in the land or waters concerned. These restrictions will generally be stricter than those which may be applied in other parts of the national territory. This means that areas with construction prohibitions or restrictions established under national land-use legislation and local zoning regulations would not normally be considered as areas set aside for conservation, even though they may well indirectly contribute to the preservation of natural features, unless additional and specific conservation-orientated measures are also applied.

Three basic factors must be taken into consideration when setting aside areas for conservation: the ownership of the land, whether public, private or sometimes common; the persons affected by the land-use restrictions, such as the landowner or occupier, the public in general and Government agencies; and the constitutional or other rights of the persons affected which may be curtailed by the conservation measures.

The interplay of these three factors is at the very basis of all legislation relating to conservation areas.

There are three basic methods which may be used to set aside land for conservation: public ownership, the exercise of the police power of the State and self-imposed restrictions. These methods are, however, often used in combination with one another.

Public Ownership

Public ownership of land is a powerful conservation tool since the State or the land-holding Government agency has the right, in its capacity as a landowner, to prohibit or restrict access and activities by third parties on the lands it owns, and to carry out any management measures that may be required. However, legislation is required to provide for the dedication of public land to conservation to the exclusion of other purposes, to bind government agencies to observe land-use restrictions and to penalise illegal activities by third parties.

Police Powers

The 'police power' is the power vested in the State to regulate human activities in the public interest, and may include regulatory measures for the conservation both of areas and species.[13]

---

[13] The regulatory powers of the State to control the exploitation of wild species are discussed in Part I, Chapter II(A) above.

In virtually all countries in the world, police powers are exercised to a varying extent to regulate such activities as forestry, mining and quarrying and the emission of dangerous substances or pollutants, to mention but a few. Such powers have also been increasingly used over past decades to control the construction of buildings and, though perhaps to a lower degree, to impose specific land-use restrictions for conservation purposes.

Police powers may be used on public land to oblige the land management agency to comply with land-use restrictions established by the law. On private land, police powers can be used to limit the rights of landowners to use their land freely. In both cases, police powers make it possible to enforce restrictions against third parties.

Regulatory measures taken in the exercise of police powers may completely prohibit certain activities; restrict a specific activity; subject any activity to the prior granting of a permit; provide that permits may be issued subject to certain conditions; or require prior notification of the intention to engage in certain activities.

Self-Imposed Restrictions

Conservation areas may also be established on a purely voluntary basis by the owners or holders of the land concerned, whether they are public agencies or private persons. In such cases, however, property rights may not always be sufficient to protect these areas against intrusions by third parties, if they are not backed by regulatory measures.

In addition, especially where private lands are involved, ownership will not preserve the land from expropriation by the State for purposes other than conservation. Finally, a voluntary decision to conserve may always be revoked or modified at any time, unless enshrined in a binding legal instrument. The perennity of the conservation area may, thus hardly ever be ensured by purely voluntary means.

As a result, a number of legal instruments have been developed in certain countries to provide some form of legal backing as well as incentives for voluntary conservation. In many nations, however, the considerable opportunities offered by voluntary conservation continue to be largely ignored.

Public Ownership versus the Exercise of Police Powers

Certain countries, mostly common law countries, rely almost exclusively on public ownership of the land, or of an interest in the land, for the conservation of natural areas, unless an agreement can be negotiated with the landowner. In other words they consider that it is not possible to impose conservation land-use restrictions on a landowner against his or her will. As a result, when land of importance for conservation is in private hands and cannot be preserved by voluntary means, it has to be acquired by the State.

In contrast, many other countries, including most European nations and Japan, are showing a marked preference for the use of regulatory measures for the conservation of private land. In these countries, acquisition of land occurs relatively rarely, although it is not completely precluded.

This difference in approach is in no way the consequence of major differences in the legal systems of the two groups of countries, as both exercise their police powers to regulate construction, for instance. The difference is apparently due instead to a greater recognition in the first group of countries that the legitimacy of conservation measures which seriously restrict the use of private property, especially for agricultural purposes, is not sufficiently well established to justify the use of the police power of the State. In the countries that favour regulatory measures, this may perhaps result from the more ingrained tradition of State control over private activities and from a long-standing reluctance to allow the State to accumulate real property.

Be that as it may, neither of the two systems can be considered as fully satisfactory. 'Public property countries' are therefore increasingly seeking ways other than acquisition to preserve natural areas, whereas 'police power nations' are resorting more and more to the purchase of land.

The advantages and disadvantages of the two systems, together with some possibilities for improving them, merit careful consideration.

B. Public Ownership

Systems of public ownership vary considerably from country to country. Public land may belong to the State and be assigned to individual agencies or belong to the agencies which manage land. Some public land forms part of the public domain, such as the seashore, the sea bed and at least the larger rivers, whilst other land is often considered to be the private property of the State, as is the case for the State forests in France. Land belonging to local communities and municipalities is often considered to be privately owned.

Where the owner or occupier of land is a conservation agency, the preservation of such land does not generally give rise to legal problems. The right of ownership associated with the will to preserve is generally sufficient. But when land is managed by an agency whose task and purpose are very different from conservation, it may not always be easy to persuade the agency to set the land aside. Ideally, land of important conservation value held by a different agency should be transferred to the conservation agency. However, this is often impossible for legal and practical reasons, unless specifically provided for by an Act of Parliament. In addition, a public agency will generally tend to resist the transfer of its lands to another department.

Rules on the alienation of public property often require that land be sold at its market price even where it is to be acquired by another public agency. Finally, there are many cases where a transfer would clearly be impossible because it would go against the very purpose for which the land is held by a particular department. Examples include land held for military purposes or along public roads.

Ways must therefore be found to ensure the conservation of ecologically important areas on public land where their transfer to a conservation agency is not possible. This can only achieved by specific regulatory measures which are binding upon the land-holding agency or by a purely voluntary commitment or agreement on the part of the agency concerned.

Regulatory measures include the establishment of statutory protected areas, such as nature reserves, or the imposition of certain land management practices. A well known example is that of wilderness areas in the United States, which may be created by Act of Congress on land held by any federal agency.[14]

One example of voluntary measures is that of Research National Areas, again in the United States, where eight federal land management agencies have decided to preserve a number of

---

[14] Discussed at Chapter IV (A)(2) below.

natural areas in an undisturbed state, which should serve as base line areas for research and monitoring purposes.

In France, the national Forest Agency, which manages State-owned forests, was very reluctant to accept the establishment of statutory nature reserves on its land, until a compromise was eventually reached. Instead of being created by Government decree, as required by the Nature Protection Act of 1976, reserves on State- or municipally-owned forest land may instead be established by Ministerial order of the Minister of Agriculture (the Minister who has forests under his or her jurisdiction).

Such orders bind the Forest Agency and third parties. However, as with the decisions creating Reserve Natural Areas in the United States, these orders are easily revocable and therefore cannot provide long-lasting protection in law. In practice, such arrangements may nevertheless be very effective, provided that the agency's will to preserve the areas concerned does not falter.

Other possibilities include leases and agreements with conservation agencies.

A land-holding Government body may agree to lease some of its lands to a conservation agency (or to an NGO) which will then manage them as protected areas. This seems to be a fairly common practice in the United Kingdom where land belonging to the Forestry Commission or to local authorities is leased to English Nature.

Legislation rarely provides for the possibility of a conservation agency entering into formal agreements with other Government bodies for the conservation of natural areas that are held or managed by the latter. However, the laws of certain Australian States, such as New South Wales and Victoria, explicitly allow the conclusion of such agreements. In Victoria, for instance, the Flora and Fauna Guarantee Act of 1988 empowers the Conservation Authority to enter into management agreements with other public bodies for the management of any species or community of flora and fauna or of any potentially threatening process.

The absence of specific legislation permitting inter-departmental agreements for the conservation of natural areas does not necessarily mean that no cooperation is possible between conservation departments and other agencies. There are many instances of informal agreements which may often be quite successful.

In the United Kingdom, it is the policy of the Forestry Commission to carry out consultations with the national conservation agencies with regard to the management of the large number of Sites of Special Scientific Interest (SSSIs) which it owns. Similarly, in the United Kingdom and in other countries as well, the protection of roadside vegetation, which may sometimes include endangered species, may be achieved by negotiations with the Highway Department. In the United States, recovery plans developed for endangered species often require the cooperation of many different agencies. This is generally done by the means of memorandums of understanding signed by the agencies concerned.

Turning to the acquisition and alienation of land by State or Government agencies, these procedures follow administrative rules which are not always well adapted to conservation needs. There is therefore often a requirement for specific rules to be developed to facilitate the acquisition of land for conservation purposes and to control the disposal of public land of conservation value.

1. The Acquisition of Land by Public Agencies for the Purpose of Conservation

a. Capacity to Acquire

It is important that conservation departments or agencies have legal powers to acquire land for conservation purposes. This is not always the case. When such agencies have to go through another department, such as the Ministry of Finance, this will invariably mean long delays and lost opportunities.

b. Right of Pre-emption

The capacity to acquire is not of itself sufficient. Conservation agencies should also have the right of pre-emption (also called the right of first refusal) over any land that comes on the market. This system exists in France, at least for land which has been specifically designated for that purpose by Government order. The beneficiaries of this right are the "départements" (the decentralised administrative units which have their own administration and budget) and the Conservatoire du Littoral et des Espaces Lacustres, a body set up specifically to acquire land along coasts and lake shores, which is discussed at section (e) below.

Other countries which have established a right of pre-emption are Hungary, in respect of land adjacent to protected areas, and Spain, over land situated within National Parks or Nature Parks. In Denmark, the Nature Management Act of 1992 incorporates the right of pre-emption which was granted to the Ministry of the Environment under the Nature Management Act of 1989. The new Italian national Protected Areas Act of 1991 gives such a right to park authorities in respect of land situated within the parks.

Under these systems, the agency which has the right of pre-emption must be informed of any proposed sale, and no sale contract can be finalised until the agency has informed the seller that it does not have the intention to exercise its right. There is, however, often a time limit after which the agency can no longer avail itself of this opportunity. This limit is often very short: in Spain, it is only 6 months. This means that money needs to be available at short notice, which is generally difficult in the context of Government spending.

A right of pre-emption may also be instituted by a contract between a landowner and a conservation agency. This procedure exists in the United States, for instance. Under such a contract and in return for the payment of a small sum of money, the landowner commits himself or herself to offer the land for sale to the agency, should he or she decide to sell it.

Another means to facilitate the acquisition of important land by conservation agencies is to allow the gift of such land to the State in lieu of death duties. This possibility exists in the United Kingdom under the National Heritage Act of 1980. To be eligible, a piece of land must be recognised by the Treasury as being of outstanding ecological value, following the advice of the competent national conservation authority.

c. Tax Incentives

Tax incentives may also make acquisition easier. In the United States, the value of land or of interests in land donated to a public agency for conservation purposes is deductible from taxable income under certain conditions.

In Italy, following the adoption of the Protected Areas Act of 6 December 1991, it is permitted to deduct from taxable income sums donated for or expenses incurred in the acquisition, protection and management of real estate which is protected under the Act on the Protection of "Natural Beauties" of 1939 or which is subject to a construction prohibition under a landscape plan adopted under the terms of that Act or the Galasso Act of 1985.

d. Compulsory Purchase

Finally, when acquisition by contract is impossible because the landowner does not want to sell, public agencies should be vested with the right to purchase the land compulsorily. It is strange, however, that although the right of compulsory purchase in the public interest exist for many purposes such as national defence or public works and is fully accepted by the public everywhere, this is not the case where the reason for the compulsory purchase is conservation. In many countries, conservation agencies are therefore unable to expropriate land.

In other countries, such a right of compulsory purchase was only recognised very recently. In Denmark, for instance, the right of compulsory purchase was only given to the Ministry of Environment in 1989 by the Nature Management Act. Moreover, even when this right exists, it is seldom exercised because it is apparently feared that public opinion would be unfavourable. Once again, the legitimacy of conservation is in question when compared to other economic or social interests.

Presumably in recognition of this situation, the laws of several countries do allow the compulsory purchase of land for conservation, but clearly specify that this should only be done in exceptional circumstances. Examples of such an approach are the Swiss federal Nature Protection Act of 1966 and the Greek Environmental Protection Act of 1986.

e. Special Financial Measures

Essential as it may be, the right to acquire land through the exercise of a right of preemption or compulsory purchase is still not sufficient. Conservation agencies should also have sufficient funds to be able to buy land when it comes on the market and be able to use flexible procedures to spend these funds when the best opportunities arise.

In a certain number of jurisdictions, special funds have now been established to finance the acquisition and management of land of conservation value. These funds are usually allocated from Government budgets. In a few cases, however, special financial mechanisms have been instituted. In France, for example. the "départements" are entitled to levy a tax on the construction of buildings, called the "sensitive natural areas tax", which may amount to up to 2% of the total construction cost. The proceeds of the tax must be used for the acquisition and maintenance of natural areas. In the Austrian Land of Vorarlberg, there is a tax which must be paid by the operator for each cubic metre of stone or gravel extracted from quarries. The funds so obtained are to be used, *inter alia*, for the acquisition of land for conservation by the government and municipalities.

An important advantage of such funds is that they permit great flexibility in the spending of public money. Whereas budgetary allocations must generally be returned to the Treasury if they are not spent before the end of the fiscal year, and are therefore lost to the Agency, the sums deposited in conservation funds may generally be carried over from one year to another. This makes it possible to accumulate a reserve which may be used to buy important land when it comes on the market. Funds may also be used as a channel for donations and bequests of money.

Another method which is sometimes used to facilitate acquisition is to rely upon certain NGOs to buy land when it comes on the market and then to buy that land from these NGOs when money is available. This is commonly done in the United States, particularly by The Nature Conservancy. This is a large organisation specialised in purchasing land for conservation, some of which it then sells back to federal or State agencies. In the Netherlands, Natuurmonumenten, a nationwide NGO, receives considerable financial grants from the Dutch Government to buy and manage natural areas. Some of these lands are subsequently sold back to the government.

Finally, it should be added that where land is acquired by a conservation agency, its preservation is generally easy from a legal point of view. The simplest way of proceeding is usually to establish a statutory protected area, such as a nature reserve, on the land thus acquired. Even where this is not done, public ownership by the agency concerned will generally be sufficient to prevent harmful activities or developments. In France, for example, the Conservatoire du Littoral et des Espaces Lacustres, a public body which was established for the exclusive purpose of acquiring land in coastal areas and along lake shores, does not usually request the creation of protected areas on its lands, as it does not believe this to be necessary in most cases.

2. Alienation

Rules governing the disposal of land held by conservation agencies are generally no different from those applicable to any other public land. However, this may not be enough to ensure that land dedicated to conservation remains in Government ownership, unless there are very good reasons for a change of ownership to take place.

In a few countries, special rules have been enacted establishing restrictive procedures for the alienation of such land. In the American State of Massachusetts, for example, the State Constitution requires that the disposal or assignment to other uses of park land be approved by the State legislature by a two-thirds majority. In France, the Conservatoire du Littoral et des Espaces Lacustres cannot sell any of its land unless authorised to do so by a Government decree and provided such disposal has first been approved by its Board of Directors by a three-quarters majority.

3. Advantages and Disadvantages of Public Ownership

Public ownership of land for conservation has many advantages. First and foremost, once land has been acquired, it will usually be permanently preserved by the agency concerned except in exceptional circumstances. In contrast, regulatory measures may often easily be revoked. The enforcement of regulations is also much easier, of course, as there are no private landowners whose activities need to be monitored and controlled. Finally, whereas regulatory measures cannot generally be used to compel landowners or occupiers to take positive management measures on their lands, the fact that a property is in public ownership allows the land-holding agency to manage the area as it thinks fit.

One of the disadvantages of public ownership is that it generally reduces the tax base for the property tax which is often the main financial resource of the local authorities concerned. This may consequently generate opposition to Government purchases of land for conservation purposes. One solution consists in compensating the municipalities or other authorities affected by this loss of income. This is done in the United States where the federal government is required to make compensatory payments to counties and municipalities for land included within the National Wildlife Refuge System.

The main disadvantage of the public property approach is obviously its cost. Confronted as they are with the rising price of land, especially in such sensitive areas as the coastal zone or prime agricultural regions, conservation agencies have tried to find other methods than "fee simple" acquisition (outright acquisition of the freehold) to preserve natural habitats. These include the use of voluntary measures, such as leases and management agreements, which will be dealt with in greater detail in Chapter VIII. Another possible method is to purchase an interest in the land rather than buying the whole range of property rights.

4. Easements

Real property is considered in certain countries, particularly common law countries, as a bundle of separate rights, such as the right to build, to cultivate the land, to drain, to fell trees, to hunt and so on. It follows that a landowner should normally be entitled to dispose of any of these rights individually, such as by selling or donating it to another person, while continuing to exercise all the other rights freely. To preserve a wetland, a Government agency or a private person, for instance a conservation NGO, could simply buy from the owner the right to drain or fill the area concerned and to clear vegetation, whilst the owner would continue to enjoy his or her property for any other purpose, such as hunting, fishing and boating. This is called creating an easement.

The original purpose of easements, in both common law and civil law countries where they are called servitudes, was to grant a benefit to one parcel of land, called the dominant tenement, by means of a burden imposed upon another parcel, the servient tenement, which had necessarily to be contiguous to the first. As an example, a right of way established through one parcel of land to allow access to an adjacent one is an easement.[15]

It is important to note that easements are made for the benefit of a piece of land rather than its owner. When the land is transferred, by sale or otherwise, so is the easement. Easements are therefore said to "run with the land" and as a result, they are binding upon successors in title. Easements are generally established by contract and are permanent unless otherwise specified.

In the form described, the easement has very little value for the conservation of natural areas because of the conditions that there must be a dominant tenement which is adjacent to the servient tenement, and there must be a connection between the two, so that the burden imposed upon the servient tenement is the counterpart of the benefit enjoyed by the dominant tenement.

In some jurisdictions, however, the requirement for a dominant tenement has been waived by legislation for the purpose of allowing the creation of conservation easements. The system has considerable advantages for both the agency acquiring the easement and the landowner. The agency will pay less than if it had to acquire the land itself, whilst the landowner will continue to enjoy many uses of his or her property and will benefit from tax reductions.

---

[15] Easements and Servitudes are described under Voluntary Conservation at Chapter VIII (A)(3)(b) below.

C. Regulatory Measures

In all countries in the world, the State has the right to enact police laws to prohibit and restrict activities which may be harmful to the natural environment. The question which arises, however, is whether compensation should be paid when a landowner is deprived of some or all of the uses of his or her land as a result of such restrictions.

Property rights are usually protected by the State constitution and just compensation is therefore usually required when private property is taken by the State for public use. Modern constitutions recognise, however, that these rights are not absolute and may be curtailed in the public interest. The most common manifestation of this new trend relates to the right to build on one's land. Land-use legislation almost universally provides for construction prohibitions or restrictions without compensation. This is hardly ever considered to constitute an unconstitutional taking of property, because the owner can continue to use his or her property in other ways.

On the other hand, where all uses are prohibited, as would be the case if a strict nature reserve were established on private land, it is clear that this might constitute an unconstitutional spoliation if no compensation were paid.

Between these two situations, there exists a grey area of uncertainty as to the extent to which compensation should be paid when some but not all activities are regulated. Two trends do seem to be emerging. Firstly, compensation is not due when the same regulatory measure is applicable to all owners of land belonging to a certain class or habitat type, such as wetlands. Secondly, compensation is not due where restrictions apply to all owners or occupiers wanting to engage in a particular activity, such as quarrying, provided that some uses of the land in question remain authorised.

The first of these conditions is relatively clear. As an example, Danish legislation provides for compensation when a landowner is affected by an individual conservation order relating to his or her land. On the other hand, no compensation is due where all permits for proposed activities are denied in respect of areas hosting a protected habitat type.

The second condition is much less clear as it is difficult to establish objective criteria to determine when compensation should or should not be paid. For example, the Swedish Nature Conservation Act provides that where prohibitions affecting private land in a nature reserve are significant, compensation must be paid to the owner. Where restrictions to property rights are particularly severe and result in substantial economic losses, the owner may require the Government to purchase the land. In the event of disputes on the interpretation of these provisions, as with very similar ones in many other national laws, it is of course up to the courts to decide in each individual case.

More specific criteria for deciding whether or not compensation is due include:

- The nature of the land itself.

- Whether the land was acquired before or after the regulatory measures came into force. After that date, it is expected that new owners will know that some activities are prohibited or restricted; they should therefore not be entitled to claim compensation if they cannot carry them out. This means that any prospective buyer of land to which such restrictions apply must be informed of that fact. This may be best achieved by a mention in the Land Register.

- Whether the prohibited or restricted activity is an ongoing or a new one; so that only actual losses be compensated and not mere expectations.

Finally, some jurisdictions consider that a proportion of the loss should, in any event, be borne by the owner.

From the Government's point of view, the advantage of regulatory measures is clearly that, at least in some cases, compensation is not payable. The exercise of police powers therefore constitutes a cheaper method of preserving natural areas. If the costs were the same, purchasing would otherwise be preferable in most cases.

On the other hand, if the Government is unable or unwilling to buy land for conservation, the fact that it may have to pay substantial compensation when applying regulatory measures may effectively deter it from protecting natural areas or denying permits for conservation reasons. Finally, if no compensation is paid, the opposition to conservation measures may become very strong.

The problem of compensation is not the only disadvantage of the regulatory system. Another major defect is that no positive management measures can be imposed by Government order, unless a voluntary management agreement is negotiated and concluded at the same time. To give an example, while it is possible for a Government agency in the exercise of its police powers to prohibit certain agricultural practices, it cannot usually force a landowner to do certain things, such as planting trees, cutting reeds or mowing grass at certain dates. There are certain exceptions whereby positive management measures may be imposed.

In addition, the monitoring of the activities of private landholders may not always be easy and enforcement measures difficult to apply.

Nevertheless, the regulatory system has also some advantages. It is generally less costly than land acquisition and much more flexible, because prohibitions and restrictions may be limited to what is strictly necessary to ensure the conservation of the natural features of the land. The system also enables emergency conservation measures to be taken quickly in the case of an immediate threat to a valuable area. In contrast, it always takes time to purchase an area for its protection, compulsorily or otherwise, and irreparable damage may occur before then unless an interim conservation order can be obtained to protect the site.

Finally, regulatory measures are the only means to control certain activities in large inhabited areas, where the optimum land use may be the maintenance of traditional and extensive agricultural or other activities, or indeed in any area where some use restrictions are preferable to outright prohibitions.

D. Conclusion: The Need for Voluntary Measures

Neither of the two systems is, as we saw, entirely satisfactory. As a result, in those countries where the system of public ownership prevails, the use of police powers to apply regulatory measures is increasing slowly but steadily. In the United States, for example, land areas adjacent to designated Wild and Scenic Rivers are mainly protected by local zoning ordinances, and the destruction or alteration of wetlands now requires a permit in many individual States as well as at federal level in certain cases.

Conversely, countries that mostly rely on regulatory measures increasingly tend to purchase land for conservation. In France, for instance, the Conservatoire du Littoral et des Espaces Lacustres was set up for the sole purpose of acquiring land in the coastal zone.

It is therefore very likely that in the future, States will increasingly try to use the two systems in combination, deploying each to its maximum potential and complementing it with the other whenever this will increase the effectiveness of conservation. Such combinations may, however, not be enough and voluntary measures will probably also be necessary, if not essential. To be successful voluntary measures will need strong incentives to ensure that the presence of important natural features on one's land should cease to be a liability, as is often the case now, and become an asset and a source of enviable income.

Nonetheless, voluntary measures will always need to be backed up by powers vested in a conservation authority, to apply regulatory measures or to resort to compulsory purchases when the need arises in order to prevent irreparable damage to important areas.

An example of such an integrated system which seems to be operating particularly well is that established under the Native Vegetation Act of 1991 in the State of South Australia.[16] This system combines the exercise of police powers with voluntary conservation measures and, to a lesser extent, acquisition by the State. The clearing of any native vegetation is subject to a permit. Where a permit is denied, a landowner who agrees to enter into a Heritage Agreement specifying his or her obligations in respect of the land in question may be paid compensation, in return for the decrease in the value of the land as a result of these restrictions, as well as incentive payments and management assistance

Where a piece of property has become economically non-viable because a permit to clear has been denied, the State may offer to buy it and may re-sell it with Heritage Agreement restrictions. Alternatively, the State may also decide to acquire it permanently and incorporate it into the State national parks and reserve system.

Similarly, the system of Sites of Special Scientific Interest in the United Kingdom also combines regulatory and voluntary methods and, to a certain extent, acquisition.

---

[16] This system is explained in Chapter V below.

CHAPTER III
CONVENTIONAL TYPES OF PROTECTED AREAS

A. Introduction

All countries in the world, with perhaps very minor exceptions, have now instituted protected areas to preserve natural ecosystems, biological processes and outstanding landscapes. Traditionally, certain specific instruments have been developed for that purpose: these are essentially national parks and nature reserves. Their main purpose is now clearly the conservation of biological diversity, although this was not always the case in the past, when national parks were often created to preserve spectacular scenery and for public recreation.

To be able to fulfil their present purposes, protected areas must satisfy two essential conditions. The protection they provide to natural ecosystems must be long-lasting and as comprehensive as possible. These objectives of perennity and integrity are now universally recognized. They are embodied in the World Heritage Convention and the associated Operational Guidelines for its implementation. They also form the basis for many legislative texts relating to protected areas.

In addition, parks and reserves often need to be managed to ensure that their objectives are effectively achieved. This means that management plans and institutions are generally required.

No protected area can be established and run without legal provisions laying down the conditions and procedures required for its creation, setting out use prohibitions or restrictions, providing for enforcement measures and penalties, instituting management bodies and determining their powers and tasks. Law is therefore an essential component of any protected areas policy. A review of the main features of relevant legislation in different countries may therefore be of interest.

Protected areas have been given many different names and their legal regimes vary from one country to another. Attempts to harmonise their names and regulations have generally failed. The IUCN definition of categories I, II and IV of protected areas provides a useful framework, however, and many existing protected areas broadly conform to the criteria laid down for these categories.

This chapter will therefore concentrate on nature reserves and national parks, that is to say mostly uninhabited areas which are to be kept undisturbed as far as possible, or which are managed for the maintenance of natural ecosystems. Some are large and some are small, some are closed except for scientific research and some are open to the public. But they all show the same characteristics from the legal point of view: most human activities that may take place in them are prohibited or strictly regulated. The legal status of these reserves and parks is therefore quite unlike that of any other part of the national territory, whether in public or private ownership, as many otherwise legitimate activities are made illegal within their boundaries.

B. Protected Area Legislation

1. General Considerations

Almost all countries in the world have some form of legislation providing for the establishment of protected areas.

In earlier times, before special Government departments were created to deal with environmental matters, protected areas usually came under the jurisdiction of Forest Departments. Provisions for their establishment were therefore usually incorporated into forestry legislation or into legislation on hunting, a matter which generally also came under forestry jurisdiction.

More recently, with the enactment of more comprehensive wildlife or nature conservation laws, a trend has gradually appeared whereby protected areas are removed from forestry jurisdiction and transferred to more specialised agencies such as Environment Departments or semi-autonomous conservation bodies. However, this process is far from being complete as Forest Services have quite naturally tended to resist what they felt to be an unwarranted encroachment upon their traditional field of competence.

As a result, certain countries now have two different networks of protected areas: one which remains governed by forest legislation and which is administered by the Forest Department, and another which is governed by the new conservation legislation and managed by conservation authorities. In parallel, other land-managing departments sometimes also establish their own networks. This is usually the case of the agencies which have jurisdiction over marine areas up to the high water mark. Finally, in federal or regionalised States, such as Italy and Spain, the federated or regionalised territorial entities are always constitutionally provided with the power to establish protected areas and have therefore also constituted their own networks.

The consequence of these jurisdictional splits has been a proliferation of separate protected area systems, each with their own basic legislation, administration, budget and management rules. Different terms are often used to designate protected areas with very similar legal status. On the positive side, however, it must be recognised that this apparently anarchic situation has often favoured the development of protected areas through competition between agencies. Moreover, in the case of federal and regionalised States, the establishment of parks or reserves by the regional government is politically much more acceptable than where this is imposed from the top by the national administration. For example, before jurisdiction on protected areas was almost completely devolved to the regions in Italy and Spain, very few national parks and reserves had been created because of regional opposition. In contrast, many regions have now enacted their own legislation on protected areas and have accordingly established many new regional parks and reserves.

Laws providing for the establishment of protected areas will in most cases have to be Acts of Parliament, as they involve restrictions of certain freedoms and impose penalties for violations. In rare cases, where reserves are created on public lands and no particular sanctions are required, they may be established by statutory instruments or by a decision of the agency concerned.

At a minimum, general enabling legislation usually contains basic provisions describing the various categories of protected areas instituted by the law and specifying the types of prohibitions or restrictions which can be imposed in respect of each of them, providing for the creation of management institutions, and laying down enforcement measures and penalties. In addition, laws provide for procedures for the establishment of individual protected areas.

In certain countries, such as Finland, Spain, Sweden and the United States, national parks can only be established by an Act of Parliament. In most countries, however, individual protected areas are created by regulations made under the main Act. Whatever the procedure, it is essential that a protected area be officially created by a formal legal instrument as prohibitions or restrictions would otherwise be deprived of a legal basis and would therefore be impossible to enforce.

Important elements, which are often and should always be included in legal texts establishing individual protected areas, include:

- a clear and accurate determination of the boundaries of the area, including, where applicable, its buffer zone, complemented if possible by a map;

- the specific prohibitions or restrictions that are applicable in the area (these may, however, be the subject of further regulations);

- the institution of a management body and a description of its duties and powers.

A certain number of important aspects of protected area legislation will now be examined in more detail, as these should be considered as the very basis upon which the operation of an effective legal system of nature conservation depends.

These elements include the legal mechanisms designed to ensure that the objectives of perennity and integrity set out for protected areas are achieved as far as possible.

2. The Perennity Objective

No human institution can be made everlasting by legislation, since what can be done by one law can always be undone by another.

The general rule which is therefore applicable to protected areas, as to almost every other type of institution, is that the same legal procedure which was used to establish the protected area must be used to abolish it or reduce its size. For example, if a national park has been created by Act of Parliament, another Act of Parliament is required to de-establish it. If a public enquiry is mandatory before a decree creating a park or reserve may be adopted, a public enquiry will similarly be required before a decree abolishing that park or reserve can be promulgated.

There are, however, legal mechanisms which may be used to make the abolition of protected areas more difficult than their establishment, and which may therefore greatly assist in ensuring their long-lasting conservation.

a. Treaties

The inclusion of an area in the list of areas protected under an international convention gives rise[17] to an obligation for the Party on whose territory it is situated to protect the area in question.

---

[17] As seen above in Part I, Chapter I(C)(2)(d).

A Party cannot, therefore, abolish the protection status of such an area or reduce its size without firstly either denouncing the convention, a rather extreme solution, or following the delisting procedure which may be provided by the convention itself or by resolutions adopted by the Parties.

The three treaties which provide for lists of protected areas benefiting from international protection are, it will be recalled, the Ramsar and World Heritage Conventions, and the Protocol on Specially Protected Areas and Wildlife in the Wider Caribbean region. Under the Ramsar Convention, Parties are free to delist sites in cases of "urgent national interest". They must, however, "as far as possible compensate for any loss of wetland resources" resulting from such action. The Wider Caribbean Protocol requires approval by the Parties before delisting may proceed. The World Heritage Convention does not entitle individual Parties to delist sites they have nominated and which have been accepted for listing by the World Heritage Committee.

The existence of an international obligation not to abolish a protected area included in a list established under a treaty was recognised by the High Court of Australia which, in the *Tasmanian Dams* case, [18] judged that there was a legal duty imposed by the Convention on each Party to protect World Heritage sites on its territory. As a result, there was also held to be an obligation not to abolish the protection status of sites so listed.

In the particular case of the Special Protection Areas instituted under the EC Birds Directive, the European Court of Justice ruled in a recent decision[19] that the abolition or reduction in size of such an area would only be lawful in exceptional circumstances, where a public interest higher than the conservation interest is at stake. This rule is now incorporated in the new Habitats Directive and will be applicable to all Special Areas for Conservation established henceforth under either of the two Directives.

b. Constitutions

Many modern constitutions now contain provisions on the conservation of the natural environment. A few provide for the protection of certain ecosystem types or of certain particular areas. Area protection resulting from constitutional requirements is clearly very strong, as any proposed change in the status of the areas concerned necessitates the enactment of a constitutional amendment to become effective. This generally entails a long and protracted public procedure which will be unlikely to succeed, unless there are indeed very good reasons to put the areas in question to other uses.

The Brazilian Federal Constitution of 1988 prohibits the alienation by Brazilian States of land set aside for the protection of natural ecosystems.

In Switzerland, the Federal Constitution was amended in 1987, following a citizen's petition and referendum to prevent the construction of a military training camp in the Rothenthurm marshes in the central part of the country. The amendment provides that marshes and marshland

---

[18] *Commonwealth of Australia v. The State of Tasmania*, n° C6 of 1983, 46 Australian Law Reports 624.

[19] Case C-57-89 (the *Leybucht* case which concerned a wetland in Germany) in which judgement was given on 28 February 1991.

areas of national interest must be protected. The Government has now listed some five hundred sites which are to benefit from this protection and other lists are in the course of preparation.

In the American State of New York, the State Constitution of 1895 provides that the forests in the Adirondak mountains shall remain "forever wild". These forests belong to the State of New York. They cannot, as a result of that constitutional provision, be sold or used in any way that may alter their wild character. They are included in the Adirondak State Park.

Many of the recent (1989) constitutions of the Brazilian States contain provisions for the conservation of certain particular areas which are specifically mentioned. The Constitution of the State of Bahia, for instance, protects certain bays, river valleys and parks which are considered to belong to the heritage of the State. The use of these areas must be subject to conditions, to be laid down by Acts of Parliament, to ensure that they are adequately managed as such. Another example is the Constitution of the State of Rio Grande do Norte, which also lists entire areas which must be specially protected by Acts of Parliament laying down conditions for the preservation and rational management of the ecosystems concerned. A similar provision is included in the Constitution of the State of Sao Paolo which, in addition, provides that all State-established protected areas shall be considered as specially protected areas and that their use shall only be authorised, again by an Act of Parliament, under conditions that ensure the preservation of the environment.

c. Provisions instituting Stricter Abolition Procedures

The laws of a relatively small number of countries require that stricter and more demanding procedures be used for the abolition or reduction in size of protected areas than for their establishment.

As early as 1933, the London African Convention already required that protected areas established by the Parties be abolished only by Acts of Parliament. This provision was taken up in the Western Hemisphere Convention of 1940 and in the African Convention of 1968 which replaced the London Convention. As a result, the laws of several African and Latin American countries also contain a similar requirement. Even where protected areas can be established by Government decree, they can only be de-established by an Act.

In Brazil this requirement is now embodied in the 1988 Federal Constitution. This provision was used recently in a case involving the construction of a highway between Sao Paolo and Santos which was to cut through a part of a protected area. As there was no legislative authority to do so, the works were unable to proceed.

Several other countries also require that stricter procedures be followed for the abolition of protected areas. The most common method is to require an Act or a resolution from the Parliament, whereas the creation of an area can merely be effected by regulations. Examples are the laws on national parks of Australia and of most Australian States, India, South Africa, and Zimbabwe. In France, there is no difference in the procedures for the establishment or abolition of national parks. In the case of nature reserves, however, a public inquiry is only necessary when at least one of the landowners concerned objects to the establishment of the reserve, whereas an inquiry is always required for its abolition.

3. The Integrity Objective

The integrity of protected areas may be threatened by activities exercised within or beyond their boundaries.

a. Activities within Protected Areas

i. Prohibitions and Restrictions

Protected area general legislation or, as the case may be, the acts or regulations establishing or regulating individual parks or reserves, always contain lists of prohibited or restricted activities. These lists can never be exhaustive, however, and there is therefore usually a need for a catch-all provision generally prohibiting activities that may be contrary to the objectives and purposes of the protected area or areas in question. In addition, such primary or secondary legislation nearly always contains lists of activities which remain authorised, although it is emphasised that these activities must be compatible with the purpose of the park or reserve.

The protected area authority is also often given some discretionary powers to grant exemptions from the prohibitions or restrictions imposed by the legislation. The exemptions to general prohibitions may be of various kinds. They may be designed to allow and promote scientific research or to facilitate the management of the park or reserve, including the construction of roads and accommodation for visitors. Other authorised activities often include the collection of fungi and other plants or forest products by local populations, as well as hunting and fishing. These activities are not generally harmful to the protected ecosystems, provided they are strictly controlled and only traditional methods are used. Indeed, it is often essential to maintain such activities if the protected area is to be accepted by local populations and integrated into the surrounding social and economic environment.

On the other hand, other authorised activities may have serious detrimental effects on the integrity of protected areas, however important they may be for the economy of the country or the region. These include mining, military activities, forestry, agriculture, skiing facilities and many others.

Mining, for instance, remains authorised in many cases, although it may be subject to restrictions or controls. For example, the Malaysian National Parks Act of 1980 authorises mining when there exists in a national park "a mineral deposit of such richness that it would be contrary to the interests of the State that it should not be mined ...". In Australia, Commonwealth legislation provides that mining rights, as opposed to other rights, are not vested in the Parks Director when the protected area is established. They remain the property of the Commonwealth. Mining can therefore be authorised by regulations. However, it must be carried out in accordance with the protected area management plan. Mining cannot be authorised in a wilderness zone.

In Zimbabwe, a permit is required for any prospecting or mining in a protected area. Mining rights automatically entail exemptions to protected area rules in respect of all activities necessary to the exercise of such rights, subject to any conditions attached to the permit, even if these activities result in the destruction of protected plants. There are many examples of this kind, relating not only to mining but to other activities as well.

In consequence, protected areas are not always well preserved by law against threats to their integrity. Perhaps this is unavoidable for economic and social reasons. What is essential, however, is that the legislation should provide for procedures to ensure that certain activities are only allowed if they are in the public interest, their public interest clearly outweighs the public interest of conservation and there are no other solutions. Such requirements do sometimes appear in protected area legislation. For example, the law of Catalonia of 30 March 1988, re-establishing the Aigues Tortes National Park, provides that in exceptional cases, for reason of public interest and provided that it has been expressly demonstrated that there is no other viable solution, prohibited activities including mining may be authorised under a permit. Mitigation measures must, however, be taken.

Before any activity which may affect a protected area is authorised, it would seem to be essential that its impact on the area be fully assessed. This should also be the case where legislation contains a catch-all provision generally prohibiting all activities liable to cause harm to the area. Without such an evaluation, it will often be impossible to determine whether or not a proposed activity is likely to be detrimental to protected area values.

This should preferably be done by using the Environmental Impact Assessment (EIA) technique. However, in most countries where EIAs are now required, the use of this instrument is generally limited to certain categories of activities or works, usually large scale industrial or transportation projects. Such projects would hardly ever be proposed in protected areas, as they would manifestly be totally incompatible with their status. Indeed, if approved, they would probably require the abolition of the protected area concerned.

Where it is proposed to undertake works in particularly sensitive areas such as national parks and reserves, there is consequently a need to subject a far greater range of projects to the EIA procedure than in the rest of the national territory. Very few laws have yet established this requirement.

Examples include Greece, where the Environmental Protection Act of 1986 provides that the decree establishing a protected area, including a nature park, may require EIAs to be carried out in respect of activities proposed within the protected area, even though such activities would normally be exempted from such a requirement. In Spain, the Decree of 30 September 1988 on EIAs makes it mandatory to perform such an assessment for open cast mines within protected areas. In addition, the Act of 27 March 1989 on the Conservation of Natural Areas and Wild Flora and Fauna provides that the Natural Resource Management Plans which must be developed under the Act for all protected areas shall determine the activities, works or installations, whether public or private, in respect of which an EIA will be required.

In the Spanish Autonomous Community of Andalucia, the Act of 18 July 1989 on Protected Areas requires EIAs to be prepared in respect of any proposed activities, other than traditional ones, to be undertaken within a particular category of reserve, called "parajes naturales".[20] Permits for such activities will only be granted when the EIA has shown that they will not jeopardise the values for which the reserve was established.

In the Canary Islands, another Spanish Autonomous Community, the Act of 13 July 1990 on EIAs contains some particularly interesting and innovative provisions, which are discussed in Chapter VII(B)(1) below. In summary, a detailed EIA must be prepared in environmentally sensitive and/or protected areas in respect of a large range of activities, many of which are not subject to the EIA procedure anywhere else in the world.

---

[20] "Parajes naturales" are reserves where traditional activities remain authorised whilst any other activity is subject to a permit from the Environmental Agency.

Finally, the new Directive of the European Community on the Conservation of Natural Habitats and of Wild Fauna and Flora of 1992 contains a provision which makes the preparation of EIAs mandatory for a wide range of projects, where these are to be undertaken in the Special Areas for Conservation established under that Directive or under the Directive on the Conservation of Wild Birds of 1979.

ii. Persons affected by Prohibitions or Restrictions

Different groups of persons may be affected in different ways by the regulatory measures applicable within protected areas.

The Public

Protected area legislation almost universally contains regulatory measures to control access by the public to parks and reserves and to regulate the behaviour of members of the public within these areas. Particular attention must be paid to ensure that certain authorised activities, particularly hunting, fishing, and the collection of wild plants and other products, are carried out according to strict management rules. Protected area authorities should therefore be empowered to regulate these activities in conformity with the conservation requirements of the area where they are exercised. For example, national hunting or fishing legislation should be replaced in protected areas by specific rules adapted to the particular needs of the areas concerned, and adopted and implemented by the protected area authorities themselves.

In Spain, for instance, special management plans are to be developed for the control of certain authorised activities in protected areas. These plans are of a regulatory nature and are therefore binding upon members of the public, as well as upon the protected areas authorities themselves.

Landowners or Occupiers

Where land included in a protected area is not Government property and certain uses of this land, such as for agricultural or forestry purposes, continue to be allowed, it is important that these activities be specifically regulated and closely monitored in order to avoid detrimental effects on park or reserve values. This is best achieved by a permit system whereby actions such as crop changes, the extension of cultivated areas, the erection of farm buildings, clear cutting of wooded areas and afforestation would only be authorised where they do not affect the integrity of the protected area. In Andalucia, for instance, permits are required for all activities exercised in the reserves belonging to the category called "parajes naturales". The only exemption is made for permitted traditional activities, provided these do not endanger the protected natural values.

Protected area authorities should also be empowered to buy or expropriate land within the boundaries of these areas and to conclude binding management agreements with landowners. They should have the right to order, or to ask a court to order, that activities undertaken in contravention of regulations be stopped immediately and that damaged areas be restored.

A different situation arises where land within a protected area belongs to a municipality, a frequent occurrence in Europe. Municipally-owned land is generally considered to be private land and is subject to all the restrictions imposed by the law. It is clear, however, that when a large part of a protected area is in municipal ownership, it will often be necessary for the protected area authority and the municipalities owning land in the park to come to some arrangement to make land-use restrictions imposed upon the municipalities more acceptable. This may be achieved by the granting of special benefits.[21] Another method, which seems to be rarely used but has proved very effective, is the conclusion of contracts.

This method was used for the establishment of the Swiss National Park. Under the terms of the contracts, the four municipalities concerned undertook to abstain from any use of their lands, including for grazing and forestry, and authorised the Swiss Confederation to maintain the area in its natural state. In exchange, the municipalities were entitled to full compensation. In addition, the municipalities agreed to waive their prospecting and mining rights without compensation, as well as their rights to use park water courses for the production of hydro-electricity. These contracts were signed as early as 1914 and are still in force today, to the satisfaction of all parties concerned.

Indigenous People

In many wild areas of the world, indigenous human populations living on and from the land by traditional means can still be considered as forming an integral part of the ecosystems concerned. Where a protected area is established on such land, it should thus have as an additional objective the conservation of the particular culture, knowledge and way of life of the indigenous populations living within its boundaries.

In certain countries, however, particularly in Africa, there has been a trend in the past decades to bring community land into Government ownership and to abolish customary rights in national parks and other protected areas. To give a few examples, the Nature Conservation Act of Zaire of 1969 provides that as a result of its having become public property, the land included in protected areas cannot be the subject of customary rights other than those which have been expressly maintained. In the Central African Republic, the elimination of customary rights in strict nature reserves and national parks is considered to be in the public interest and no exercise of these rights is allowed.

In many other countries, traditional activities by inhabitants of protected areas are allowed, subject to certain conditions and controls.

In Mauritania, for instance, the Decree of 24 June 1976 establishing the Banc d'Arguin National Park authorises traditional subsistence fishing by local populations, provided that any alteration in the traditional means that are used is first approved by the park authority.

The Decree of 16 March 1977 of Colombia lays down the principle that a special regime may be established for the benefit of indigenous populations living in a national park, respecting their rights to harvest natural resources with techniques compatible with the park objectives.

In Canada, the Parks Canada Policy adopted in 1979 provides that certain traditional subsistence resource uses may continue to be permitted and that the treaty rights of Indian peoples will be honoured. Such rights can only be terminated by mutual agreement with the people concerned.

---

[21] See section (D)(2) of this Chapter.

In Australia, the federal National Parks and Wildlife Conservation Act of 1975 allows Aborigines in parks, reserves and conservation zones to continue their traditional uses of land or water for hunting or food-gathering, otherwise than for the purpose of sale, and for ceremonial and religious purposes, subject to any regulations made for the purpose of conserving wildlife.

The Kakadu National Park in the Northern Territory is now in Aboriginal ownership. It is leased to the Director of Australia's National Parks and Wildlife Service for a period of one hundred years.

The importance of preserving indigenous culture and knowledge has now been formally recognised in international law by article 8 of the Convention on Biological Diversity, which deals with *in situ* conservation. Each Contracting Party shall, as far as possible and as appropriate:

> "subject to its national legislation, respect, preserve and maintain knowledge, innovations and practices of indigenous and local communities embodying traditional lifestyles relevant for the conservation and sustainable use of biological diversity.... and encourage the equitable sharing of the benefits arising from the utilization of such knowledge, innovations and practices."

Public Agencies

The State and public agencies often behave as if protected area legislation did not apply to their own activities. This means that permits are sometimes granted by agencies other than park or reserve authorities for activities which may result in considerable damage to protected areas. Even if this constitutes a violation of the law, conservation agencies are often powerless to prevent it, as the public interest of the proposed activity will be considered as outweighing the importance of preserving a natural area. Moreover, these agencies may sometimes not even be consulted or informed of the proposed activity and will then be faced with a *fait accompli*.

There are no easy legal solutions to the problem which is, of course, largely of a political nature. Some legal mechanisms may, however, be of help.

Protected area legislation does not generally have priority over other laws charging other agencies to carry out certain works, such as the construction of dams or the exploitation of forests for the production of timber. These laws are usually silent as to how to resolve conflicts that may arise when these activities are exercised in protected areas. Development agencies feel that they do not have to comply with legal provisions running counter to a mandate which they hold under legal provisions of the same rank and force. It may therefore be important to specify in the protected area legislation that in the case of a conflict, the latter will prevail. In this way, the Australian Northern Territory Fish and Fisheries Act of 1979 specifies that in respect of aquatic parks, regulations may prohibit or restrict any activity under any law of the Territory including an Act, notwithstanding that such a law may expressly provide for the activity in question.

Another possible legal mechanism is to state in the law on protected areas that it is applicable to all Government departments or agencies. In certain countries of the British Commonwealth, this is expressed by the sentence : "This Act binds the Crown". Several laws of the Spanish Autonomous Communities contain provisions requiring all public agencies to exercise their powers in such a way that the values of protected areas will not be jeopardised. In Mexico, the Decree of 16 January 1986 establishing the Sian Ka'an Biosphere Reserve provides that all federal agencies which undertake activities or investments in the Reserve must do so in accordance with the objectives laid down in the Decree, and must abstain from any action contrary to these objectives. The Minister of Finance is not allowed to make funds available for programmes or activities in contravention of the objectives of the Decree.

It may also happen that the public agency contravening the legislation will be the protected area authority itself, especially when the law has provided it with a considerable degree of discretion in the granting of exemptions to statutory prohibitions or restrictions.

Parks and reserve authorities are not immune to political pressures and abuse of their discretion is always possible. This is particularly likely to occur when the discretionary powers of an agency are not limited by clear rules. For instance, where the law merely states that only those activities which are compatible with the objectives of a protected area may be authorised, it is clear that the determination of whether an activity is or is not compatible will often be a matter of personal judgment on the part of the authority concerned, especially if no EIA is required. On the other hand, it is also essential for an agency to have a certain degree of flexibility in the way it may exercise its powers.

One solution could be to bind a protected area authority not to grant exemptions in certain well-defined cases. For example, the Parks and Wildlife Act of 1975 of Zimbabwe gives an exhaustive list of the powers of the Minister concerned in relation to activities in national parks. By inference, an activity that does not fall within these powers is consequently illegal.

Mechanisms binding public agencies can only be really effective, however, if agency decisions can be subject to judicial review. Without this sanction, nothing really prevents a Government department or other public body from exceeding its powers and no legal remedies are possible when this occurs. The scope of judicial review is very wide, however, as it encompasses any decision by a public body which may affect the natural environment. It will therefore be dealt with in some detail in the General Conclusion to this paper.

Nevertheless, it may be of interest to point out here that certain laws relating to protected areas give members of the public the right to institute proceedings in the courts, to ensure that public agencies comply with the legislation and, where necessary, to seek the invalidation of illegal administrative decisions. In Spain, provisions to this effect appear in certain regional laws, such as the one of Andalucia on Protected Areas, and in some but not all of the Acts establishing certain national parks.

Even when judicial review is possible, there is still a need for procedures to suspend the effects of a decision affecting the protected area, until such time as a court has had the opportunity to decide on the merits of the case. Without such procedures, there is a serious risk that unlawfully-authorised works will be completed by the time the final judgment has been pronounced, which would make their demolition very difficult to achieve.

To give a recent example, the Conseil d'Etat, the supreme French administrative court, recently judged that a permit from the Western Pyrenees National Park Authority was illegal. The grounds were that although the law gave a considerable degree of discretion to the Authority for the issuing of permits, a permit that allowed the construction of large buildings and parking lots for the development of winter sports in the core area of the park violated the spirit of the National Parks Act and should therefore be declared void. However, as the project had been completed in the meantime and the court has no powers to order its demolition at this stage, this decision is of little more than moral significance.

b. Outside Activities affecting Protected Areas

Most protected areas are vulnerable to the effects of activities exercised outside their boundaries, in particular to these affecting the quantity or quality of the waters flowing into them. The traditional approach to this problem has been to establish buffer zones around parks or reserves, in which activities that may affect the integrity of the core area may be prohibited or restricted. However, this approach may not be sufficient where harmful activities originate at great distances from a protected area.

i. Buffer Zones

Many laws provide for the establishment of buffer zones around protected areas. Their purpose is to institute a transition zone between the core area of a protected area, where most if not all human activities are prohibited, and the outside world where these activities may be freely exercised, subject to any general controls applicable to the whole of the national territory. In other words, buffer zones are areas where special controls may be applied for the purpose of preserving the integrity of a protected area.

National laws, however, differ widely on the interpretation of this concept. In Cameroon, for instance, the prohibitions applicable in buffer zones are the same as those in core areas. In Spain, buffer zones around national parks are areas where construction is in principle prohibited and where all activities that may have an ecological impact on the parks can be strictly controlled. In the buffer zone of the Aiguestortes Park in Catalonia, only traditional activities are allowed without a permit and all others must first be authorised by the park authority. In Kenya, the Wildlife Act of 1976 provides that the Minister may designate protection areas around national parks or reserves and specify the activities that will be prohibited or restricted in these areas.

Buffer zones are, in general, established by the law creating the protected area itself and are usually considered as forming an integral part of that area, albeit with a different legal regime. Alternatively, they may be established by the creation, around the boundaries of an existing protected area, of another protected area belonging to a different legal category. This system is used in Spain, where the Autonomous Communities have established regional natural parks as buffers around some of the national parks created by national legislation.

In those countries where it is not possible to restrict land uses on land that is not in public ownership, it may not be easy to establish buffer zones on private land surrounding national parks or equivalent reserves. In most cases, buffer zones will accordingly have to be created within the protected area itself. This could mean that the core was comprised of a strict nature reserve, in which all activities except scientific research are prohibited, whilst the rest of the protected area was considered as a buffer zone for these core reserves.

The buffer zone concept is bound to gain in importance as biosphere reserves, in which the establishment of buffer zones is an essential element, are increasingly recognised as a valuable tool for the creation and management of protected areas.

ii. Beyond the Buffer Zone

Beyond the boundaries of the buffer zone, or when there is no buffer zone, the control of activities that may be harmful to protected areas may be much more difficult.

Ideally, provision should be made to control any outside activity which may have an adverse impact on such areas. This would require the enactment of specific rules subjecting a wide range of activities to permits; requiring that no permits be issued unless it has been shown that proposed activities will not harm a protected area; providing for conditions to be attached to permits to ensure that damage will be avoided; establishing procedures making it possible to stop an activity immediately whenever unforeseen damage is observed; and requiring that such damage be repaired and providing for the deposit of a security or performance bond to ensure that the damage is effectively repaired.

No law goes as far as that. However, some laws do establish general principles for the protection of parks and reserves against outside activities. The German Nature Conservation Act of 1976, for instance, provides that all actions which may lead to the destruction of, cause damage to, or induce changes in a nature reserve, a national park or natural monument shall be prohibited, "subject to more specific provisions to be adopted". However, it is unclear to what extent this general provision is effectively applied in practice to control harmful outside activities. The Law of the Spanish Autonomous Community of Valencia of 1988, on the establishment of the reserves called "parajes naturales", prohibits any action or omission resulting in the deterioration of any wetland included in such reserves or in the pollution of the waters or of the environment of these areas.

Another mechanism consists of establishing a third protection zone around the buffer zone: biosphere reserves should in principle be surrounded by such transition zones. This method has been used in respect of the Doñana National Park in Spain. In that zone, called the "zone of influence", special measures may be taken for the protection of the waters flowing into the park. The Government may, in particular, order the suspension of any activity liable to affect the quantity or quality of these waters pending the adoption of corrective measures. Special rules may be applied to the use of pesticides. By way of example, an order requiring that phyto-sanitary treatments be effected against a rice pest specifically exempts rice fields situated in the zone of influence from this requirement. In that zone, only certain types of pesticides can be used under controlled conditions.

It is also possible to regulate certain activities, either within a certain radius around protected areas or wherever they occur.

For example, the Law of Luxembourg of 1982 on the Protection of Nature and Natural Resources prohibits any construction within 30 metres of nature reserves. In Sweden, under the Water Act, permits relating to the use of waters upstream of a national park or nature reserve must be denied where the proposed activities will result in damage to these areas.

Finally, perhaps the most effective way to preserve protected areas from harmful outside influences, although not necessarily the easiest one, may be to enact special physical planning rules to be applied in those areas where certain development projects are most likely to have undesirable effects on parks or reserves.[22]

In all these cases, the difficulty lies in the fact that it may not be easy to determine whether or not a proposed activity will affect a protected area, especially if it is to take place a long distance away. It is therefore necessary to require EIAs for all activities which may have harmful effects on parks or reserves.

---

[22] See section (D)(2) of this chapter, dealing with the integration of protected areas into their physical, economic and social environment.

This requirement is included in the new European Community Habitats Directive of 1992. According to article 6 of the Directive which deals with measures for the protection of Special Areas of Conservation,

> "any plan or project not directly connected with or necessary to the management of the site but likely to have a significant effect thereon, either individually or in combination with other plans or projects, shall be subject to appropriate assessment of its implications in view of the site's conservation objectives..."

The competent national authorities may only agree to the plan or project after having ascertained that it will not adversely affect the integrity of the site concerned. Where appropriate, the opinion of the general public must also be obtained.

There is no express requirement under the Directive for Member States to ensure that the authorities responsible for the protection and management of these areas are closely associated in the assessment procedure. In addition, there are no criteria to determine when a project or plan is "likely" to have a significant effect on a protected area. This lack of precision may well result in uncertainties and disputes in the future.

Whatever the requirements that are imposed for the control of activities outside protected areas, there is an equal need for the availability of judicial review of administrative decisions that may affect the integrity of protected areas, including decisions as to whether or not an EIA is required.

Finally, no legal measures adopted within a particular jurisdiction can preserve a protected area against the impact of activities carried out in a neighbouring country or federated entity. The Law of the Spanish Autonomous Community of Valencia on the reserves called "parajes naturales" is perhaps unique in that it acknowledges this problem and directs the Regional Government to seek the cooperation of the other Autonomous Communities concerned, in order to avoid damages to "parajes naturales" caused by activities taking place on their territory.

Possible solutions to this problem lie in the conclusion of specific agreements between the federated or regionalised territorial entities concerned, or of international conventions when the damaging activity is carried out in a neighbouring country. In this context, it will be recalled that the World Heritage Convention requires its Parties not to take any deliberate measures which might directly or indirectly damage World Heritage Sites situated on the territory of other Parties.

Mention should finally be made of the Convention on Environmental Impact Assessment in a Transboundary Context, signed in Espoo, Finland on 25 February 1991. The Convention requires the making of an EIA in respect of all proposed activities listed in its Appendix I, which are likely to have a significant transboundary impact, including upon transboundary waters. In addition to potentially polluting activities, the Appendix includes large dams and reservoirs and the deforestation of large areas. The Convention also provides that where the Parties so agree, other activities likely to cause a significant transboundary impact shall be treated as if they were listed in Appendix I, but this effectively depends on one Party taking the initiative for this purpose.

Appendix III sets out general criteria for the determination of the environmental significance of activities not listed in Appendix I. Most interestingly, one criterion relates to the location of the activity and covers proposed activities in or close to an area of special environmental sensitivity or importance, such as wetlands designated under the Ramsar Convention. This criterion does not of course extend to more distant activities, which may still be capable of adversely affecting Ramsar sites.

The very fact that such protected areas are characterised as liable to benefit from special treatment for the purposes of the Espoo Convention is of great importance, including for the effective implementation of the Ramsar Convention. However, the Parties concerned are under no obligation to grant this preferential treatment to Ramsar sites, nor does the Espoo Convention establish any mechanism for the conclusion of such agreements. There is obviously nothing to prevent the Ramsar Convention from recommending such an agreement to the countries concerned on the basis of article 5 of that Convention.

C.   The Management of Protected Areas

1.   Introduction

The primary purpose of protected areas is the conservation of natural ecosystems. However, many parks and reserves also have as an object the provision of a certain number of uses, such as scientific research, recreation, and other permitted activities, even though these are generally subordinate to their main purpose. Protected areas must therefore be managed to ensure that such uses are only authorised where they do not run counter to the conservation objectives of the area, and that, where authorised, their degree and extent remain compatible with that goal. It is also also necessary to prevent or punish unauthorised uses.

This form of management may be called regulatory management. To be effective, it must clearly be based on scientific research and monitoring and take into consideration all the interactions between human uses, the various components of the protected ecosystem and the environment as a whole. The management of a protected area, and this is even truer of a whole system of protected areas, is thus a complex operation where the results of research should govern the policy to be followed, the regulations which are to be adopted, and the measures that are necessary to enforce them in each individual area.

In addition, most protected areas also need ecological management, namely the performance of certain manipulations of the ecosystem, including restoration of degraded habitats, where these are necessary to safeguard, rehabilitate and sometimes enhance park or reserve values.

Finally, a protected area is almost always an administrative unit with its own staff, buildings, services and budget, and must accordingly be administered as such.

For all these reasons, many laws now provide for the establishment of management bodies and the development of management plans.

2.   Management Bodies

There are considerable differences between countries as to the nature, powers and functions of the bodies in charge of the management of protected areas and it is therefore not possible to examine existing systems in detail. There are, however, some general trends which are worth considering.

a. Centralised Management

A first approach is to manage all protected areas from a central administration. This may for instance be the Forest Service, as in for the national parks in Greece and Turkey and in many African countries, or a specialised Government body, such as the National Parks Service in the United States, Parks Canada and the Australian National Parks and Wildlife Service. Park or reserve directors and staff are directly appointed by the central body. There is sometimes a central advisory committee or council.

In Sweden, national parks are managed by an authority appointed by the Crown, whereas nature reserves are established and managed by the counties. Reserve administrators are appointed by the counties after consultation with the National Environment Protection Board, the State administration in charge of environmental matters. In Malaysia individual parks are administered by a committee appointed by the Minister and composed of representatives of the Federal and State Governments. The Minister is advised by a National Parks Advisory Council.

The system of a centralised administration for protected areas is generally used when these areas have been established on public land.

b. Decentralised Management

Many countries, on the other hand, have opted for a more decentralised system of management, providing for a closer association with local authorities and populations. This would seem to be the case particularly where protected areas are mostly created by regulatory measures on private or municipally owned land. Where this is the case, it is clear that the success of the conservation measures taken depends on the involvement and cooperation of the local authorities and populations concerned. The institutionalisation of that cooperation through the establishment of management bodies, in which local people can make themselves heard and often participate in the decisions is, therefore, all the more important.

There are two main approaches to decentralised management. The first, which is widely used in Spain both at national level and by the Autonomous Communities, consists of the appointment of a Board for each park which is composed of representatives of the Government, local authorities, scientific institutions, local interests and conservation NGOs. The Board's functions are to oversee compliance with the park regulations, promote and implement management measures and give its advice on the management plan and work programme.

National parks, which are always established by the central government, are administered by ICONA, the Government institute for nature conservation, a public body under the Ministry of Agriculture. Board chairmen are appointed by the Government. Similar systems exist in many of the Spanish Autonomous Communities for protected areas created under their jurisdiction.

This system is therefore of a mixed nature. The parks enjoy only a limited degree of autonomy and remain closely controlled by the central administration. The Boards, however, play an important role in that their advice is required by law before many decisions can be taken, including the adoption of the management plan. In a few cases, certain decisions are subject to the prior agreement of the Board. In Andalucia, for instance, any activity in certain nature reserves which is not provided for by the management plan must be first approved by the Board.

The second approach to decentralised management consists in providing the parks with a high degree of administrative autonomy, by establishing them as corporate bodies with legal personality. This method is used, for instance, in Algeria, in France and in certain Italian parks. The Boards and park directors continue to be appointed by the central administration, but they are vested with considerable powers. In Italy, powers are conferred on the Boards for such matters as the determination of park policies, the approval of management plans, the issuing of permits and the adoption of action programmes. Each park is legally represented by the chairman of its Board.

In France, park boards lay down general principles relating to park management and regulations and supervise the action of the director. The director is the legal representative of the park. He is vested with police powers and, as a result, may make regulations binding upon third parties. In particular, he may make any regulations that may be required to implement the prohibitions or restrictions laid down in the decree establishing the park. Even more important, he has the power to substitute himself for the mayors of the municipalities holding land in the park in the exercise of their jurisdiction over the management of municipally-owned land and the policing of the park.

As corporate bodies with legal personality, parks are naturally entitled to buy or lease land and to conclude contracts. They may also exercise activities outside their boundaries. In France, for instance, they can be the managing body of nature reserves established in the surrounding region.

3. Management Rules

This section only deals with regulatory management as ecological and administrative management do not generally require the enactment of specific legal provisions.

a. Zoning

The first aspect of regulatory management relates to the zoning of protected areas. Indeed, zoning is increasingly considered as an essential management tool, as it allows for the fine tuning of regulations to meet the particular requirements of the various types of areas included in a park or reserve. In addition, as the protected area concept evolves to include areas where some uses of the land may continue to be pursued, zoning also becomes necessary to separate these zones from those in which stricter rules apply.

As a result, protected areas no longer appear as monolithic entities where all land was subject to the same prohibitions or restrictions, but rather as mosaics of individual smaller protected areas each with a different legal regime. These regimes range from strict nature reserves or wilderness areas, where almost all human activities are prohibited, to buffer zones where only certain activities are controlled.

The different types of zones which may be established are generally defined in the basic protected areas legislation. This method is lacking in flexibility, however, as it makes it difficult to adapt the zoning system to the particular requirements of each individual area. Increasingly, therefore, legislation tends merely to provide for the power to zone a protected area and to determine in each case whether the area should be zoned and, if so, what should be the functions of each zone and the rules applicable therein.

To give an example, the Fisheries Act of 1976 of the Australian State of Queensland listed six types of zones which could be established in marine parks. These provisions were repealed by the Marine Parks Act of 1982, and replaced by a new text which simply confers the power to divide a marine park into zones and to make specific regulations relating to each one of them.

b. Management Plans

Management plans, or regulatory instruments for the management of protected areas, are a legal requirement only in a small number of countries. However, they have many advantages and deserve, it would seem, to be more widely recognised as an important management tool. They do constitute a unique method of combining, in a legally binding instrument, the establishment of management objectives, based on the results of scientific research and monitoring, with regulatory measures aiming at achieving these objectives.

Management plans of this type are adopted by a formal procedure and are binding on the protected area authorities, all Government agencies and private persons. They may provide for the establishment of zones and for specific regulations particular to each zone or type of activity and may also contain general indicative provisions relating to such matters as land acquisition, the construction of necessary infrastructures, the restoration of degraded habitats and other general management objectives.

The Australian and Spanish legislation on management plans are of particular interest as examples of this approach.

The Australian federal National Parks and Wildlife Act of 1975[23] requires that a Plan of Management be prepared in respect of each park or reserve. These plans may provide for the zoning of such areas and must describe the manner in which it is proposed to manage the protected area. In particular, where a Plan of Management provides for the mining of minerals or other works, the plan must set out any conditions that are applicable. Criteria and objectives are laid down for the preparation of plans. If the plan provides for the division of the park or reserve into zones, the conditions under which each zone shall be kept and maintained must also be set out.

Plans of Management must be submitted to the public before they are approved. Representations made by interested persons must be given due consideration. The Plans are then approved by the Minister and laid before both Houses of the Parliament. If either House of Parliament disallows the Plan, a new Plan must be prepared. Plans of Management, once approved, are binding upon the Director of Parks and Wildlife. They are made for a maximum period of ten years and must therefore be replaced by new plans after this period has elapsed.

In Spain, management plans called "Planes Rectores de Uso y Gestion" must be established in respect of all national parks and nature parks. They must be approved by the national Government for national parks, and by the competent regional body for nature parks. They must lay down general rules for the use and management of these protected areas and must be periodically updated. An important feature of Spanish management plans is that they constitute mandatory land-use regulations, which prevail over any other normally binding land-use plan. In the case of a conflict between a management plan and any other land-use plan, the latter must be revised accordingly. Management plans are binding on public agencies and private persons.

---

[23] This Act is only applicable to land owned or held on lease by the Federal Government, and to marine areas under federal jurisdiction. Individual Australian States have their own protected area legislation, which may or may not require the adoption of management plans.

Management plans provide for the zoning of each protected area, set out management objectives for each zone and lay down binding rules accordingly. More detailed objectives and regulations may be established by special plans dealing with particular activities.

To give an example, the Law of the Valencia Autonomous Community on "parajes naturales" provides that a management plan must, at a minimum, lay down management objectives, describe the area, establish zones and their level of protection, provide for the necessary instruments for the carrying out of the plan, regulate activities in the area, contain an operational programme including the development of economic and financial studies, and provide for monitoring and follow-up measures.

D. The Integration of Protected Areas into their Physical, Economic and Social Surroundings

It is commonly said that no park is an island. Yet, by their very nature and purpose protected areas must be preserved from harmful influences coming from the outside world as much as possible. On the other hand, if all links are severed between a protected area and the surrounding communities, the area will soon be considered by them as a foreign body and rejected as such.

Some form of integration with the outside world is therefore essential if protected areas are to survive. As article 5 (a) of the World Heritage Convention puts it, States must consequently endeavour

"to give the natural and cultural heritage a function in the life of the community and to integrate the protection of that heritage into comprehensive planning programmes".

1. Protected Areas and Physical Planning

Physical planning may be an important instrument for the preservation of protected areas from harmful outside influences. This is recognised by the Ramsar Convention which requires Parties to formulate and implement their planning so as to promote the conservation of the wetlands included on the List of Wetlands of International Importance.

Physical planning policies which take the needs of protected areas fully into account could be effectively used to locate certain developments, such as highways and other infrastructures, housing projects and factories, in areas where they are least likely to cause harm to parks or reserves through increased visitor pressure, water use, pollution or other factors. In keeping with its size or vulnerability, a protected area should therefore be able to trigger the adoption of specific planning rules and restrictions in the surrounding environment and particularly within the whole of the watershed concerned.

No law goes that far, although a few do contain certain provisions which are going in that direction.

One example is the Spanish Act of 1989 on the Conservation of Natural Areas and Wild Flora and Fauna. The Act provides, in particular, that water planning for each river basin must provide for the conservation and restoration of the natural areas they contain, especially wetlands. The same Act institutes a special and very novel, kind of planning instrument, called Natural Resources Management Plans ("Planes de Ordenacion de los Recursos Naturales").

195

The purpose of these Plans is to apply the principles of the World Conservation Strategy to the management of natural areas and species requiring protection. Such a Plan must establish indicative criteria for the formulation and implementation of all sectoral policies affecting the area concerned and determine the activities, works or facilities, whether public or private, which will require EIAs. The Plan is a binding instrument, which means that no other instrument of physical or land-use planning can prevail over it or amend any of its provisions in any way in respect of any matter regulated by the Act. The provisions of the Plan in all other matters are merely indicative.

Another interesting example is that of the American State of Montana, wherein the Major Facilities Siting Act of 1983 deals with installations, such as thermal power plants. For each potential site for a new plant, an impact zone must be determined. If the predicted impact of air pollution on the vegetation of natural areas, including parks and other protected areas, is considered unacceptable, the plant must be located elsewhere.

Finally, the Convention on Biological Diversity lays down the broad requirement that Contracting Parties should, in accordance with their particular conditions and capabilities:

"integrate, as far as possible and as appropriate, the conservation and sustainable use of biological diversity into relevant sectoral or cross-sectoral plans, programmes and policies." (article 6)

2. Protected Areas and their Social and Economic Environment

Protected areas must be accepted by the populations living in their surroundings. At a minimum, there should be peaceful coexistence between the two. Where the existence of a park or reserve is considered by local people as an asset from which they derive tangible benefits, the chances of success of the protected area concerned clearly become much greater, especially when the local people are invited to participate in its establishment and management.

Many laws now require that local inhabitants be consulted before a protected area is created. This is of course essential when a park or reserve is established on private land. Even where only public land is concerned, consultation would nonetheless seem to be important, because neighbouring populations are bound to be affected in one way or another by the creation of the protected area.

Some examples, amongst many, of legislation requiring such consultation include an Order made under the 1982 Hungarian Nature Conservation Law-Decree, whereby the opinion of all interested parties and citizens must first be obtained before a protected area is established; the Law of Morocco of 1934 which requires that information about the intention to create a national park be made known by town-criers and posters and that the observations of the public be recorded; and the Norwegian Nature Conservation Act of 1970 which provides that municipalities and counties must be consulted and that all persons who may be affected by the proposed conservation order be able to make their views known.

It is also important to provide for some representation of local populations on the management boards of protected areas, where these exist. This is best achieved, as seen above, by the appointment of representatives of local authorities, such as the mayors of the municipalities concerned, and of various local interest groups. When there are no appointed management boards, some laws, such as that of Australia, require that the public be consulted when management plans are prepared.

Another important aspect of the integration of protected areas into their environment is the continuation of traditional activities in the park or reserve by local people, provided that these do not run counter to protection objectives. Many laws provide for this possibility, at least in certain categories of protected areas. As an example, the laws on protected areas of Algeria, France and Greece state that all traditional activities that are compatible with the purposes of these areas may be maintained.

The importance of traditional activities is moreover rightfully emphasised by the three Protocols on Specially Protected Marine Areas which have been concluded so far under the Regional Seas Conventions. To quote the Kingston Protocol of 1990,

"each Party shall, in formulating management and protective measures, take into account and provide exemptions, as necessary, to meet the traditional subsistence and cultural needs of its local populations".

The fact that when traditional activities are allowed, the right to exercise them is usually limited to neighbouring populations which consequently excludes outsiders, may also help to make protected areas more acceptable to local communities. Finally, provision may also be made to make certain benefits available to the municipalities concerned, particularly in the form of development projects and improved community services, as compensation for the restrictions resulting from the establishment of a protected area.

In Algeria and France, peripheral zones have been created around national parks. Their main purpose is to serve as the seat of specific social, economic and cultural projects. In Spain, some park laws provide for the establishment of a third zone, called the zone of influence, around the buffer zone surrounding a national park. In that zone, there are no particular restrictions on human activities, but special grants are available for the construction of infrastructure projects and the carrying out of many types of community development projects. In addition, the municipalities concerned have a preferential right to benefit from concessions for the provision of services in the park itself.

E. Conclusion

The traditionally rigid concept of protected areas, which mostly applied to uninhabited zones where all human activities were prohibited or severely restricted, is now gradually evolving towards a much more flexible approach where, around a closed core area, other zones are being established in which activities compatible with the conservation requirements of that area are authorised or encouraged. This approach is exemplified by the evolution of the biosphere reserve concept, which now emphasises the need to ensure the greatest possible degree of integration of protected areas into their social and economic environment.

However, another approach is steadily gaining momentum whereby, instead of starting from an uninhabited area and allowing compatible uses, the reverse process is actually encouraged, namely the delimitation of an inhabited area, in which certain no-use or limited use zones are established, and activities incompatible with the purposes of the area are eliminated. This new form of protected area is generally called "nature park" and will be discussed in detail in Chapter VI below.

There are also a certain number of other legal conservation instruments which have been developed over the past decades for the conservation of natural areas, and which are characterised by a greater degree of flexibility in the ways they can be used and adapted to particular conditions and circumstances.

CHAPTER IV
INNOVATIVE SITE-SPECIFIC INSTRUMENTS

A. Wilderness Areas

Wilderness areas can be defined as large roadless areas of pristine vegetation where certain developments, in particular the construction of roads, tracks and other means of access, and certain activities such as the use of motor vehicles, are prohibited. They generally remain open to the public and hunting, trapping and fishing are usually permitted. Wilderness areas cannot therefore be considered as protected areas in the classical sense of the term, but rather as a particular kind of land-use instrument prohibiting certain developments and activities in natural areas which meet certain conditions. In practice, as many destructive activities are impossible when there are no roads, the degree of protection provided by wilderness areas can be very high.

As far as it can be ascertained, wilderness areas are always established on public land. There is nothing, at least in theory, which would seem to prevent the use of this instrument on private land as long as the construction of roads and the use of off-road vehicles can be prohibited. Most of the remaining wild areas of the world are, however, in some form of Government ownership.

1. The Wilderness Concept

Wilderness is both a philosophical and a legal concept.

As a philosophical concept, it probably has its roots in the recognition in the United States, in the late 19th and early 20th centuries, that wild areas had values which needed to be preserved, coupled with the awareness that these were being destroyed at an ever increasing rate. The first national parks were then established, at least partly as a result of this new awareness. At that time, the conservation of species and ecosystems was not a major concern and parks were mainly set aside because of their spectacular scenery and wilderness values.

This concept was crystallised in the Convention on Nature Protection and Wildlife Preservation in the Western Hemisphere which was signed in Washington in 1940. The Preamble of this treaty refers to the protection and preservation of

"scenery of extraordinary beauty, unusual and striking geological formations, regions and natural objects of aesthetic, historic or scientific value and areas characterized by primitive conditions...".

In addition, provision is made for the establishment of "strict wilderness reserves" which are defined as regions

"under public control characterized by primitive conditions of flora, fauna, transportation and habitation, wherein there is no provision for the passage of motorized transportation and all commercial developments are excluded".

However, the protection of scenery and regions meeting these conditions was achieved by the establishment of protected areas in the conventional sense of the term, and it is clear that whenever possible the latter continue to constitute the most appropriate instrument for the preservation of wild areas. Indeed, most of the protected areas of wilderness throughout the world are within the boundaries of national parks or equivalent reserves.

2. Wilderness Legislation

The legal concept of wilderness appeared for the first time with the enactment by the Congress of the United States of the Wilderness Act of 1964. The Act established the National Wilderness Preservation System in response to the increasing acknowledgment by public opinion that wild areas had recreational, spiritual, aesthetic, ecological and other values and should be kept intact

"to secure for the American people of present and future generations the benefits of an enduring resource in wilderness".

Centred as it is on a "no roads, no motor traffic" approach, the wilderness legal concept is distinct from the protected areas method of wild land preservation, in that it may be used in any area meeting the required conditions, particularly that of roadlessness, whether in an already established protected area or not.

Pursuant to the Act, wilderness areas have now been created on a variety of federal lands under the control of different federal agencies, such as the Forest Service, the Bureau of Land Management, the National Parks Service and the Fish and Wildlife Service. The objective is to include some 400,000 sq. km of federal roadless lands into the system.

It is clear that in areas which are already protected under other Acts, such as national parks and national wildlife refuges, the new concept adds little which could not have been achieved by other means. In such cases, the wilderness status is merely superimposed upon another protection status pursuant to which roadless and vehicle-free zones are established. However, the concept has the advantage of being a legislative requirement which is binding on the agencies concerned and to which no exceptions may be allowed. Outside conventional protected areas, the new instrument obviously has enormous advantages in practice, as it permits the establishment of a new kind of protected area.

The concept has been taken up by a certain number of US States, which have adopted wilderness legislation applicable to State-owned roadless lands, as well as by a few countries where there are still large areas of wild lands in public ownership.

The State of Michigan, for instance, adopted a Wilderness and Natural Areas Act in 1972 which provides for the establishment of wilderness areas. These must be of at least 3000 acres in size or be on an island. Prohibited activities include mining, the destruction of vegetation, permanent or temporary roads, and the use of motor vehicles.

At federal level in Canada, wilderness areas may be established within national parks. In addition, several provinces have enacted their own wilderness legislation. Newfoundland, for instance, has "wilderness reserves" which are large areas where no roads, no vehicles and no tree felling are allowed. Ontario has "wilderness parks" which are provincial parks or zones within such parks where motor vehicles are not allowed.

In Australia, wilderness areas may be established in national parks under federal legislation and under the laws of several States.

The State of New South Wales adopted a Wilderness Act in 1987 which empowers the competent Minister to designate a wilderness area on publicly-owned land, subject to the conclusion of a wilderness protection agreement with the Government department or agency or with the local authority owning or controlling that land. Designations so made cannot be revoked except by an Act of Parliament. Wilderness areas must be managed so as to protect the unmodified state of the area and its plant and animal communities, preserve the capacity of the area to evolve in the absence of significant human interference, and permit opportunities for solitude and appropriate self-reliant recreation.

In New Zealand, wilderness areas may be established on public land under the National Parks Act, the Reserves Act and the Forest Act. Such land may consequently be not only in statutory protected areas but also on land owned by the Forest Department.

Under the Forest Act of 1989 of the Republic of South Africa, wilderness areas may be established in State forests. In such areas, all activities which are incompatible with their conservation objectives are prohibited.

In Europe, the recent Wilderness Act of 1991 in Finland has designated a total area of about 1.5 million hectares of land as wilderness areas in Lapland. Mining and permanent roads are prohibited and so is the construction of buildings, other than for traditional uses by the Laplanders or for recreation. These areas are of particular importance for the preservation of the traditional way of life and culture of the local inhabitants, especially reindeer herding. The establishment of wilderness areas in Northern Norway and Sweden has also been under consideration for some time.

In Italy, the Region of Abbruzi has a law which prohibits quarrying, building and the construction of roads in the mountains above the 1,600 metre line.

Sri Lanka has a National Heritage Wilderness Act of 1988 which empowers the competent Minister to establish wilderness areas on any Government land, subject to confirmation by Parliament. Access to these areas is prohibited without a permit.

Japan also has legal provisions allowing for the establishment of wilderness areas on land owned by the central Government as well as by local authorities. These areas are protected and managed as strict nature reserves.

This short review of wilderness area legislation clearly shows that the institution is currently used for two specific purposes. The first is to establish zones of stricter protection within national parks or equivalent reserves. The second is to provide for the protection of pristine areas on public land which is owned or managed by agencies other than the statutory conservation agency, generally the Forest Department, and to assign to these agencies the responsibility of conserving and, where required, managing these areas.

While the first objective may be achieved through appropriate provisions in protected areas legislation, the second would, in the absence of wilderness legislation, require the transfer of land and management responsibility in most cases to the agency in charge of protected areas. The wilderness area solution circumvents the predictable difficulties which are inherent in such transfers. Although the result will often be the same, this is nonetheless conceptually far from the philosophical definition and purposes of wilderness areas.

B. Conservation Orders

Conservation orders may be defined as administrative decisions prohibiting or restricting certain activities or uses on specified tracts of land for conservation purposes. They are therefore regulatory measures imposed in the exercise of the police power of the State, which may generally be taken in respect of both public and private land.

Areas protected by conservation orders differ from nature reserves in that there is no official reserve designation and that prohibitions or restrictions are adapted in each individual case to the specific requirements of the area or feature concerned.

In certain countries, conservation orders may be made for the preservation of any area. In others, they may only concern certain particular habitat types, landscape features or the habitats of endangered species.

Denmark and the Netherlands are perhaps the best examples of countries that use conservation orders extensively for the preservation of natural areas. In Denmark, the Nature Protection Act of 1992[24] empowers County Conservation Planning Commissions to request special quasi-judicial bodies, called Conservation Boards, to issue conservation orders to preserve individual areas. The contents of each order vary from site to site. An order may, for instance, prohibit the ploughing or cultivation of an area or the use of pesticides. Several thousand of these orders have been made so far, covering more than 3.5% of the total area of the country. Some relate to very small areas or to groups of trees, other apply to larger tracts of land and may cover hundreds and even thousands of hectares. The orders are very seldom revoked and exemptions from their provisions are granted very sparingly. Compensation is paid to the owners.

In the Netherlands, the Nature Conservation Act of 1967 provides for the designation of natural monuments on both public and private land. The damaging of a natural monument is prohibited. Private landowners may, however, apply for permits for activities that may affect the natural beauty or scientific value of a designated natural monument. When a landowner has suffered economic loss as a result of the designation of his or her land as a natural monument or of the denial of a permit or the attachment of conditions to a permit, compensation is payable.

Legislation may also provide for the preservation by administrative order of specific areas containing certain particular habitat types. Many countries thus provide for the designation of protection forests on both public and private land. In a protection forest changes in land-use and clear cutting are generally prohibited. The purpose of these designations is usually not the protection of natural areas as such, but rather the preservation of water catchments and other vulnerable zones such as sand dunes. The designation may therefore apply to artificial forests as well as to natural ones.

However, where protection forests are established in natural areas, the degree of protection they provide may be quite high. In the Republic of South Africa, for instance, the Minister in charge of forests may declare mountain catchment areas and order the protection of the natural vegetation in any part of such areas. Such orders are binding on all owners of the land concerned and their successors in title (Mountain Catchment Areas Act of 1970).

In a few countries, protection forests may be specifically established for conservation purposes. In the Republic of South Africa again, the Minister may make orders protecting groups of trees on private land for the purpose of maintaining the natural diversity of species or to preserve tree-dominated biomes (Forest Act of 1984). In France, the Nature Protection Act of 1976 specifically provides for the designation of protection forests for ecological or conservation purposes.

---

[24] Replacing the Nature Conservation Act of 1969, as amended in 1972, 1978, 1983 and 1991.

A few countries empower nature conservation authorities to make conservation orders for the preservation of specific areas belonging to certain other habitat types.

In the Austrian Land of Upper Austria, for instance, such orders may be made in respect of particular peatlands or marshes. In the United Kingdom, under the Wildlife and Countryside Act of 1981, special orders may be made for the preservation of limestone pavements. These are flat outcrops of rock with deep crevices in which grow unusual and rare communities of plants. Limestone Pavement Orders prohibit the removal or disturbance of limestone in the areas they designate. In the American State of Massachusetts, the Commissioner of Environmental Management may make orders prohibiting or restricting the destruction, alteration, filling or dredging of specified wetlands.

Conservation orders may also apply to certain particular landscape features such as trees, spinneys or hedgerows. In the United Kingdom and the Republic of Ireland, Tree Preservation Orders may be made to preserve single trees, lines of trees or groups of trees. The Forest Act of the Republic of South Africa, mentioned above, also allows for the protection of individual trees or groups of trees. In Switzerland, the laws of certain Cantons empower the local government or the municipality to designate protected trees, spinneys or hedgerows.

Finally, a certain number of laws provide for the possibility of making conservation orders to protect the habitat of protected species. These include the French "arrêtés de biotope" described in Part I, Chapter VII (A) above. Similar orders may be made in certain Swiss Cantons for the preservation of the habitat of protected species.

In a few countries, conservation orders may only be made as an interim measure, to safeguard a site pending acquisition by the State or the conclusion of a management agreement with the landowner.

In the Australian State of New South Wales under the National Parks and Wildlife Act of 1974, as recently amended, the Minister in charge of conservation may make interim protection orders to prohibit or restrict activities which may affect the preservation of a particular area and its fauna or flora. Such orders are valid for a period of twelve months and cannot be renewed. They may apply to any area of natural, scientific or cultural significance.

The Australian State of Victoria also permits the making of interim conservation orders under the new Flora and Fauna Guarantee Act of 1988 to preserve the critical habitats of endangered species.

In the United Kingdom, Nature Conservation Orders may be made to safeguard Sites of Special Scientific Interest of particular importance for a period of twelve months pending the conclusion of a management agreement or expropriation.

Under amendments to the Finnish Nature Conservation Act adopted in 1991, where the compulsory purchase of land for conservation purposes is proposed, conservation orders may bemade for a maximum period of two years to prohibit any activity which may adversely affect the values for which the land is to be expropriated.[25]

---

[25] The laws of the State of Victoria, the United Kingdom and Finland are also discussed at Part I, Chapter VII above.

C. The Granting of a Special Legal Status to all Areas of Conservation Importance

The purpose of this type of instrument is to provide for a certain degree of protection to all areas meeting certain conditions. Whereas conservation orders are made in a discretionary way in respect of individual sites, this particular kind of regulatory measure applies automatically to all sites in a certain category. It is the fact that they belong to that category which generates their legal status. As the instrument is site-specific, however, the sites to which it will apply must first be identified and designated.

The best, if not the only, example of this type of measure is that of the Sites of Special Scientific Interest (SSSIs) in the United Kingdom. The SSSI concept was initially included in the National Parks and Access to the Countryside Act of 1949 in respect of activities requiring planning permission, and was extended by the Wildlife and Countryside Act of 1981 to all other activities liable to affect the natural values of areas thus designated.

Where English Nature is of opinion that any area of land is of special interest by reason of its flora or fauna, or its geological or physiographical features, it must notify the fact to the local planning authority, to the Minister competent for the region concerned (i.e. England, Scotland or Wales) and to the owner or occupier of the land concerned. Such a notification has a number of legal consequences which differ according to whether the proposed activity which may have adverse consequences on the designated site is or is not subject to planning permission.

When the activity requires a permit, the local planning authority must first consult with English Nature before coming to a decision. The authority is not obliged to follow the English Nature's advice, although it will of course frequently do so. Where it proposes to grant planning permission against that advice, the Secretary of State for the Environment has the power to call in the planning application and to make the decision himself, following the holding of a public inquiry.

A recent judgment of the High Court[26] has confirmed these discretionary powers to grant planning permission even where this will result in the destruction of a site of major importance for conservation. The case concerned a housing development in an SSSI which constituted the habitat of a number of protected species. The local city council had granted planning permission for the development, after the Minister had refused to call in the application for his own determination.

The outcry following this judgment has now led to a decision that the Minister will normally call in planning applications which are likely to affect sites of international importance or of recognised national importance. New Planning Policy Guidance Notes (PPG) on Nature Conservation have since been issued to planning authorities after consultation with nature conservation authorities and major NGOs: these are not legally binding but are nevertheless highly persuasive. Planning applications for activities in areas within a radius of approximately 2 km around SSSIs should henceforth be subject to consultations with English Nature, which should also be consulted where the proposed activities are likely significantly to affect internationally or nationally important sites.

---

[26] *R v Poole Borough Council ex parte Beebee and others* (1991) Journal of Environmental Law, Vol 3, no.2, p.295: judgement given on 21 December 1990.

When no planning permission is required, as is the case for agricultural and forestry activities, the system works in a completely different way. When the owners or occupiers of land designated as an SSSI are notified of the designation, the notification must specify the "potentially damaging operations" which are likely to damage the fauna, flora or other features for the conservation of which the designation was made.

If the landowner or occupier wishes to carry out any of these activities, he must first inform English Nature of his intentions and may not engage in the proposed activities for a period of four months. Failure to comply with this requirement constitutes an offence. During these four months, English Nature will try to negotiate and conclude a management agreement with the owner or occupier. Management agreements commit the owner or occupier not to engage in the activities that will harm the site, in return for a sum of money in the form of a lump sum or, more usually, in annual payments.

If the negotiations fail, the Secretary of State may make a Nature Conservation Order prohibiting the proposed activities for a further period of nine months. Stricter penalties apply if the landowner breaches this prohibition. If no management agreement has been reached by that time, the land may be compulsorily purchased but this is a costly and cumbersome procedure which has only been invoked once for an SSSI.

The SSSI system is somewhat paradoxical. No compensation is due where planning permission is refused, whereas the only way to prevent harmful activities for which no such permission is required is to provide for compensatory payments through management agreements. Compensation is guaranteed under a management agreement if the owner states an intention to undertake the destructive activity. It is not always easy to know whether such a claim is genuine or not, and the system has been criticised for that reason. On the other hand, the entitlement to compensatory payments necessarily contributes to the market value of the land concerned. The system is therefore a rare example of a conservation measure that does provide direct benefits to landowners.

By the end of 1992, a total of 5,854 SSSIs had been notified in the United Kingdom covering more than 1,819,451 hectares, of which some 40% were in public ownership.

Finally, France has recently completed an Inventory of Zones of Ecological, Fauna and Flora Interest (ZNIEFF). The designation of such zones does not, however, have any legal consequences except that, as a general rule, this should be taken into consideration when municipal or other land-use plans are prepared and when decisions are made for the grant of building permits. Failure to do so may result in the invalidation of land-use plans or building permits by the administrative courts.

CHAPTER V
NON-SITE-SPECIFIC REGULATORY INSTRUMENTS

Certain countries have enacted legal instruments designed to protect particular types of natural or semi-natural habitats or landscape features. There are also laws that regulate specific activities which may have adverse effects on the natural environment.

A. Habitat Types and Landscape Features

1. Forests

Forests the world over have been and often continue to be considered only as a source of raw materials and revenue with little regard for other values. Over the past decades, some of the multiple functions of forests have begun to be recognised. The role of forests for the regulation of the water regime, the prevention of erosion and public recreation is thus now largely acknowledged, and concepts such as that of "protection forests" were developed accordingly. However, these functions may often be performed by modified or artificial forests as well as or almost as well as by natural forests. As a result, there continues to be little incentive to preserve natural undisturbed forests, as their importance for the preservation of biological diversity and of certain ecological processes such as the decomposition and recycling of vegetable matter continues to be largely ignored.

Public forests are managed by Forest Departments according to rules and objectives that they have established and which are essentially aimed at the production of timber and woodchips. Stringent legal provisions protect these forests from encroachments by third parties, while the departments concerned enjoy full and uncontrolled discretion, unhampered by legal limitations, to determine their policies and management objectives and methods as they wish.

With regard to privately-owned forests, on the contrary, forest legislation often considerably restricts the freedom of landowners. In many countries, for example, private persons are not allowed to cut their own trees except under a permit from the Forest Department. The purpose of these laws, some of which were enacted a long time ago, is usually not the preservation of natural forests, however, but rather to ensure the optimum exploitation of timber resources and to control erosion, whether the forest be natural or man-made.

With the increasing recognition of the ecological role of forests, and often under pressure from public opinion and the conservation movement, legislative reforms have recently been introduced in certain countries with a view to limiting the discretionary powers of forest administrations to a certain degree. In consequence, recent laws officially acknowledge that the ecological functions of forests are as important as their other roles and that multiple uses should be encouraged as much as possible.

In the United Kingdom, an amendment in 1985 to the Forestry Act of 1967 imposes a duty on the Forestry Commission to seek to achieve a reasonable balance between timber production and nature conservation. The new Swedish Forest Act of 1979 and the Spanish Natural Areas, Flora and Fauna Conservation Act of 1989 contain similar provisions. The Spanish law, in particular, specifies that the technical management of forests must be in accordance with their ecological, forestry and socio-economic characteristics.

Another example is the new Mexican Forest Act of 1986 which, in laying down the objectives of the law, places conservation and exploitation of forests on the same footing.

All these obligations are, however, of a very general nature and may therefore not be sufficient to guide, still less to bind, forest departments in their day to day activities.

A few recent laws contain more specific provisions. In the United States, the National Forest Management Act of 1976, which applies only to federally owned forests, requires management practices designed to maintain the multiple functions of the National Forests. The objectives of the US Forest Service now include the preservation of viable populations of all plant and animal species and their existing pattern of distribution, as well as the promotion of the recovery of endangered species. Forest plans must now be drawn up for each National Forest using an interdisciplinary approach which takes into consideration, *inter alia*, the protection of watercourses, their banks and wetlands. Clear cutting will be authorised only after all the effects on the environment have been evaluated and found to be compatible with the protection of soils, waters, fish and wildlife, recreation and the landscape. These plans must be prepared with the participation of the public. However, their development seems to be taking place rather slowly.

In Europe, where the restrictions on property rights imposed upon private forest owners seem to be well accepted in view of their acknowledged public interest, recent legislation now tends to provide better protection for natural forests.

For example, the laws of several Spanish Autonomous Communities restrict the clear cutting of indigenous woodlands. Where permits to clear cut are granted, replanting must be with native species. In the Community of Cantabria, permits may be denied for ecological reasons. Clear cutting and thinning, when permitted, give rise to an obligation to replant the same or at least similar species. Preference must be given to natural regeneration. In the Community of La Rioja, permits are to be denied in areas with an exceptional fauna or flora. In Navarra, 2% of the area of municipally-owned forests must be preserved intact and 5% within nature parks. In the Canary Islands, a permit from the nature conservation authority is required for the cutting of any indigenous tree, in addition to any permit already required from the Forest Department. Restrictions on the clearing of indigenous forests or the felling of native trees also appear in the recent legislation of several Italian regions.

In France, wooded areas may be protected under land-use legislation and all changes in land-use prohibited accordingly.

In Switzerland, the Forest Act of 1902 already laid down the principle that the forest area of the country could not be diminished, that any clearing of forests or wooded area required a permit, and that where permits were granted, an area of equivalent size had to be afforested or re-forested as compensation for the clearing.

The new Forest Act, adopted in 1991, contains many far-reaching provisions, which are applicable to both public and private forests. One of its stated purposes is the preservation of forests as a natural environment. Any clearing of forests is consequently prohibited in principle. Exceptions will only be granted for important reasons. The financial interests of the owner do not justify the grant of a permit. Compensatory reforestation is to be carried out in the same general area and must be equivalent to the cleared forest or woodland in both size and quality. It is possible, however, to replace in-kind compensation by measures for the conservation of nature or the landscape. Forests are to be managed in such a manner that all their functions are maintained. Clear cutting is prohibited. Any felling of trees within forests is subject to a permit from the forest authorities. The use of fertilisers and pesticides is, in principle, prohibited.

2. Wetlands

Wetlands are now considered as one of the most threatened types of ecosystem and have accordingly been singled out by an increasing number of countries as requiring specific conservation measures.

In a few cases, the obligation to preserve wetlands or certain types of wetlands is enshrined in the constitution. For example, the Brazilian Constitution of 1988 declares that the Pantanal, an immense wetland area in the south western part of the country, belongs to the national heritage and that any utilisation of that area must be approved by an Act of Parliament under conditions ensuring the preservation of the natural environment. Similarly, many of the constitutions of the Brazilian States, adopted in 1989, list permanently protected habitat types which include river banks, lakes, springs, estuaries, lagunas, mangroves and deltas, although this varies from one State to another.

In Switzerland, as mentioned in Part II, Chapter III(B), an amendment to the federal Constitution was adopted by referendum in 1987 to the effect that marshes and marshy areas of particular beauty and national interest must now be safeguarded.

In most cases, however, the principle of wetland preservation is established by legislation. The most commonly used method is to establish a permit requirement for activities that may destroy or alter wetlands or wetland functions.

In some jurisdictions, the law only protects certain wetland types. In the Austrian Land of Upper Austria, for instance, the permit requirement applies to marshes, peatbogs, natural or man-made ponds and alluvial forests. Permits may only be issued if the proposed activity will not harm the natural balance of the plant and animal communities concerned or affect the landscape in a manner that would be contrary to the public interest. Permits may also be granted when the public or private interest of the activity is considered to be higher than the public interest in the conservation of the natural environment or the landscape.

In Australia, the Philippines and Thailand and in the American State of Florida, mangroves are protected by legislation and permits are required for the cutting of mangrove trees.

In Venezuela, a far-reaching Decree dated 18 April 1991[27] also provides extensive protection for mangrove areas. The Decree requires any person or legal entity intending to carry out projects, activities or building works liable to affect mangrove ecosystems and the associated environment to obtain a permit from the competent administrative authorities. An environmental impact assessment must be conducted, and the permit may only be granted if the technical assessment satisfies four conditions. The Ministry of the Environment must confirm that there is no other practicable location for the proposed activity; the project should involve as little damage as possible to the ecosystem; interestingly, the natural flow of sea and river water must not be interrupted; and by way of mitigation, the contractor must guarantee to correct and minimise any damage caused to the environment.

---

[27] Decreto Sobre Protección de los Manlares y sus Espacios Vitales Associados, n° 1544, replacing the Order of the Minister of Agriculture of 1972.

Specific activities are banned in mangrove ecosystems under the Decree. These include the use of pesticides except in case of epidemics, the construction of houses on stilts or floating houses, the replanting of mangrove swamps with alien species, the discharge of building refuse or liquid effluent into mangroves, and any other activity which, in the opinion of the Ministry of the Environment, could damage mangroves or their associated or dependent species.

There are two exceptions to the Decree. Firstly, "special administrative zones" given over to permanent forestry may be exempted from these requirements, provided that management plans have been prepared for the area in question. Also excluded are indigenous populations (including small-scale fishermen) which depend on such ecosystems for their subsistence, provided that their activities do not degrade the mangroves concerned.

Although these measures are of general application, the Decree empowers the Minister of the Environment to enact more rigourous site-specific regulations for mangrove swamps and the surrounding biotopes in need of greater protection. Such additional restrictions would effectively convert the areas in question into nature reserves.

In other countries, the law either does not make a distinction between various types of wetlands or lists all or almost all the wetland types occurring in the country as protected.

The Tunisian Forest Code, for instance, prohibits the filling or draining of wetlands, defined on the basis of the Ramsar Convention definition, except under a permit which may only be granted for over-riding reasons of national interest. In Zimbabwe under the Natural Resources Regulations of 1975, it is prohibited to cultivate or destroy any natural vegetation on wetlands or to dig up, break up, remove or alter in any way the soil or surface of a wetland without a permit. In Luxembourg, the Nature Protection Act of 1982 forbids the destruction or alteration of ponds, marshes and reedbeds and establishes a permit requirement for any works liable to modify the level or flow of waters or to have an adverse effect on aquatic fauna and flora and on the quality of the habitat. Draining, dredging and water extraction are explicitly covered by this requirement.

In Spain, the Water Act of 1985 lays down the principle that any activity affecting a wetland requires a permit or an administrative concession. Several Spanish Autonomous Communities have also adopted their own legislation to protect wetlands. In an Act of 1990, the Madrid Autonomous Community prohibits any construction, and any other activity that may affect the naturalness of the waters or the ecological or landscape values of these areas, in all wetlands and in a surrounding fifty metre-wide buffer zone.

In Denmark, which is probably the country which has the most comprehensive legislation protecting habitat types, all kinds of wetlands, including watercourses, are protected by a permit system.

In France, the new Water Act of 3 January 1992 lays down as one of its objectives the preservation of aquatic ecosystems and wetlands. However, it remains to be seen what type of permit requirements will be imposed under this very general provision, by means of the regulations for its implementation which have yet to be adopted.

In the United States, section 404 of the federal Water Pollution Control Act of 1972, as amended by the federal Clean Water Act of 1977, establishes a requirement for a permit from the Army Corps of Engineers for the discharge of dredged and fill materials into the waters of the United States. This provision has been interpreted broadly by the courts to include almost all categories of wetlands. On the other hand, a large number of destructive activities, such as drainage, flooding, excavations, clearing of vegetation and upstream alteration of the water regime, are outside the scope of section 404 and therefore remain unregulated. There are also many exceptions to the permit requirement. Moreover, as the Corps clearly does not have the capability to examine each individual case, it has developed a system of general permits covering certain categories of wetlands or activities which in fact amounts to the establishment of further exceptions.

Regulations made by the Corps of Engineers lay down criteria for the issue of permits. No permits may be granted where a wetland performs functions which are important to the public interest. Exceptions to this rule may only be made when the benefits of a proposed project outweigh the damage that it will necessarily cause to a wetland. In evaluating a proposed wetland alteration, the Corps must consider whether the project is water-dependent or not, and whether feasible alternatives exist. Conditions may be attached to permits including mitigation requirements.

An interesting and unusual feature of the scheme is that the Environmental Protection Agency is empowered to veto the issue of a permit by the Corps. So far, this right has been used very sparingly.

Wetlands are also protected in the United States by the laws of many of the individual States. The earliest one was adopted by the legislature of Massachusetts in 1963. All thirty coastal States, including these bordering the Great Lakes, have now enacted legislation providing some degree of protection to coastal wetlands. At least thirteen of them have instituted a permit system. Some fifteen States only regulate inland wetlands.

The requirements laid down by State wetland laws are of course additional to those established by federal legislation. This means that two separate permits are usually necessary when a wetland or an activity is covered by both laws.

In practice, State wetland protection laws generally apply to those wetlands and activities which are not covered by the section 404 requirements. State legislation therefore constitutes an essential complement to federal legislation. Draining, dredging and other destructive practices are, in particular, usually regulated by State laws.

3.   Other Habitat Types and Landscape Features

A small number of jurisdictions have legislation instituting a permit system in relation to types of habitats other than wetlands.

The country that has enacted the most comprehensive legislation in this regard is certainly Denmark. The Danish Nature Protection Act of 1992 imposes a general prohibition on all activities which would result in changes to the beds of watercourses and lakes or in the equilibrium of peatbogs, salt marshes, marshlands, wet meadows and dry grasslands, as well as earth or stone walls which were traditionally built to separate pieces of land belonging to different owners. This prohibition replaces the very wide permit requirement set out in the Nature Conservation Act of 1969 as amended, which covered all natural and semi-natural habitats existing in the country with the exception of forests. However, certain exemptions are still permitted under the 1992 Act and the difference is probably one of emphasis.

The Danish law also makes it clear that the prohibition applies to all changes in land-use, including the cultivation, plantation and afforestation of the areas concerned. This has been interpreted broadly to include, for instance, the application of fertilisers.

Another law that provides for the preservation of most natural habitat types is that of Germany. The Federal Nature Conservation Act of 1976 was amended with effect from 1987 to provide a long list of protected habitat types, including inland and coastal wetlands, dunes, heathlands, dry grasslands, cliffs, rocky shores and alpine meadows. However, the implementation of this provision is left to the individual Länder. In Bavaria, for instance, natural meadows and dry grasslands cannot be destroyed without a permit.

In Switzerland, the federal Nature Protection Act of 1966, as subsequently amended, lays down the general rule that river banks and lake shores, wetlands, rare forest associations, dry grasslands, spinneys, hedgerows and other habitats must in principle be preserved. Where, after taking all interests into consideration, it is impossible to avoid damage caused by human intervention to habitats worthy of protection, the person causing the damage is obliged to take specific mitigation measures.

The implementation of this general provision is the responsibility of the Cantons, i.e. the federated territorial entities. The Cantons may use the method of their choice for that purpose. For instance, the Canton of Aargau protects hedgerows through their inclusion as protected natural objects in municipal land-use plans. Where such plans have not yet been developed, the destruction of hedgerows is prohibited except under a permit which may only be issued in exceptional cases.

The new Landscape Protection Act of 1993 in France has now made it possible to preserve certain landscape features through municipal land-use plans. These plans will be able to identify landscape elements which must be preserved for aesthetic or ecological reasons and to provide prescriptions for their protection. Any works which may result in the destruction of a landscape element protected in this way will be subject to a permit.[28] The Act also makes it possible to preserve single trees, rows of trees and hedgerows as specially-protected wooded areas ('espaces boisés classés') within municipal land-use plans.

Furthermore, article 1 of the Act empowers the State to adopt binding 'directives' for the protection of the landscape in areas which are remarkable for their landscape values. These 'directives' will prevail over any land-use plan and will be binding on any authority competent to grant building, forest clearance or any other land-use permits. The 'directives' must be prepared in consultation with the local authority concerned, as well as with approved environmental or landscape NGOs and the relevant professional organisations.

The vegetation along river banks is protected in a relatively large number of countries. This is generally achieved by the establishment of a protection strip along the banks where the felling of trees or the destruction of vegetation is prohibited. The width of the strip generally varies according to the width of the watercourse concerned. As an example, in Brazil, the federal Forest Code of 1965, as amended in 1989, gives permanent protection to forests and other forms of natural vegetation along river banks according to the following scale: for rivers less than 10 metres wide, the width of the protected strip is equal to 30 metres; it is extended to 50 metres for rivers that are 10 to 50 metres wide, and to 100 metres for these that are 50 to 200 metres wide, to 200 metres for those that are 200 to 600 metres wide, and to 500 metres for those wider than 600 metres.

In Switzerland, the federal Nature Protection Act of 1966, as subsequently amended, prohibits the destruction of all forms of riparian natural vegetation, including reed beds and alluvial formations. Permits from the fisheries authorities are required for any alteration of river

---

[28] Article L.442 of the Town and Country Planning Code, inserted into the Code by the new Act.

banks and clearing of their vegetation under the Water Act of 1991. The Cantons must as far as possible ensure that river banks remain covered by sufficient vegetation, or at least that the conditions which are necessary for the development of vegetation are maintained. The elimination of river bank vegetation may, however, be authorised when it is necessary to carry out projects which do not contravene water legislation and cannot be located elsewhere.

Certain countries sometimes protect very specific habitat types. A particularly interesting example is that of caves and karstic areas. In Hungary, the Law-Decree of 1982 on Nature Conservation lays down the principle that all caves are protected, except those which have been specifically exempted by regulations. This constitutes perhaps a unique example of the use of a negative list to protect natural areas. As a result, any newly discovered cave is automatically protected until the Nature Conservation Authority decides whether its continuous preservation is necessary. The discovery of new caves must be immediately reported. Permits are required for any alteration or utilisation of protected caves.

In several Austrian Länder, legislation provides for permit requirements for activities which may adversely affect caves. In the Italian region of Liguria, an Act of 1990 prohibits the destruction, damaging or obstruction of caves, the discharge into caves of solid or liquid waste, changes in the water flow of underground streams, the removal of concretions and many other activities. Access to caves may be prohibited to safeguard scientific values. Karstic areas of hydrological, environmental or landscape value must be identified and listed by regulations. Listed areas will be subject to special restrictions.

The laws of several Austrian Länder also provide for specific measures for the protection of glaciers. In Carinthia, for instance, any change or alteration of glaciers is prohibited, with a small number of limited exceptions subject to a permit system.

In rare cases, the permit system has been made applicable to the general clearing or alteration of native vegetation. The Italian autonomous province of Bolzano, for instance, prohibits changes in the cultivation of pastures and alpine meadows, as well as any substantial alteration of the natural vegetation in those areas, without a permit.

The Australian State of South Australia is, it would seem, the only jurisdiction in the world which has instituted a permit requirement for the clearance of native vegetation, wherever it occurs. The Native Vegetation Act of 1991, replacing the Native Vegetation Management Act of 1985, has as one of its purposes

"the conservation of the native vegetation of the State in order to prevent further reduction of biological diversity and further depradation of the land and its soil".

The Act lays down a number of principles which should be followed when considering whether a permit should be granted or denied[29]. In particular, native vegetation should not be cleared when it comprises a high level of diversity of plant species; has significance as a habitat for wildlife; includes rare, vulnerable or endangered plant species or contains a plant community which is rare, vulnerable or endangered, or is a significant remnant of vegetation which has ben extensively cleared, or is growing in, or in association with, a wetland environment, where clearance is likely to contribute *inter alia* to soil erosion or salinity, the deterioration of water quality or flooding.

---

[29] The management and financial provisions applicable under the Act in the event of a permit being denied are discussed in section (B) below.

In summary, the Act firstly requires an assessment of the vegetation to be prepared. Permits may only be granted on the basis of this assessment and will generally be denied if clearing would infringe the principles laid down by the Act. The Act therefore provides a very interesting example of the practical implementation of the precautionary principle.

B. Discussion of the Permit System as applied to Habitat Types

There are many difficulties in the implementation of a permit system for activities affecting habitat types.

The first such difficulty relates to definitions. If the system is to work satisfactorily, it is essential that land owners or occupiers, as well as the permit-issuing authority itself and the enforcement personnel, know exactly to what areas it applies. In many laws, either no definition at all is provided or the definition is so general that it will be of little use in solving practical difficulties.

In other laws, especially those relating to wetlands, very technical definitions are made which, albeit probably scientifically correct, are impossible for the average farmer or enforcement officer to understand. There is no easy solution to this problem. One possibility may consist in listing some plant indicator species that are generally known to non-scientists. Another is to provide an understandable description of the habitat types, together with the vernacular names that are commonly used to designate them.

In Denmark, detailed descriptions of the habitat types concerned have been included in Government circulars and do not seem to have given rise to serious problems. In the United States, on the other hand, there continues to be considerable difficulty in determining which wetlands are covered by section 404 of the Federal Water Pollution Control Act, as amended. Several definitions have been tried and a manual for the identification and delineation of "jurisdictional wetlands" has been developed by the federal agencies concerned. However, the more detailed the definition, the more it seems to be open to controversy and its scientific basis challenged. Moreover, suggested definitions based on numerical criteria, such as the number of days in a year during which an area is saturated with water, may be difficult to use in the field, as proof of the duration of the saturation period in the past may not be easy to obtain.

On the other hand, if it is impossible to arrive at generally accepted definitions of habitat types, the only alternative is to revert to site-specific protection systems. This can be achieved through the drawing up of inventories of all the important remaining areas of any given habitat types by competent scientists and their subsequent notification to the owners, with possibly a mention in the land register.

This is the system which will be generally followed in Switzerland. After the Nature Protection Act was amended in 1988 to require the conservation of biotopes of national importance, the drawing up of inventories was started for certain habitat types such as peatbogs, marshes and alluvial forests. The identification of sites is to be based on lists of indicator plant species established by regulations. Once an inventory is completed, other regulations will be adopted listing all sites considered to be of national importance. Conservation measures will then have to be taken by the Cantons. They are free to decide on the most appropriate method to be used in each case, although the Act advocates the use of management agreements concluded with the landowners. Nevertheless, acquisition, conservation orders or the use of the planning system are also possible. Regulations have now been issued in respect of peatbogs and alluvial areas, and over five hundred sites are listed.

A second difficulty arises from the necessity to allow for exceptions to the permit requirement. There is first a question of size. It would clearly be unrealistic to apply the permit system to all areas occupied by a particular habitat type, as for practical reasons this would make the system impossible to implement and enforce in most cases. As a result, most laws institute size thresholds below which no permits are required. These vary widely from one jurisdiction to another without apparent justification.

Useful as they may be to make the law politically more acceptable and to relieve permit-issuing agencies from excessive administrative burdens, thresholds ignore the cumulative effects of a large number of small destructive projects. This became apparent in Denmark where the threshold set for lakes did not permit the conservation of small ponds that are essential to the survival of amphibians. That threshold has now been lowered from 1000 sq. metres to 500 in 1983 and to 100 sq. metres in 1991. For all other habitat types, the minimum size under which a permit is not required has been set at 2,500 sq. metres by the 1991 amendments. This constitutes a very considerable lowering of the thresholds laid down earlier, which were respectively 5000 sq. metres for bogs, 5 hectares for heathlands and 3 hectares for salt marshes. For newly protected biotope types, i.e. wet meadows and dry grasslands, the threshold is set at 2,500 sq. metres.

Other exceptions relate to certain categories of activities. Exempted activities vary from one jurisdiction to another. As will be recalled, an extreme example is that of section 404 of the Federal Water Pollution Control Act of 1992, as amended, which only applies to the discharge of dredged and fill materials into wetlands and exempts all other activities.

Exemptions may concern the nature of the activity such as grazing, recreation or the erection of small structures, the intensity of the activity or the time at which it may be exercised. The most important exceptions relate to agriculture and forestry. As these activities are the primary cause of the destruction of natural habitats, the exceptions defeat the purpose of the law. Yet this exception appears in the legislation of many jurisdictions, probably out of political necessity. Nevertheless, Denmark and Luxembourg are examples of countries where the agricultural exception does not apply.

Government agencies generally enjoy a considerable degree of discretion when issuing permits. There is a growing tendency, however, to provide some criteria or guidance in the legislation to assist the authorities concerned in making their decisions. In a few cases, the law binds the permit-issuing agency and prohibits the grant of permits in particular circumstances, such as where a proposed activity would destroy the habitat of an endangered species.

The refusal of permits may also in certain cases be the result of a policy decision by the agency concerned in the exercise of its discretionary powers. For example, in Denmark, the Nature Conservation Agency does not grant permits for activities that would affect Ramsar sites or Special Protection Areas established under the EC Birds Directive. Pursuant to a growing number of laws, the agency is required to carry out a balancing test to weigh the importance of a proposed project for the community against the value of conserving a natural area.

When a permit is denied, the question of financial compensation to the owner may arise. In most countries where a permit system has been instituted, restrictions on property rights for conservation reasons are deemed to be in the public interest and do not give rise to a right to compensation. This approach permits the owner to continue with the former use of his or her land, since only the freedom to change the use of the land is affected by the refusal of a permit.

As a result, compensation is usually not payable as no actual loss has been incurred. This has been generally accepted by the courts.

There are signs, however, that in certain countries, particularly in the United States,[30] conservation restrictions without compensation are becoming less acceptable to the public.

A recent case, *Lucas v. South Carolina Coastal Council* (Supreme Court, 1993), dealt with the scope of the Fifth Amendment to the American Constitution which states that private property shall not "be taken without just compensation". A developer sued for compensation when his coastal property was designated by the State's Beachfront Management Act as a 'critical area' two years **after** his purchase, and his application for a permit to develop the property was accordingly refused. The developer claimed that this refusal rendered his property valueless[31] and therefore amounted to a total "taking" of his property, a claim which was accepted by the lower court.

The State's Supreme Court refused compensation on appeal, however, holding that a taking did not occur if the purpose of the Beachfront regulation was to prevent serious public harm and was reasonably calculated to achieve this purpose.

This ruling was overturned by the majority decision of the federal Supreme Court, which held that a prohibition on all beneficial use of land could not be newly legislated without compensation, simply because an activity was in effect held to be a noxious use and therefore outside the scope of the Fifth Amendment. Otherwise, all limits to the police power would be removed. Instead, such a prohibition might only arise from restrictions that background principles of the State's law of property and nuisance already placed on land ownership, since property owners were not entitled to assume that conducting a nuisance was part of their property right.

The case leaves unresolved several questions. "Nuisance" is defined in common law as "a substantial and unreasonable interference with uses of other property or the rights of the public". This is not a sufficiently precise concept to leave property owners certain of where they stand, nor is it as separate from legislation as the Court inferred: the interpretation of nuisance is shaped by regulations defining what is harmful in the area of public health and safety. There has been criticism that this ruling gives courts too great a scope to sidestep legislation, whereas in fact the allocation of the benefits and burdens of regulation should more properly be for political and legislative judgment.

Given the ready assumption by the Supreme Court that economic injury has been caused to a landowner, the case is also certain to cause concern to local authorities, the primary regulators of land use in most States. These authorities will probably be more vulnerable in the future to individual judgments requiring them to award compensation.

An intermediary solution which seems to operate satisfactorily is the system instituted under the Native Vegetation Act of 1991 in South Australia. Where permits for the clearance of native

---

[30] Environmental statutes in the US which limit the ways people can develop their property include the Endangered Species Act, section 404 of the Federal Water Pollution Control Act, as amended, various Historic Preservation Acts and the Coastal Zone Management Act which encourages States to pass laws to protect barrier beaches.

[31] No final ruling has been made on this factual question, as the case has been remanded back to the State court for decision on this point.

vegetation are denied, which happens in the large majority of cases, the Minister may enter into a heritage agreement with the owner of the land. This agreement is binding upon successors in title.

The agreement may include payment by the State to the landowner of an amount in respect of the decrease in the value of the land resulting from the agreement, as well as an amount as an incentive to enter into the agreement. The landowner may also apply for financial and other assistance for:

- the management of the land, its native vegetation and animals living on or visiting the land;

- the preservation or enhancement of native vegetation on the land;

- the establishment of native vegetation on the land;

- the undertaking of research in relation to the preservation, enhancement or management of native vegetation on the land or of animals living on or visiting the land.

The system therefore promotes the carrying out of positive management or restoration measures. This is rather unique and is made possible only by the conclusion of heritage agreements following the denial of a permit.

More usually, regulatory protection of habitat types only involves prohibitions and does not provide for the possibility of active management measures.

It is not easy to evaluate the success of the permit system as it applies to habitat types. It would seem, based on the scant information available, that permits are more frequently granted than refused.[32] On the other hand, potential applicants are clearly discouraged from asking for permits when they have manifestly no chance of success. Statistics on the number of permits granted or denied have therefore little meaning. Furthermore, the fact that permits may be granted subject to conditions will often enable the conservation authority to negotiate an acceptable compromise. Without a permit system, this would be much more difficult to achieve.

---

[32] In contrast, it is said that the rate of permit refusals in South Australia, represents some 95% of the total area in respect of which applications for clearing had been submitted. Between 1985 and 1989, 225,000 hectares had thereby been safeguarded. This degree of success is probably due to the availability of financial and other assistance under the heritage agreement scheme.

CHAPTER VI
PROTECTED LANDSCAPES AND NATURE PARKS

A. Introduction

Protected landscapes, or nature parks as they are often called, are a new type of conservation instrument which is developing rapidly, particularly in certain Mediterranean countries such as Italy and Spain. Their primary purpose is the conservation of man-made landscapes, together with the natural and semi-natural areas they contain. They are inhabited areas, often with a relatively high population density, and are sometimes located close to urban centres. They frequently attract a considerable number of visitors.

As initially conceived, nature parks were essentially designed to preserve the natural beauty of the countryside, to afford opportunities for open-air recreation and to serve as physical planning instruments to assist in the development of less favoured rural areas unsuitable for modern agriculture.

The conservation of the natural environment was, therefore, only a secondary or incidental objective. As a result, regulatory measures applicable within these areas were limited to controls on construction and other works and agricultural and forestry activities remained unregulated. Indeed, it was believed at the time that the first protected landscapes or nature parks were established that sprawling urbanisation, second homes, open-air advertising and the like were, and would continue to be, the major evil and that these could be remedied by the imposition of appropriate planning controls.

Agriculture and forestry, on the other hand were considered as activities that contributed to the preservation of the rural landscape. It was felt that prosperous agriculture would facilitate both the maintenance of rural populations and the conservation of the environment. At that time, it was perhaps impossible to foresee that the intensification of farming methods, which was becoming increasingly necessary if farm incomes were to be raised, could also lead to the destruction of the very landscapes that nature parks were designed to preserve. As a result of agricultural subsidies and other incentives, landscapes and natural environments therefore continue to disappear, even in the less favoured areas. The ploughing and fertilisation of natural grasslands, the drainage of wetlands and the afforestation of heathlands and moors consequently continue unchecked in many nature parks.

However, there has been a growing recognition in recent years of the important role that protected landscapes and nature parks can play in the preservation of wild species and natural and semi-natural ecosystems. This recognition has led to the taking of specific measures to assist in safeguarding natural values within these areas through voluntary agreements and, in the new generation of nature parks that are now established in certain countries, through the imposition of much stricter land-use controls.

Nature parks are particularly suited to those areas of the world where nature has been under the influence of man for centuries and where the rural landscape forms an integral part of local culture. So far, the institution has mostly been developed in Western and Central Europe, in Japan and in Korea. In the United States there are two important examples of nature parks, both in the eastern part of the country: the Adirondaks State park in the State of New York and the Pinelands

National Reserve in New Jersey. There is of course no reason why it could not be used in other parts of the world where similar conditions prevail, particularly in South-East Asia.

B. Nature Park Legislation

From the legal point of view, protected landscapes or nature parks are areas which have been officially designated as such, the boundaries of which have been delimited and where special planning rules, which are more restrictive than those applicable elsewhere, may be used. They differ, however, from conventional protected areas such as national parks in that many human activities remain authorised.

There are considerable differences in the legal regime of nature parks and protected landscapes between countries, and sometimes even within the same country where several categories of these areas have been established. This is the case in the United Kingdom where there are National Parks (which correspond in spite of their name to the nature parks of other countries), Heritage Coasts and Areas of Outstanding Natural Beauty (AONBs).

At one end of the scale are those areas whose designation has no or very few legal consequences. The British AONBs, for instance, are designated by the Countryside Commission, subject to confirmation by the appropriate Minister, and are administered by the local County Councils. The latter may, in principle, take measures to preserve and develop the natural beauty of those areas, for example through the denial of building permits. There are, however, no particular restrictions or planning rules which are specifically applicable to those areas.

Regional nature parks in France have, until recently, also enjoyed no statutory protection at all, simply being areas where the conservation of nature, along with other matters, should receive priority attention. They are established at the initiative of the regions and in agreement with the local authorities concerned by the approval of a charter. There are at present 26 regional nature parks covering more than 7% of the national territory.

The charter was not formerly a statutory instrument but a contract between the authorities concerned which did not bind third parties or establish any land-use restrictions. This purely voluntary system had some advantages in that it relied on a consensus between the local authorities concerned to work towards common objectives, and did allow for management measures to be drawn up and implemented. Unlike the British AONBs, regional nature parks have always been administered by joint commissions, on which the local authorities and other bodies concerned are represented. The fact that there existed an administrative structure, a director, a staff and limited funds mades it possible to act by persuasion, sometimes quite successfully.

On the other hand, the total absence of rules binding the local authority to observe even a minimum of constraints, especially with regard to the conservation of the natural environment, was considered to be a major weakness. The French environmental lawyer, Jean Untermaier, described regional nature parks as nothing more than areas where it had been decided to apply nation-wide conservation legislation better than elsewhere!

However, the new Landscape Protection Act of 8 January 1993[33] has enshrined the protection of regional nature parks in primary legislation for the first time. The Act adds a new

---

[33] Loi sur la protection et la mise en valeur des paysages.

article, L.244-1, to the Rural Code, affirming the role of the parks within national policies for environmental protection, land use, social and economic development and public education and training, as well as their importance for the preservation of the landscape and the natural and cultural heritage.

The charter has finally been given statutory force. Although it will continue to be drawn up by the Region in agreement with all local authorities concerned, it will subsequently have to be approved by Government Decree which will establish the park for a period of ten years. The charter will set out prescriptions for the protection and development of the park, as well as measures for their implementation. It must include a plan based on a heritage inventory which divides the park into different zones according to their varying objectives, as well as a document setting out the fundamental principles for the protection of landscape components within the park.

If the Act is fully implemented by Government Decree, these prescriptions and measures will be binding on the State and all the local authorities concerned. For the first time, all land-use plans will have to be compatible with the provisions of the charter.

Where the terms of the charter are not respected and destructive activities authorised or promoted, the Minister of the Environment is empowered to withdraw the label. This happened before the new law came into force, the regional nature park label of the Marais Poitevin Park being withdrawn in 1991 after a large part of the remaining marshes had been drained and converted into maize fields.

In Belgium, nature parks are created by statutory instruments made by the regional executive bodies and are administered by regional or local authorities. Each park must have its management plan and a management commission. The commission must give its opinion on any development project which is proposed to be carried out within the park. However, permits are not issued by the administrative body but by the competent national or regional administration. So far, only one park has been created.

The National Parks of England and Wales (there are none in Scotland) are the first instance of nature parks for which some powers were conferred on the body entrusted with their administration.

Under the National Parks and Access to the Countryside Act of 1949, National Parks, like AONBs, are designated by the Countryside Commission and confirmed by the relevant Minister. Each park is administered by a National Park Authority. For two of the parks, the Peak District and Lake District parks, the management authorities are known as Joint Boards which are autonomous bodies with legal personality. They have the power to make the park land-use plan, to control development by issuing permits, to buy land and to conclude management agreements with landowners. As a result, these parks are autonomous planning units.

In the other parks, the Authority is known as a National Park Committee which, unlike a Board, does not constitute a local authority in its own right. Powers with regard to planning and permits are retained by the relevant County Councils which may, however, delegate some of them to the Committee. It was confirmed in 1992 by the Minister of the Environment that new legislation would be introduced, probably in 1993, to enable all parks to be administered by Joint Boards in the near future.

Members of both Boards and Committees are appointed by the appropriate Minister (one-third of members) and by the County Councils (two-thirds). Every Park authority must appoint a National Park Officer and prepare a park management plan. Central Government contributes 75% of the budget of national parks.

There are no particular restrictions established by law on activities in National Parks, except in respect of those activities for which "planning permission" is universally required in any event. Park authorities have few powers beyond the limits of their statutory planning functions.

Agriculture and forestry, including afforestation, cannot be directly regulated, although the Minister of Agriculture may designate for special protection certain areas of moorland or heath in any National Park which have not been used for agriculture within the last 20 years. Pursuant to such a temporary 'stop order', it is an offence to carry out specified agricultural or forestry operations likely to affect the character or appearance of the area.[34]

However, it is possible for a Park Authority to oppose the grant of an agricultural subsidy to a farmer where the proposed activity may have adverse effects on the landscape or the natural environment. In such a case, the Authority must propose a management agreement to the farmer concerned. Grants may also be made to farmers for the preservation of natural areas or landscape features.

All ten existing National Parks were established between 1951 and 1957. In 1988, the Norfolk and Suffolk Broads Act was enacted to establish a new protected landscape covering an area of rivers and marshes in the Eastern part of the country called the Broads. Although not technically a National Park, the Broads has an autonomous management authority with similar functions and some additional powers. It must be notified of proposed operations damaging to marsh, reed and woodlands, but it cannot stop these operations except by negotiations. It may close parts of the area to navigation for nature conservation purposes, and a representative of the navigation service is a member of the Broads Authority.

The legislation awaited in 1993 is also expected to confer special status on the New Forest in southern England, with some 200 square miles to be designated "an area of national significance". Following the precedent set by the Broads Authority, the existing New Forest Committee will probably be given statutory recognition as the management authority of the Forest.

A greater degree of control over activities in nature parks appears in the legislation of such countries as Japan, Portugal and Spain.

In Japan, National Parks and "Quasi" National Parks are designated by the Director General of the Environment Agency. These parks are zoned with different rules applying in each kind of zone. In the "special areas" of these parks many activities, in addition to construction, are subject to a permit from the Director (or from the provincial government in the case of Quasi National Parks). These include the cutting of trees, modification of the water regime, the filling or drainage of wetlands and the clearing of vegetation. In the "special protection areas", this requirement extends to other activities such as the plantation of trees and bamboos, grazing and the use of off-road vehicles. "Ordinary areas" serve as buffers to the special areas. Proposed activities in those areas must be notified to the Director of the Environment Agency (or to the provincial Governor) who may, during a period of thirty days, prohibit or subject to conditions the exercise of the activity concerned. There are no individual managing bodies for the parks and no activities are actually prohibited by law.

In Spain, jurisdiction over the establishment of nature parks has now been transferred to the Autonomous Communities. Several of them have enacted their own park laws, whilst others base themselves on the national Act of 1989 on the Conservation of Natural Areas and of Wild Flora

---

[34] Section 42, Wildlife and Countryside Act 1981.

and Fauna, which is a framework law. The Act merely provides that any use of natural resources which is incompatible with the objectives of a park must be prohibited. The Act also allows the Autonomous Communities to establish protected landscapes and any other form of protected areas they may wish. A Plan of Use and Management must be drawn up for each park. Those plans have the authority of planning orders and prevail over any other land-use plan, including municipal or local plans. When the latter are incompatible with a park plan they must be revised accordingly.

In recent years, many Autonomous Communities have started to establish nature parks, often in considerable numbers. This is the case for instance of Andalucia, Cataluña and the Madrid and Valencia Communities. In Andalucia, 23 nature parks have now been created, covering almost one and a half million hectares, more than 15% of the total area of the region.

Nature parks are generally zoned, with certain zones benefiting from very strict protection rules. Agricultural and forestry activities may be prohibited or made subject to permits. Detailed rules applicable to each kind of zone and to various activities are laid down by the management plans or other regulations.

The parks are usually established by an Act of the regional parliaments and the Plans of Use and Management by decrees. They are administered by the environment department of the regional Government, which has the right of pre-emption over land situated in a park and may also undertake the compulsory purchase of such land. Each park has an advisory body composed of representatives of various Government departments, local authorities, scientific institutions, conservation NGOs and local interests. There is usually a Director appointed by the regional Government.

One example of a regional nature park is that of the Cuenca Alta de Manzanares in the Autonomous Community of Madrid. It was established by a regional Act of 1985 and covers an area of 37,500 hectares. The area is divided into zones which comprise a strict nature reserve on 18% of the area and an "educational nature reserve" occupying another 17% of the park surface. Next comes a zone where only traditional land-uses which contribute to the maintenance of the conservation status of ecosystems are allowed. Forage crops and artificial grasslands and afforestation with species that are not indigenous to the park are prohibited. In the other zones, intensive agriculture is generally prohibited except under a permit. In all zones modifications of the water regime are prohibited. Agriculture is regulated by annual plans. Forest exploitation is subject to a permit and hunting is restricted to a small number of species. All these activities are of course prohibited in zones designated as nature reserves.

In Italy, jurisdiction over nature conservation, nature reserves and nature parks was largely transferred to the Regions in 1977. As in Spain, there has consequently been a considerable development of regional legislation providing for the establishment of nature parks.[35] A regional framework law is generally enacted to deal with all categories of protected areas, including nature parks, nature reserves, and other types of areas such as natural monuments or areas of particular environmental importance. These categories vary from one Region to another, but all Regions provide for the establishment of nature parks and reserves and more than 70 regional nature parks have now been established. Individual protected areas are then created by special legal instruments, which are also usually Regional Acts. Specific regulations are subsequently developed for each individual area designated in this way.

---

[35] Regional legislation is now governed by national rules on protected areas, summarised later in this section.

The law usually requires the preparation of a park plan for each park, called the "territorial co-ordination plan" (TCP), which must be formally adopted by another Regional Act. The TCP is simultaneously a master management plan and a planning instrument. As a planning instrument, its main feature is that it legally prevails over all the planning instruments of those local authorities which have part of their territory within the park. This means that in the event of any discrepancy, it is the local plans which must be modified in order to bring them into conformity with the park plan.

Park management bodies may also develop sector plans covering particular activities or conservation problems within a park. These plans provide general guidance to the authorities concerned on specific activities and may contain rules that are binding, at least upon the administrative authorities concerned.

In contrast with Spain, where parks are managed by the national or regional conservation authorities, Italian parks are autonomous planning units. They are managed by a Board which is an independent public body with legal personality, composed of representatives of the region and of the local authorities concerned. There is usually also a scientific advisory committee on which conservation NGOs are represented. The Chairman of the Board is often empowered to issue or deny the permits required under the Regional Act or the TCP and to inflict administrative penalties. Parks may also acquire property, if necessary by compulsory purchase.

As in the Spanish system, parks are divided into zones including strict or managed nature reserves. Agricultural and forestry activities are regulated with varying constraints according to zones. In certain parks in Lombardia, for instance, there are strict limits on the number of animals allowed per unit of surface to prevent intensification of stock raising.

An interesting feature of the Italian system is the development of fluvial nature parks in several regions. The Ticino Park in Lombardia, established in 1974, and the corresponding park on the other bank of the river in Piedmont, are interesting examples of this new kind of nature park.[36] More than 450,000 people live in the Park which stretches for 85 km along the river. The Park Plan, which was approved in 1980 by the Regional Parliament, provides for the control of water pollution and the water regime. There are several kinds of zones, including strict and managed nature reserves, which encompass all the remaining natural areas along the banks of a river and cover a surface of 18,000 hectares. In the other non-urban zones, agriculture and forestry are authorised subject to various constraints.

The Plan lays down guidelines and criteria relating to the preparation of local land-use plans. The grant of building permits in the Park is subject to the presentation by the applicant of a statement of environmental compatibility in respect of the proposed works. This statement must be supported by an affidavit from an expert certifying that the legislative and regulatory provisions made by the State, the Region and the Park Board are effectively complied with. In addition, in most of the Park zones, the Park Board can veto a building permit if the project is incompatible with the protection of the environment.

The Plan also provides for the preparation of sector plans relating to water quality, hydraulic works, forests, protected habitats, open-air recreation, roads, quarries, hunting and fishing. These sector plans are adopted by the Park Board and approved in the form of regulations by the

---

[36] The Regions of Liguria, Emilia-Romagna, Lazio and Veneto have also established river parks of this kind.

Regional Executive. Certain general rules applicable to particular activities are included in the Park Plan itself.

For example, with regard to the water regime of the river, the Plan specifies that the corresponding sector plan must regulate the use of water and floodable areas. Flood protection works may only be authorised to preserve important installations located near the river. The construction of houses in floodable areas is prohibited, as is the extraction of sand and gravel from the river bed and in the reserves. A minimum flow of three cubic metres per second must be maintained downstream of any point of water extraction. Motorboat traffic on the river is also regulated and may be prohibited in certain zones. Hunting is prohibited in the reserves and all along the river. Angling is prohibited in most of the reserve zones and is regulated elsewhere according to the relevant sector plan.

At national level, the framework Protected Areas Act of 6 December 1991 now sets out a number of general rules to provide a basis for the harmonised establishment, planning and management of regional nature parks. These are defined as areas of natural and environmental value constituting homogeneous systems characterised by their natural components, their landscape and aesthetic values and the cultural tradition of the local populations. Most innovatively, they may now be established in adjacent marine areas as well as on land, rivers and lakes.

Within a period of twelve months from this Act coming into force, regional legislation must be amended to conform to national rules set out in the Act. These cover issues which include the participation of local authorities in park creation and management; the evaluation of the effects of the establishment of a park on the regional territory; the prohibition of all activities which may affect the conservation of the landscape and of the natural environment; and significantly, the express prohibition on hunting in all regional parks.[37]

The park management body is now empowered to order the suspension of any works undertaken in contravention of the legislation and to order the restoration of the damaged area to its initial condition. It is also given standing to institute proceedings in the administrative courts to request the annulment of illegal administrative decisions affecting the parks.

Regions remain free to choose the administrative structure that they feel best suited to their parks. Park plans are now deemed to be landscape plans in the sense of the Galasso Act of 1985[38] and replace any other existing landscape or territorial planning instrument at any level. Each park must also develop a long-term economic and social plan for the promotion of activities compatible with its conservation purposes, to be formally approved by the Region. The State, the Region, local authorities and any other bodies concerned may all contribute to the financing of the implementation of the plan. Provision is also made for the establishment of inter-regional

---

[37] Hunting is at present authorised by regional legislation in many zones in existing parks which are not designated as nature reserves. To minimise opposition to the new ban, the new Hunting Act of 11 February 1992 has extended the twelve-month time limit laid down by the 1991 Act up to 1 January 1995, as the deadline for translating the hunting prohibition into regional legislation. Regions may also modify park boundaries to exclude those areas in which they wish to continue to authorise hunting, although they are encouraged to give these areas the status of buffer zones. This will allow the Regions to control most damaging activities whilst continuing to authorise hunting.

[38] Discussed in Chapter VII (A)(2) below.

nature parks by common agreement between the Regions concerned. These parks must be managed in a unitary way.

Finally, the type of regional nature park which results from the new national Act is probably closer to the conventional national park concept than the type of park developed by regional legislation until the entry into force of the 1991 Act. Nevertheless, regional parks will still continue to be areas containing human settlements and in which only those activities which are incompatible with the purposes for which the parks have been created are prohibited or otherwise regulated. It would seem, however, that there is now less difference between national and regional parks in Italy, as defined and regulated by the new Act.

Turning now to the two known examples of nature parks in the United States, the Adirondak Park and the Pinelands National Reserve are both autonomous planning units with their own administration vested with considerable powers to regulate land-uses.

The Adirondak Park was established by an Act of 1971 of the State of New York. The Park includes a large area of State-owned lands which, as already mentioned, must remain "forever wild" under the State Constitution. The Adirondak Park Agency was established by the Act as an autonomous planning agency to prepare and implement Park Plans to regulate both public and private land. For each of the six types of zones into which private land is classified, the Plan provides specific housing density restrictions and compatible uses. All important development projects must be approved by the Agency. Middle-sized projects must be approved by the local governments of the municipalities situated within the Park, provided they have adopted their own land-use plans which have been approved by the Agency. Otherwise, these projects must also be approved by the Agency. The Agency has also imposed specific restrictions relating to watercourses and wetlands.

The Pinelands National Reserve was established by an Act of Congress in 1978 and an Act of the New Jersey legislature in 1979. The Pinelands Commission was established under these Acts as a "planning entity" to regulate development in the Reserve. The Commission is composed of persons appointed by the counties concerned and by the Governor of New Jersey, and of a representative of the federal Secretary of the Interior. The Commission, as required by the Acts, prepares a comprehensive management plan for the Reserve. County and municipal land-use plans must be compatible with this plan and must also be approved by the Commission.

The Reserve is zoned according to "land capability" types. Land-use in each zone is regulated by a distinct set of rules. The core of the Reserve is largely publicly owned, and an acquisition programme is under way to purchase a number of special habitats and to link areas which are already public property.

In the other zones, the objective of the plan is to promote compatible agricultural and recreational uses and to prohibit incompatible development. The permissible density of housing units varies from one zone to another. An innovative scheme allows a specialised financial institution to purchase and sell building rights. This allows the relocation of development towards the less environmentally sensitive areas of the reserve. The plan also establishes high water quality standards in all zones.

Successful as the system is beginning to be in certain parts of the world, the extent to which it could be adapted to most developing countries is still a matter for conjecture, since it has only been tried out on a limited scale in a relatively small number of these countries.

Sophisticated land-use instruments as used in some western countries for nature parks may not be workable. However, the simpler biosphere reserve concept, as recommended by the Man and Biosphere programme, may be a useful model. It amounts to a simplified form of nature park, containing a core zone and a buffer zone which may be inhabited and in which traditional activities compatible with the objectives of the park and the integrity of the core should be allowed and encouraged. In addition, as will be recalled, there should also be a third zone, called the transition zone, beyond the buffer zone, in which specific cooperative links with the local population should be developed and maintained, but where no specific land-use or activity restrictions should be imposed.

Examples of protected areas of this type in developing countries are still few, but their number is increasing.

Mexico has enacted specific biosphere reserve legislation, namely the Ecological Balance and Environmental Protection Act of 1988. Biosphere reserves must have a core and a buffer zone, but no transition zone is required. One of the best known Mexican biosphere reserves is Sian Ka'an in the Yucatan Peninsula. This has three core zones and a large buffer, in which traditional activities for local inhabitants are allowed.

Venezuela has national park legislation[39] very similar to that of the Spanish nature parks, which provides for the establishment of areas subject to a special management regime. These include national parks and other protected areas. Management plans and regulations concerning permitted uses are adopted by Presidential decrees. These establish different categories of zones, each with different rules, including strict nature reserves, wild areas, managed reserves, natural restoration areas, recreational areas and buffer zones. Authorised activities for each type of zone are listed in the decree.

In Zaïre, a Mangrove Nature Reserve was established in 1992, with a core area completely closed to the public in which all activities are prohibited, and a surrounding area of wet savanna and coastal strip in which traditional and other activities are allowed, provided that they do not disturb the natural environment.

There is generally little provision made for participation by local authorities or populations. The Sian Ka'an biosphere reserve has a Joint Management Committee with representatives of federal Government, State government and the municipalities concerned.

In Zaïre, nature conservation authorities must contribute to the socio-economic development of human populations neighbouring the Mangrove Nature Reserve, by the construction of schools and other development infrastructure and the maintenance of roads.

In Togo, the Environment Code set out in the Act of 3 November 1988 lays down the legal basis for the establishment of 'Areas of Protected Environment'. Such areas may be established, *inter alia*, to preserve landscapes, geological formations, soils, sea shore, hydrological systems, forests, animal and plant populations, their habitats and the ecosystems of which they form a part.

Within these areas, the Minister in charge of the Environment may prohibit, restrict or otherwise regulate all activities incompatible with the objectives of the area, and may implement programmes for the restoration of the natural environment. He or she may also adopt management plans to achieve the objectives of the area.

Areas of Protected Environment may be established after an evaluation of the social and economic effects of imposing restrictions on property rights and access to natural resources, and after consultation with local authorities and populations. It is possible for compensation to be

---

[39] Planning Act of 11 August 1983 (Ley organica para la ordenacion del territorio).

paid. Once the Area has been established, landowners, whether public or private, may be required to plant and maintain trees or other plant formations.

This is a rare case of legislation in a developing country specifically empowering conservation authorities to establish areas akin to nature parks.

C. Conclusion

Certain protected landscapes or nature parks constitute an almost unique example of a legal instrument which has the potential to integrate planning and conservation law. This is achieved by the establishment of the areas concerned as autonomous planning units and the development of park plans which are binding on all public and private persons and to which all local plans must conform. Zoning allows for the fine-tuning of regulations with a view to limiting development, even in agriculture and forestry, to those activities which are compatible with the conservation of the natural environment and landscape features in each zone. The regulation of specific activities, such as off-road motor vehicle traffic, hunting or fishing, through sectoral regulations complements the zoning system.

The key to the success of nature parks seems to be the designation of a Government agency or the establishment of an administrative body as specifically responsible for the area, with the jurisdiction to prepare park plans and regulations and to issue permits or at least to advise on their issue. When these matters are left to the Government agencies which would otherwise normally be in charge of them, experience has shown that there tend to be many implementation and enforcement difficulties.

It is therefore essential to make these transfers of jurisdiction acceptable to the Government departments and local authorities concerned. This may be achieved by associating them closely with the management of the park, as a member of the body in charge of its administration.

Nature parks must of course also be accepted by their inhabitants. The law should therefore also allow for their representation on the managing bodies of the parks, either through their elected officials or by means of spokesmen from particular interest groups.

In developing countries, parks need to increasingly based on the biosphere reserve model, with at least a core and a buffer zone in which traditional activities are allowed, as well as preferably a transition zone. There is a need to involve local populations as much as possible in park management and to provide specific benefits for these populations in return for their acceptance of restrictions on their activities, especially in the core areas.

Nature parks and protected landscapes are areas where specific land-use controls are applied to protect the natural environment in areas of particular value. However, land-use controls may also be used for the maintenance of natural components of the countryside in general, outside any specifically designated area.

CHAPTER VII
PLANNING CONTROL INSTRUMENTS

Planning instruments are of two kinds. They may apply to particular areas and regulate various land-uses in such areas. Alternatively, they may apply specifically to certain activities, whatever the area in which these are to be conducted.

A. Area-Based Planning Instruments

1. Introduction

During recent decades, many countries have developed systems of land-use controls which have continually grown in sophistication and complexity. Planning law now constitutes a distinct branch of administrative law with its own corpus of rules, abundant court cases, and specialised scholars. It cannot be examined here except in the form of a brief overview.

The basis of the system consists of developing the regulation of certain activities, essentially construction and other works, within a specific area. This is achieved by means of a plan drawn up for an entire administrative territorial unit, usually a municipality. In most cases, the plan divides the territory concerned into zones and lays down certain rules for the control of development for each of these zones. These include prohibitions, restrictions, or permit requirements. In this way, there may be zones where no construction is allowed and others where a certain density of construction, varying from one zone to another, is authorised, subject in each individual case to the grant of a permit.

At higher levels of Government, other plans are drawn up to serve as guidance to local plans. These may be binding on local authorities. Where this is the case, local plans must be compatible with them.

Where no local plan has been made, as is still a common occurrence for small rural municipalities, there is usually still a permit requirement for controlled activities.

The object of the planning system, at least when it was initially conceived and developed, was to ensure the harmonious development of urban areas and, more recently, of the countryside. Economic, social and landscape considerations, rather than conservation, were therefore at the basis of the system. Agriculture and forestry are almost universally excluded from the system and may consequently not be the subject of prohibitions or permit requirements.

Construction restrictions are now well accepted as a social obligation in most countries where planning controls have been instituted, notwithstanding the limitations on property rights that these imply. No compensation is generally due when land is included in a no-construction zone or when a permit is denied.

The planning system has recently began to evolve, albeit timidly, towards a better preservation of natural areas. One sign of this evolution is the emerging trend to interpret planning rules in the light of the conservation duties of the State, insofar as these are embodied in the constitution or a conservation Act. For example, French administrative courts have on several occasions invalidated local land-use plans on the grounds that important natural areas should have been included in the zones benefiting from the highest degree of protection. In many countries, building permits are denied because the proposed construction would affect a natural habitat or the habitat of a rare or endangered species.

In certain countries, recent changes in planning legislation have been enacted for the specific purpose of preserving the natural environment. Two possible methods may be used to that end. The first is exclusively area-based. The second consists in bringing under the ambit of planning law those operations or developments which were initially exempted from planning permission. This makes it possible to strengthen considerably the degree of protection provided to individual areas.

2. Areas

In France, legislation now provides that a building permit may be denied or granted subject to special conditions where the proposed works, by reason of their location or object, may have adverse effects on the natural environment. It is also possible to designate wooded areas in local land-use plans in which clearing and changes in land-use are prohibited.

In Sweden, central Government may designate natural areas of national interest. If a local plan does not provide a sufficient degree of protection for any of these areas, the municipality concerned may be required to modify its plan.

In the German Land of Bavaria, where all remaining natural habitats have been mapped, the law requires the preparation of Landscape Plans. These plans must identify all natural habitats, particularly wetlands and dry grasslands which must be protected. Landscape Plans are binding on all Government agencies and municipalities. The latter are therefore required to protect the areas thus identified in their local plans.

In Switzerland, watercourses, lakes and lakeshores, landscapes of particular beauty or interest for science and the habitats of animal and plant species "worthy of protection" must, *inter alia*, be preserved by the local plans. These federal rules must be observed by the Cantons when making their own land-use legislation. Thus, in the Canton of Zürich all wetlands, dry grasslands and the habitats of rare species or plant associations must be included in zones where no construction is allowed.

In certain countries, special planning rules have been enacted to apply to coastal or mountain areas which are particularly vulnerable to development pressure. In France, for instance, construction is now prohibited within one hundred metres of the coastline, outside urbanised areas, and roads built parallel to the coast are no longer allowed within two kilometres of the sea shore. Exceptions may, however, be made in special cases.

Of particular interest is an innovative provision which requires that local plans must preserve those environments which are necessary to the maintenance of "biological balance" or which have an ecological value. The law provides a non-exhaustive list of these habitats, which include dunes and coastal heaths, beaches, coastal wooded areas, uninhabited islets, marshes, mudflats and other wetlands, mangroves and coral reefs (in the French Overseas Départements), sea-grass beds, spawning areas and the habitats of the birds protected by the EC Birds Directive. When no local plan has been made, building permits must be denied where the proposed works would have detrimental effects on these habitats.

In Italy, the Galasso Act of 8 August 1985 prohibits any changes to many different kinds of areas without a permit. The categories of sites specified are protected automatically, and there is consequently no need to list them by individual orders. These areas include a 300 metre-wide strip of land along the coastline and lake-shores, a 150 metre-wide strip along river banks, mountains above 1,600 metres in the Alps and 1,200 metres in the Appenines, and islands, glaciers and glacial cirques, volcanoes and forest areas even when the latter have been damaged by fire.

These protection measures are only temporary, pending the adoption by the Regions of landscape or territorial land-use plans which specifically take landscape and environmental values into consideration, and impose land-use restrictions for all the categories of area to which the Galasso Act applies. The Regions were given a relatively short period of time, up to 31 December 1986, to adopt these plans, failing which the State may prepare such plans in place of the defaulting Region.

A few Regions have in fact adopted such plans, examples being Emilia-Romagna, Liguria, Marche and Veneto. However, the State has not yet exercised its powers of substitution and is unlikely ever to do so, since the issue is highly sensitive from the political point of view. The Constitutional Court has ruled that this is only justified where a Region has failed to adopt the necessary plans and such intervention is necessary to achieve the essential aims of the protection regime established by the Act.[40]

Most Regions which have now drawn up their landscape or territorial land-use plans have preferred to do so for the entire territory of the Region, and not only for those types of areas for which such plans are required under the Galasso Act. The result is a comprehensive plan which takes into account the need to preserve natural areas throughout the Region, by means of appropriate zoning and the restriction of activities varying from one zone to another. The plan prevails over all other land-use instruments, such as municipal zoning plans, which means that where the latter do not comply with the regional plan, they must be amended accordingly. This new approach is tantamount to extending the nature park system to the whole territory outside urbanised areas.

In Portugal, certain areas of particular vulnerability or ecological value are to be designated as part of the National Ecological Reserve. The system was established in 1983 and modified in 1990. Its purpose is to maintain the ecological functions and potential, together with many of the economic, social and cultural values of the areas concerned. The system covers beaches, dunes, cliffs, estuaries, coastal wetlands, river beds, flood plains, riparian wetlands, erodible areas and escarpments with a slope of more than 45 degrees. Subdivisions, urbanisation, the construction of buildings and other works, including roads and hydraulic works, excavations and the destruction of vegetation cover are prohibited in these areas. Permits may, however, be granted for action of recognised public interest where it has been demonstrated that there is no other economically acceptable alternative. Afforestation and forest exploitation are amongst the projects which must be approved or authorised by the Directorate of Forests. All the areas subject to these restrictions must be delimited as such in all land-use plans. The plans of neighbouring municipalities will have to be coordinated accordingly.

Other laws lay down special planning rules to preserve coastal areas from excessive development.

---

[40] Constitutional Court, judgement n° 151 of 27 June 1986.

In Denmark, it has been prohibited since 1937 to construct buildings or structures within one hundred metres from the coast-line. In Portugal, under new legislation enacted in 1990, stricter planning rules apply to the coastal strip defined as a two kilometre-strip from the highest tide line. The Spanish Coastal Act of 1988 lays down special restrictions on construction and other activities within one hundred metres of the coast-line. There are many other examples.

In coastal areas, another means to preserve the natural environment may be to extend the public maritime domain to cover coastal inland habitats. This was one of the purposes of the new Spanish Coastal Act, which provides that beaches, dunes, coastal wetlands and other habitats form a part of the public domain of the State.

Legislation may also provide non-binding guidelines for the preparation of local land-use plans.

In Luxembourg, for instance, landscape plans or "green plans" are drawn up for rural municipalities. These plans include an inventory of natural ecosystems located within the territory of the municipality concerned, proposals for their protection and management and a list of priority measures for their conservation or restoration. These plans are not binding, but municipalities may take them into consideration when they make their local land-use plans.

In Germany, landscape plans list measures which should be taken to preserve natural habitats and landscape features as well as the habitats of wild animals and plants, particularly those which are legally protected. These plans are generally not binding on the municipalities, although they may use them to develop their local plans, as in Luxembourg. However, such plans are binding in Bavaria, as mentioned above.

3. Operations

Planning legislation is almost universally limited to the control of construction. Many activities which may be damaging to the natural environment are consequently excluded. There is therefore a need to extend the application of the legislation to certain operations or developments which have hitherto been exempt from planning permission.

As an example, the construction of golf courses is not considered as an operation falling within the scope of the law in a certain number of countries. A recent case in Ireland showed that it was impossible to prevent the destruction, as a result of the development of a golf course, of an important habitat for the natterjack toad, *B. calamita*, a species protected by the Berne Convention on the Conservation of European Wildlife. It is envisaged that Irish legislation may be amended to remedy this problem, but no measures have yet been introduced.

A very small number of countries have incorporated provisions into their planning laws aimed at controlling agriculture and forestry in environmentally important areas.

The Swiss Canton of Zurich, as seen above, requires that certain habitats be included in the no-construction zones of local plans. This obligation is binding on the municipalities as well as on the Canton when it approves local plans. However, as the strongest protection under a local plan may not be sufficient to preserve these areas, the law provides that it must be supplemented either by additional regulatory measures applicable to the areas concerned, or by the conclusion of management agreements, especially where active management measures are required.

In the Netherlands, planning legislation provides that a permit may be required for any operation that may affect the status of land as determined by its inclusion in a particular zone.

This rule may apply to agricultural activities if they are liable to harm natural areas or landscape features which are protected as such by the plan.

It is, perhaps, in Spain that the integration of conservation into land-use planning has reached the most advanced stage, as evidenced by a certain number of very recent regional laws, such as those of Aragon, the Canary Islands and Navarra.

In Aragon, for instance, the municipal planning regulations for the province of Teruel provide for the establishment of "nature zones" in the local plans and prohibits in these zones all uses which would affect their ecological values. Existing agricultural uses and traditional grazing and wood cutting remain authorised unless they are incompatible with the conservation objectives of these zones. Construction is prohibited in a 300 metre-wide strip of land around "nature zones".

In the Balearic Islands, an Act of 1991 provides for the designation of Nature Areas of Special Interest which are subject to particularly strict planning controls, including a prohibition on the construction of golf courses and yacht harbours, and the institution of a density rule of one house for 20 hectares. In certain parts of these designated areas, such as a one hundred metre-wide strip along the shore, dunes, cliffs, wetlands, and oak, juniper and wild olive forests, no new construction is allowed and all operations are prohibited except for the maintenance of existing buildings and public infrastructures which cannot be located elsewhere.

In the Canary Islands, special plans have been made to preserve certain areas where in addition to construction and quarrying, the intensification of agriculture, stock-raising and all activities liable to affect the hydrological balance of these areas are either prohibited or subject to a permit. Permits will not be granted where proposed activities may result in an irreversible alteration of the areas concerned. Specific provisions apply to coastal areas, river banks and wetlands. The draining of wetlands and changes in their water supply are prohibited.

In the Community of Navarra, the Planning Act of 1987 establishes a zoning system applicable to all municipal rural local plans. Zones range from strict nature reserves to agricultural areas. For each type of zone, the Act lists those activities which are prohibited, subject to a permit, or authorised. In some of these zones, agricultural and forestry activities are prohibited or restricted. Special planning rules apply to forest areas, water courses and wetlands.

Two trends seem to emerge from these very recent Spanish laws. It is possible to establish special protection areas, distinct from nature parks, where planning prohibitions and restrictions are much stricter than elsewhere, or to control any kind of activity or operation through the more conventional local land-use plans. In both cases, the innovative aspect of these institutions is that agriculture and forestry are brought under planning control in those zones where they are liable to cause damage to the natural environment.

In addition, Spanish national legislation provides for the preparation of plans for the conservation and management of certain natural resources. The Water Act of 1985, for instance, requires the development of basin-wide plans which must, *inter alia*, reserve waters for the preservation of the natural environment. As mentioned earlier, the Act on the Conservation of Natural Areas and Wild Flora and Fauna of 1989 also requires the preparation of plans for the preservation of species and ecosystems, called Natural Resources Management Plans.

Both Hydrological Plans and Natural Resources Plans establish general and specific limitations on human activities in the area concerned, or in different zones of that area, and determine the activities and projects for which impact assessments are required. Both are binding instruments which prevail over all other land-use plans. When the latter are incompatible with the Plans, they must be modified accordingly.

B. Activity-Based Planning Instruments : The Role of Environmental Impact Assessments (EIAs)

Regardless of the particular location where they are exercised, many activities are regulated by special legislation in a large number of countries. These activities generally include mining and quarrying, public works of all kinds, the construction of buildings and the abstraction of water. Such activities accordingly require a permit even when there is no land-use plan in the area where they are to be carried out. Rules relating to activities that are subject to permit requirements vary considerably and cannot be examined in any detail here. It must be emphasised, however, that there are many activities which may have considerable detrimental effects on the natural environment which continue to be unregulated in many countries, particularly in the fields of agriculture and forestry.

One such operation is afforestation which can be an extremely damaging operation when it involves the plantation of large numbers of exotic trees in natural or semi-natural habitats. Despite this, it seldom requires a permit, although there are a few exceptions. In the Republic of South Africa, it is prohibited to plant trees for commercial or industrial purposes on any land, whether public or private, without a permit from the Forest Department.

When an activity is subject to a permit, however, it is essential that the permit-issuing authority be as fully informed as possible of the effects that the activity will have on the natural environment before it takes its decision to grant or deny the permit or to issue it subject to particular conditions. The recognition of this requirement has led to the establishment in an increasing number of countries of a legal requirement to prepare Environmental Impact Assessments (EIAs).

The primary object of an EIA is, therefore, to inform the Government agency responsible for the approval of a project of its foreseeable consequences on the environment. Equally important, although this may not always be fully realised, an EIA may often be the only way to determine whether a proposed project complies with environmental protection legislation.

Four particular aspects of EIAs are of special importance for the preservation of the natural environment and will therefore now be examined in some detail.

1. Field of Application

If important habitats are to be preserved and endangered species saved, it is of course essential that the impact of any proposed project on these habitats and species be thoroughly evaluated before it is approved. On the other hand, EIAs are costly and time-consuming and it would be physically impossible to require them in all cases. In consequence, legislation always limits the field of application of EIAs to certain categories of projects.

Four different methods can be used for limiting the field of application. Firstly, a list of projects may be drawn up for which an EIA is required. This can include size or other thresholds, where necessary, to eliminate small projects. This method is used in the EC Directive on EIAs and by most European countries. By and large, only major projects are covered by this method.

Secondly, the negative listing approach is used by France, which was one of the first countries in Europe to adopt EIA legislation.[41] All projects are, in principle, covered by the EIA requirement if their cost exceeds twelve million francs. This threshold remained at 6 million francs for many years, but has recently been updated by a Decree of 1993. There are a number of exceptions which are specifically listed in the legislation. There are also certain types of projects for which an EIA is always required, even when their cost is lower than twelve million francs.

The third method, which is the one used in the United States, for example, consists in requiring an EIA wherever a proposed project will have a significant impact on the environment. This method first requires a determination on the part of the Government agency concerned as to whether or not such an impact is likely to occur.

The fourth method is entirely discretionary. It allows the Government department designated for that purpose to decide in each individual case whether or not an EIA will be carried out.

Whatever the method used, many projects are likely to be excluded from the EIA requirement. This is particularly true of projects which, because of their nature, size or cost, are presumed not to have significant impacts. The cumulative effects of such projects may nonetheless have very serious consequences for the environment.

Although such difficulties are unavoidable, at least to a certain extent, it should be possible for countries that use the positive or negative listing systems to require EIAs for a wider range of projects than has been the case to date.

Another possibility is to enact special EIA rules in respect of certain areas in which stricter precautions than elsewhere should be taken before projects may be adopted, because of their conservation importance or vulnerability. A few countries have already included provisions along those lines in their legislation.[42] These include the Greek Act on Environmental Protection of 1986, which provides that the decree establishing a protected area may require EIAs in respect of projects to be carried out in the protected area which would not normally be subject to such a requirement, as well as the Spanish Act on the Conservation of Natural Areas and of Wild Flora and Fauna of 1989.

The Environmental Impact Assessment Act of 13 July 1990 of the Canary Islands Autonomous Community is particularly noteworthy in this respect. The Act lays down a special regime for "ecologically sensitive areas". These areas include the four national parks that have been established in the archipelago and their buffer zones, as well as any other area designated

---

[41] Article 2 of the Law on the Protection of Nature of 10 July, 1976 (n° 76-629) and Decree of 12 October 1977 (n° 77-1141).

[42] EIA requirements for protected areas in both national and international law have already been mentioned in Part II, Chapter III, which deals with Conventional Types of Protected Areas. As will also be recalled, the Espoo Convention of 25 February 1991 on Environmental Impact Assessment in a Transboundary Context makes it possible for EIAs to be required in respect of proposed activities located in or close to areas of special environmental sensitivity or importance, such as wetlands designated under the Ramsar Convention.

for that purpose by an Act or by the general land-use plans which must be made for each of the islands. In all these areas, a detailed assessment is required for a large number of projects, many of which are rarely subject to EIAs anywhere else in the world. These include: pasture improvements, crop changes in areas of more than three hectares, rodent control measures, the breeding of and trade in live exotic animals, applications of pesticides in areas of over 50 hectares if certain products are used, cattle farms over a certain number of breeding animals, any changes in land-use affecting an area over 25 hectares, fish farms and many other activities.

A further aspect of EIAs which needs to be developed is that few laws so far require the assessment of policies, plans and programmes, that is to say the evaluation of the potential impacts on the environment of policy and political decisions of which individual projects are generally only a consequence. In that respect, it would seem to be particularly important that land-use plans, whether at municipal or higher level, be subject to EIAs: their effects on the natural environment may be very far-reaching, because of the zoning system adopted for individual areas covered by the plan.

In France, although no formal EIA is required for municipal land-use plans, the law lays down an obligation to prepare a "presentation report" which must, *inter alia*, analyse the initial status of the environment in the area covered by the plan, the effects of the plan on its evolution and the measures provided for its enhancement. Some plans have been invalidated by the administrative courts because they had failed to analyse the status of the local environment in the "presentation reports".

Another example is that of the American State of California where, under the California Environmental Quality Act, EIAs are required for the enactment and amendment of zoning ordinances.

The EC Commission has prepared a proposal for a Directive with a view to extending the EIA requirement to policies, plans and programmes, although it may be a long time before such a Directive is approved. However, the Habitats Directive of 1992 does establish an EIA requirement for any plan or project likely to have a significant effect on Special Areas of Conservation established under this Directive or Special Protection Areas established under the Birds Directive. This is supported by the existing provisions of Annex II to the EC EIA Directive, which lists those projects which shall be made subject to an assessment only where Member States consider that their characteristics so require. Such projects must always be subject to an assessment when they are located in or are likely to affect the conservation potential of a Special Protection Area.

Article 14 of the Convention on Biological Diversity, which deals with "Impact Assessment and Minimizing Adverse Impacts" requires Parties, "as far as possible and as appropriate" to introduce procedures for

"environmental impact assessment of those proposed projects that are likely to have significant adverse impacts on biological diversity with a view to avoiding or minimizing such effects and, where appropriate, allow for public participation in such procedures."

In addition, the environmental consequences of its programmes and policies likely to have an adverse impact on biological diversity must be taken into account by the Party concerned. Curiously, this requirement does not extend to plans.

Where the adverse impact of activities may affect the territory of other States or areas beyond national jurisdiction, the Party responsible is required to exchange information and consult with the States concerned. The conclusion of bilateral, regional or multilateral agreements is encouraged in this context. Where such extra-territorial damage is imminent or grave, the Party responsible must immediately notify the States liable to be affected and take measures to prevent or mitigate such damage.

Finally, legislation may also determine that certain potential effects are always significant, so that an EIA must always be prepared wherever these effects may occur. The California Environment Quality Guidelines, for example, provide that an EIA is required when a project has the potential to cause a fish or wildlife population to drop below self-sustaining levels or threatens to eliminate a plant or animal community or to reduce the number or restrict the range of a rare or endangered plant or animal.

2. The Content of EIAs

One essential aspect of the EIA system is that it always requires a description of the site liable to be affected by a project and of its wild fauna, flora and habitat components. However, the methods to be used to make this analysis are seldom specified in the legislation. There is therefore a serious risk that important elements will be missed. For instance, a survey of flora which is conducted in the winter time in Europe will not be very informative, yet this has been known to happen.

These problems may be avoided if the legislation provides a more detailed description of the factors that must be taken into consideration when making a site analysis. For example, Italian law lays down technical standards for the making of EIAs which list in considerable detail the various factors which must be taken into consideration, including plants, vertebrates, invertebrates, ecosystems, biological diversity and the vulnerability of these different elements. The EIA must also show the relationship between the proposed project and the territorial and sectoral planning and programming instruments, as well as any other legal restrictions on land-use which apply in the project area. The accuracy of all documents submitted must be certified by a declaration under oath on the part of the experts who have prepared the impact study.

The identification of the endangered species or habitat types which may be affected by a proposed project is also of particular importance. By way of example, the "biological assessment" provisions of the federal Endangered Species Act of the United States in respect of the critical habitats of species listed as endangered or threatened have already been discussed in Part I, Chapter VII of this paper.

3. The Review of EIAs

If the permit-issuing agency is to be sufficiently well informed of the impacts of a proposed project on the natural environment, it is essential that the EIA should provide a correct assessment of the effects that are likely to occur. It would therefore seem appropriate to establish a mechanism to review the EIA and to require its author to provide any additional information that may appear to be necessary or to prepare a new EIA if it has not been correctly prepared.

Two conditions are required if a proper review system is to operate. The reviewing body should have a large degree of independence and it should be well qualified from the scientific point of view to be able to pass judgment on the content of EIAs. Indeed, experts are necessary even to detect what important information has not been made available.

These conditions are hardly ever fulfilled. In most cases, the permit-issuing agency is simultaneously the reviewing agency and is therefore both judge and jury in its own causes.

The EC EIA Directive does not provide for any review mechanism other than a requirement to make the EIA available to the Environment authorities for advisory purposes.

Nevertheless, a few laws not only require the national Department of the Environment to be consulted, but also empower that body, if it is dissatisfied with the EIA, to request additional information, to call upon expert advice, or even to require a new EIA.

In Italy, for instance, a Commission on Environmental Assessments has been established in the Ministry of the Environment to review EIAs. The Commission may request the advice of any public body and may request from the project proponent any additional information that it may require. The Minister of the Environment must then state whether, in his or her opinion, the proposed project is environmentally compatible. Where the Minister in charge of the project disagrees with that finding, the matter must be referred to the Council of Ministers for a final decision.

In Switzerland, the Environment Protection Department must verify that the information included in the EIA report is complete and correct. If it finds that this is not the case, it may request the department concerned with the project to ask for additional information from the project proponent or to call upon experts.

In the Canadian Province of Ontario, EIAs must be submitted for review to the Minister of the Environment. No permit may be issued unless the Minister has accepted the assessment and has given approval to proceed with the undertaking. There is also an Environmental Assessment Board, which is a quasi-judicial body. Any person may request a hearing by the Board about any project subject to an EIA, its assessment or review. However, the request must be submitted to the Minister who has the discretionary power to refuse to put it to the Board. In addition, the Minister is not obliged to comply with the decisions of the Board, although it would seem that this seldom happens in practice.

4. The Effects of EIAs

The result of an EIA is never binding, as the purpose of this instrument is simply to inform public agencies of the potential consequences of their decisions. It may be, however, that an EIA will show that the carrying out of a proposed project will result in the violation of legal requirements in respect of the conservation of nature or the environment and that, as a result, the project cannot be approved unless it is modified in such a way as to make it compatible with the law.

In Switzerland, the legislation explicitly specifies that where the EIA shows that a project does not meet environmental legal requirements, the permit-issuing authority may only issue the permit subject to those conditions or restrictions which are necessary to ensure that the legal requirements are met.

In a few jurisdictions, the law allows the reviewing authority to veto a project if the EIA shows that it is incompatible with conservation needs. This is the case in Ontario, where, as seen above, the Minister of the Environment must approve all projects before they can proceed. In the Spanish Autonomous Community of the Canary Islands, the reviewing authority must make an "Ecological Impact Statement" on the basis of the environmental assessment before a project can be approved. This rule applies most particularly to designated "Ecologically Sensitive Areas". When the Statement is unfavourable in respect of an individual project, the proponent is required to revise the project accordingly.

CHAPTER VIII
VOLUNTARY CONSERVATION

Necessary as they may often be, regulatory measures are not sufficient in themselves to preserve the natural environment. There are political limits on the extent to which the State can control land-use. In addition, statutory controls cannot oblige a landowner to manage a particular site in a specific way, except in rare cases. Voluntary conservation measures are therefore an essential complement to the direct conservation role of the State.

It should accordingly be incumbent upon the State to assist private landowners to preserve natural areas. There are many ways in which this can be achieved. These include the elimination of legal obstacles to conservation and of financial and fiscal incentives for the destruction of the natural environment, the development of legal tools to facilitate voluntary conservation, the payment of subsidies, the grant of tax incentives and the provision of advisory conservation services.

A distinction must, however, be made between individual landowners, to whom certain incentives must be provided to encourage them to conserve nature on their land, and conservation NGOs which are in need of assistance from the State to purchase, preserve and manage land for conservation.

A. Conservation by Individual Landowners or Occupiers

Voluntary conservation may result from a unilateral decision on the part of a landowner or occupier, or from a contract concluded between a Government agency (or a private body) and the landowner or occupier. In both cases, the provision of financial incentives may assist in the decision to conserve.

1. Unilateral Commitments

Landowners are often willing to preserve valuable natural habitats on their land, but may not always be informed of their existence or importance for conservation. In some jurisdictions, therefore, the conservation agency has developed a notification scheme for that purpose. Admittedly, it has sometimes been argued that disclosure of the location of an important habitat would invite disaster. However, experience to date does not seem to be negative in this respect. Notification often builds up pride and a genuine wish to conserve, and it also eliminates the risk of inadvertent destruction. It should be added that notification alone has no legal consequences and owners remain free to destroy the sites concerned if they so wish.

In several States of the United States, legislation provides for the establishment of a State Registry of important natural sites. Landowners may commit themselves voluntarily to protect registered sites on their land. In some cases, they are also obliged to inform the State conservation authorities of any proposed changes they intend to carry out on registered sites. There are, however, usually no penalties for non-compliance and the commitment is purely moral. In return for the owner's commitment, the conservation department sometimes presents him or her with a plaque and, where necessary, provides management advice.

In California, "Significant Natural Areas" are identified by the Fish and Game Department. That agency then tries to ensure the preservation of these areas, wherever possible by a voluntary commitment on the part of their owners. At federal level, there is also a system of National Natural Landmarks which has been developed since 1963 by the Department of the Interior for the purpose of encouraging the preservation of areas illustrating the ecological and geological characteristics of the United States. The Landmarks are designated by the Secretary of the Interior, after notification has been made to the landowners and other persons, including the relevant local authority and the Governor of the State concerned. Landowners are merely invited to sign a voluntary commitment to preserve the sites in return for a certificate and a plaque. They do not relinquish any of their rights over the land and may terminate their commitment at any time. On the Government side, the presence of Landmarks must be taken into consideration when EIAs are prepared. The National Parks Service which administers the Landmarks system advises the owners, at their request, on recommended conservation measures.

In the Republic of South Africa, there also exists a network of National Heritage Sites. In return for their voluntary commitment, landowners may benefit from grants to manage the areas concerned.

None of these instruments is binding and landowners may therefore withdraw their commitment as they wish.

In contrast, there is also another type of instrument which enables an owner to accept voluntary restrictions, irrevocably and in perpetuity, on the use of his or her land. This is called a covenant. The obligations thus created are real obligations[43] for the beneficiary of the covenant, which is usually a Government department or an institution like a Conservation Trust. Covenants are enforceable in the courts in cases of non-compliance. For a landowner, covenanting has the advantage that the land will be preserved for posterity whilst he continues to own, and if so desired, to occupy that land. For the Government Department or Trust, it is an inexpensive way to preserve important habitats as no payments are due to the owner. Owners may, however, benefit from assistance to manage the land, as their rates or land taxes may be lower due to the decrease in the value of the land as a result of the covenant.

Another form of unilateral commitment is called dedication. The concept was developed in the United States to preserve important habitats in perpetuity. Once again, this amounts to a unilateral and irrevocable commitment on the part of a landowner to preserve a particular area. Once accepted and registered by the appropriate State authority, the "articles of dedication" become binding on the owner.

2. Incentives and Disincentives

Voluntary conservation may also be encouraged by various systems of financial incentives or disincentives, whether or not there has been any kind of formal or informal unilateral commitment.

---

[43] Meaning that they run with the land.

The first aspect consists of the removal of incentives to destroy the natural environment. Such incentives may include unduly high land taxes on land which is not farmed, special tax reductions for the cultivation of undeveloped land and the grant of subsidies for the same purposes.

In many countries, the tax system still heavily favours the development of agriculture. Unfarmed arable land is often taxed at a high level as a deliberate incentive to encourage landowners to develop it. A common method is to tax the land as if it were producing agricultural crops. There is consequently no incentive to conserve these lands, as even extensive agriculture or stock-raising would be unprofitable after these taxes have been paid.

On the contrary, with the special tax benefits and subsidies which are generally available for that purpose, the only profitable alternative is conversion to intensive agriculture. Changes in the land tax system are long overdue but tend to meet with the resistance of treasury departments and local authorities which generally benefit from the proceeds of the land tax.

Some changes are, however, gradually taking place. In France, a provision of the tax law which granted a twenty-year land tax exemption to farmers who had drained wetlands was repealed in 1990. In addition, a review of the land tax system is under way with a view to reducing the tax level on land reserved for conservation. There are some examples, such as in the American States of Minnesota and Indiana, where substantial tax reductions have been implemented to encourage the preservation of certain habitats, such as wetlands and the remnants of the Great Prairie of Minnesota. In the Canadian Province of Ontario, there is a complete exemption from land tax for "significant" wetlands, and "areas of natural and scientific interest". In return, the landowners must comply with guidelines to maintain these areas in their natural state.

Another promising form of tax exemption consists of granting an exemption from death duties in respect of areas which have an exceptional value for nature conservation. In the United Kingdom, land deemed to be of "Outstanding Scenic or Scientific Interest" (LOSI) may be thus exempted if formal aproval if given by the Inland Revenue. The Countryside Commission has a statutory duty to advise the Government whether the land is appropriate for such exemption, and employs a National Heritage Adviser to assess the properties concerned for this purpose. Over 60,000 hectares of land in England are now exempted from inheritance tax, much of which is made up of SSSIs. The landowner and heirs must agree to manage the area according to a management plan approved by English Nature and/or the Countryside Commission, as appropriate. If the provisions of the plan are not complied with, the tax in question becomes payable immediately.

Expenses incurred in the conversion of natural areas, particularly wetlands, to agriculture have traditionally been tax deductible or eligible for accelerated depreciation. In the United States, most of these incentives have now been eliminated by the Tax Reform Act of 1986. In Australia, farmers were formerly allowed to deduct fully the costs of clearing and draining their land to put it under cultivation, until these provisions of the Income Tax Assessment Act were repealed in 1983.

Special subsidies often continue to be paid for the conversion of natural areas to cultivation. Nevertheless, in the United States, the Presidential Executive Order of 1977 prohibited federal agencies from encouraging the destruction of wetlands, with the result that many of these subsidies, including low interest loans, have been removed. In other countries such as Denmark, grants made for the conversion of wetlands have also been brought to an end.

Free or low-cost flood insurance is a form of disguised subsidy that encourages agricultural and other developments in areas that would be otherwise unsuitable, because of the frequency of the floods that affect them. In the United States, the Coastal Barrier Resources Act of 1982 removed these incentives, with a view to preserving certain barrier islands on the east coast of the country together with their adjacent wetlands and aquatic habitats. In addition, the use of federal funds to construct roads, bridges or any other structures in these areas is also prohibited.

The Coastal Barrier Improvement Act of 1990 more than doubled the size of the area covered by the System, the total surface of which is now 520,000 hectares. All federal subsidies, including disaster relief and water treatment grants which might be used to develop undeveloped shorelines and coastal barriers included in the system, are prohibited. It is expected that the System will be extended to the Pacific Coast in the near future.

Considerable disincentives to the use of highly erodible land and the conversion of wetlands for agriculture were instituted, once again in the United States, by the Farm Act of 1985 and substantially maintained by the Food, Agriculture, Conservation and Trade Act of 1990. These Acts deny a large number of important federal subsidies to farmers who have converted wetlands or highly erodible land for crop production. Benefits thus denied include price support payments, crop insurance and disaster relief.

Finally, an increasing number of laws now empower conservation agencies or other Government bodies to provide financial assistance to landholders who manage their land for conservation. As an example, the Wildlife and Countryside Act of 1981 empowers English Nature and its sister bodies to make grants or loans for any activity contributing to nature conservation. Another example is the Australian State of Victoria where grants, loans or other forms of assistance are available to landowners to achieve the objectives of conservation legislation. The provision of financial assistance is, however, usually linked to the conclusion of a management agreement.

3. Contracts

a. Leases

Leases are often used as a substitute to acquisition. In the United Kingdom, for instance, many national nature reserves have been established on land leased by English Nature or its sister bodies. NGOs also frequently lease land from private and sometimes public landowners. However, this form of land tenure cannot guarantee long-term conservation of the areas concerned, as leases are generally concluded for relatively short periods of time and may not be renewed if the landlord wishes to use the land for other purposes. When this happens, past conservation efforts will often have been in vain. Nevertheless, it is often possible to conclude long term leases, for instance for a duration of 99 years.

Leases may also be a convenient method for a Government agency or landowning NGO to contract with another person for the management of land held for conservation. The method has the advantage that the tenant pays rent to the owner. This provides a useful income to the agency or organisation concerned, which would otherwise often have to pay labour to perform essential management tasks. However, it is important that the terms of the lease impose strict conditions on the tenant so that farming practices are compatible with conservation requirements. This means that rents will frequently be set at a lower level than would be the case if farming practices were unrestricted.

In the United Kingdom, for instance, the National Trust's income as a landowner is derived mainly from farm rents. Working with English Nature, the National Trust now routinely inserts covenants into all its tenancy agreements to promote good conservation management. Some 35,000 hectares are now covered by these positive and restrictive covenants, which bind those renting the land to preserve its natural amenity.

Nevertheless, there are often serious legal obstacles that may prevent or hamper the use of leases as a management tool for natural areas. The rules applicable to public property may prohibit long-term leases. The rules governing farm leases may forbid the insertion in the contract of clauses restricting the freedom of the tenant to farm the land as he or she wishes. In France, for example, the Rural Code does not allow the lessor to prohibit his or her tenant to eliminate hedgerows or to plough natural grasslands. Any provision to that effect in a lease would be null and void.

b. Easements or Servitudes

The legal nature of easements has already been briefly examined in Part II, Chapter II (B)(2)(d)). However, since easements are established by contract and constitute the basis of management agreements where these run with the land, it is appropriate to consider them in more detail here.

As seen above, the purpose of the institution from its very origin in Roman law was to confer a benefit on one piece of land, called the dominant tenement, by imposing a restriction or an obligation upon the adjoining parcel of land, called the servient tenement. Easements may be negative, in the sense that there are some things which cannot be done on the servient tenement, such as the construction of a building that would obstruct the view from the dominant tenement. Easements may also be positive where they allow certain things to be done on the servient tenement to the benefit of the dominant tenement. Examples would be the extraction of water or the establishment of a right of way. As a general rule, however, an easement cannot require the owner of a servient tenement to carry out any positive measures to the benefit of the dominant tenement.

The value of this institution for conservation is limited for several reasons. An easement contract can only be concluded between two landowners. The dominant and servient tenements must be contiguous, or at least very close to each other. The restrictions imposed on the servient tenement must provide a benefit to the dominant tenement.

In a relatively small number of jurisdictions, legislation has been enacted, usually very recently, to encourage the use of easements for conservation by means of eliminating these obstacles.

This is generally achieved by removing the need for a dominant tenement. For example, the Countryside Act of 1968 in the United Kingdom provides that management agreements concluded between English Nature or its sister bodies and the owners of SSSIs, which impose restrictions on the exercise of property rights, confer the same rights on English Nature and its sister bodies to enforce these restrictions as if they were the owners of adjacent land. As there is no dominant tenement, the requirement that the easement must serve the needs of that tenement is consequently also eliminated.

Similarly, a large number of American States, perhaps more than forty, have enacted legislation allowing the creation of "conservation easements" in respect of which the contiguity requirement has been dropped. Under its "Reinvest in Minnesota" Programme, the State of Minnesota has even instituted a system that enables farmers to receive payments in return for accepting permanent easements on their wetlands.

In Canada, a recent study on the role and use of conservation easements[44] shows that an increasing number of Provinces have now also waived the contiguity requirement, at least in certain specific cases.

In civil law jurisdictions, on the other hand, conservation easements remain almost universally impossible, as the double requirement of contiguity and benefit to the dominant tenement continues to be solidly entrenched in legislation. The major exception to this rule is Switzerland, where any person is entitled to conclude an easement contract with a landowner for any purpose.[45]

Even where conservation easements may lawfully be created, there are often other restrictions that may limit the extent to which the institution can be effectively used.

In a number of American States, conservation easements cannot be established in perpetuity, which means that they cannot be binding upon successors in title.

The owner of a conservation easement should be entitled, like the owner of a conventional easement, to transfer his or her rights, by gift or sale, to another person. As an example, if the owner is a Government agency, it should be able to transfer the easement to a conservation NGO and vice versa. However, the law does not always allow for this possibility, which may make the system unduly rigid.

Finally, restrictions are often placed on the categories of persons entitled to enter into easement contracts. Such persons will generally be Government agencies or autonomous public bodies or foundations which have been established for that particular purpose. In the Canadian Province of Ontario, for instance, the right to create conservation easements is vested in the Ontario Heritage Foundation. The Foundation is entitled to assign an easement that it has acquired to any other person. The right of conservation NGOs to conclude easement contracts is recognised by law in a certain number of American States, but not in all. In contrast, the right of any private person to do so is very rarely recognised.

Easements are, in principle, enforceable in the courts at the suit of their holders. This obviously implies that the beneficiaries are able to monitor the sites on which they hold easements, as well as to bring the matter to court where damage occurs or is imminent as a result of the landowner violating the terms of the easement. If the site has already been damaged, the court should order its restoration or, when this is not possible, provide for the payment of monetary damages to the easement holder.

In the American State of Massachusetts, there exits a unique system whereby conservation easements must first be approved by the State before they are valid in law. Once this has been done, the State Attorney General, the municipality concerned or any group of ten residents of the State are entitled to bring the matter to court whenever the conditions of an easement have not been complied with.

One of the advantages of easements is that they may have important consequences with regard to the payment of certain taxes.

---

[44] O. Trombetti and K. Cox : *Land and Wildlife Conservation*, published by Wildlife Habitat Canada in 1990.

[45] This possibility also exists in other countries, such as Germany, but does not seem to be extensively used.

As far as property tax is concerned, the amount of tax due may be reduced in proportion to the decrease in the value of a parcel of land as a result of the creation of the easement. This form of compensation has been expressly recognised by the legislation of several American States. Moreover, as easements are considered as interests in land, they may be subject to the same rules as those which are applicable to any transfer of real property. In the United States, as a result, exemptions from property transfer taxes are also applicable when easements are transferred to charitable organisations, including conservation NGOs. Income tax deductions in respect of gifts made to such organisations also apply to the gift of easements.

Finally, another advantage of the easement system is that a Government agency or an NGO that has acquired land may sell that land with an easement attached to it. This reduces the cost of land acquisition and management for conservation.

In South Australia, by way of example, where a permit to clear native vegetation has been denied under the Native Vegetation Act of 1991 and a parcel of land becomes economically non-viable as a result, the Government may offer to buy the land from the landowner. It may then re-sell the land with a Heritage Agreement (which is an equivalent of an easement) attached to the land.

In the United States under the Farm Act of 1985 and the Food, Agriculture, Conservation and Trade Act of 1990, land acquired by the Farmers Home Administration as a result of loan foreclosures or voluntary conveyances may be resold with perpetual conservation easements.

In the same way as regulatory measures, however, easements cannot generally be used to require landholders to carry out active management measures. This objective may only be achieved through the conclusion of management agreements.

c. Management Agreements

Management agreements are contracts between a public authority (or a conservation organisation) and an owner or occupier of land, under the terms of which the latter undertakes to manage his or her land in a specified way in return for regular payments or, more rarely, a lump sum paid once and for all.

Management agreements usually provide for both negative and positive obligations. For example, an owner may agree not to drain a wetland, plough a natural meadow or use fertilisers or pesticides and also to mow the grass after a certain date, to remove woody vegetation but only by certain means, or to have the area grazed by cattle in an extensive way. The payments made under these agreements may be considered both as compensation for profits foregone and as remuneration for activities carried out in the public interest.

From the legal point of view, management agreements belong to two main categories. Some run with the land and are binding upon successors in title. They should therefore be considered as easements to which a number of positive obligations may be added. The other category encompasses purely personal contracts concluded with individual landowners, generally for a specified period of time. These expire at the end of that period unless they are renewed, or when the owner dies or sells the property.

Management agreements running with the land are used extensively in the United Kingdom, particularly in respect of national nature reserves and SSSIs. These agreements may include positive obligations since, under the Countryside Act of 1968, they

"may provide for the carrying out of such work and the doing of such other things as may be expressly for the purpose of the agreement".

The law therefore authorises the creation of positive obligations binding upon successors in title, although this remains a rare occurrence.

In the Republic of Ireland, the appropriate Minister may also conclude agreements with landowners that run with the land. Whereas management agreements in the United Kingdom are automatically binding upon successors unless they provide otherwise, this rule only applies in Ireland if the parties to the agreement have expressly decided that this should be the case.

In Australia, there are many examples of agreements running with the land. In New South Wales, under the National Parks and Wildlife Act of 1974 as amended, "conservation agreements" may be concluded for the preservation of "conservation areas". These agreements may include positive obligations, require the implementation of management plans and commit the relevant Minister to provide financial and technical assistance. These obligations are binding on both parties and run with the land. However, conservation agreements may be modified or terminated by mutual consent. The Minister may also terminate the agreement unilaterally if he or she is of the opinion that the area is no longer needed for, or is no longer capable of being used to achieve, any purpose for which it was entered into.

In Victoria, the Conservation, Forests and Lands Act of 1987 empowers the Director General of Conservation to conclude management agreements with private landowners and other Government departments. The agreements may provide for positive obligations and can be made binding upon successors in title. The Act contains a number of particularly innovative provisions. If the agreement so provides, the Director General may be required to reimburse the property tax due in respect of the area which is subject to the agreement. Alternatively, the Minister may recommend to the appropriate rating authority that the whole or part of the rates payable with regard to the land in question be remitted. The Minister may then reimburse the sums thus remitted to the rating authority concerned.

These agreements are also binding upon third parties to a certain extent: the Act provides that if an agreement prohibits a landowner from doing certain things on his or her land, this prohibition also applies to third parties unless they can prove they did not know of the existence of the agreement. Finally, the Director General is empowered to institute proceedings for breach of an agreement. The court may award punitive damages and order the landowner to restore the land to its prior condition.

In South Australia, under the South Australia Heritage Act of 1978, the Minister may enter into Heritage Agreements with landowners. The terms of such agreements may include restrictions on the use of land, and may require owners to carry out works or to permit the Government to carry out works for the preservation and enhancement of the site. Heritage Agreements are binding on both landowners and their successors in title.

As a counterpart, the Agreement may contain terms which bind the relevant Government authority to provide financial assistance and technical advice and assistance. Heritage agreements are available to landowners where permits to clear native vegetation under the Native Vegetation Act of 1991 have been denied.

In Switzerland, agreements binding upon successors in title may be concluded with landowners in some of the Cantons. In the Canton of Aargau, for instance, legislation provides for this possibility as regards the protection of dry grasslands and tall grass meadows which are the habitat of endangered species.

By and large, however, there are relatively few jurisdictions whose legislation specifically authorises the conclusion of management agreements running with the land. Nevertheless, this type of agreement has many advantages, especially when concluded in respect of areas of considerable natural value, such as SSSIs in the United Kingdom. Such agreements may be the only means to secure long-term commitments for the conservation of important features of the environment, short of imposing regulatory measures or acquiring the land. Where agreements do not run with the land, changes of tenure may result in the refusal by the new owner to renew the agreement. On the other hand, and quite understandably, farmers are usually reluctant to be tied down permanently by an agreement and fear that, should they wish to sell their land, they will have difficulties finding a buyer who will accept to be bound by the terms of the agreement, unless the agreement is sufficiently attractive.

This is presumably the reason why systems of personal contracts have developed recently in many countries: in some cases, agreements running with the land are not legally possible. Contracts are therefore only binding upon the parties and consequently expire when the property changes hands. They are also often concluded for a specified period of time but may, of course, be renewed if the two parties so wish.

Management agreements of this kind were initially used mainly for the conservation and management of certain habitat types. However, their use has been extended in recent years to the extensification of agricultural production on certain lands. During the past two years, innovative types of agreements have been developed in the United Kingdom, particularly for the restoration or creation of natural habitats, which will be dealt with in greater detail at the end of this section.

An example of agreements relating to specific types of habitats are those concluded in the United States under the federal Water Bank Program, which was established pursuant to the Water Bank Act of 1970. Under this programme, private landowners receive annual payments in exchange for a commitment not to drain, fill, level or destroy wetlands and to maintain vegetation on adjacent land. The contracts are concluded for a period of ten years and are renewable. They are transferable when land is sold. They may be terminated unilaterally by the farmer before the ten years are up, but all previous payments must then be reimbursed. The programme has been used extensively for the conservation of prairie potholes in the American States of Minnesota, North Dakota and South Dakota.

In Switzerland, federal regulations adopted in 1989 provide for the payment of special subsidies to farmers who agree to maintain the natural vegetation of dry grasslands and tall grass meadows. These payments are considered to be a remuneration for ecological services performed by landowners in the public interest. The regulations lay down general conditions which must be complied with. In particular, the natural vegetation must not be damaged by the use of fertilisers, irrigation, draining, grazing or any other means. Individual agreements lay down the conditions which must be respected for each particular site.

Similar systems have been developed in several Swiss Cantons, either to complement the federal regulations relating to these two particular habitat types or to make it possible to conclude agreements for other categories of habitats.

In Lichtenstein, an Act of 1988 also provides for the payment of subsidies for the conservation of dry grasslands and wet meadows. These payments include a lump sum in compensation for the lower agricultural yields resulting from the extensification of production, an allowance for the inconvenience this entails, and a premium for the consequent increase in biological diversity.

In the German Land of Bavaria, management agreements may be concluded for the conservation of wetlands and dry grasslands.

In Bavaria once again and also in the Netherlands, special management agreements may be signed for the protection of certain ground nesting birds, such as the Ruff and the Black-tailed godwit. Under these agreements, farmers undertake not to mow or graze their meadows before a certain date.

In parallel to systems of management agreements aiming at the conservation of specific habitat types, a few countries have now developed new forms of agreements which are designed to contribute to the extensification of agricultural production and a resulting decrease in agricultural surpluses, through the maintenance of semi-natural habitats. These agreeements make no particular reference to particular types of environments.

The origin of this new system was a European Community Regulation of 12 March 1985 (n 797/85) on the improvement of the efficiency of agricultural structures, now replaced by a new Regulation of 30 June 1992 (n 2078/92). Certain provisions of the earlier Regulation authorised Member States to provide financial assistance to farmers who committed themselves for a period of at least five years to introduce or maintain agricultural practices which were compatible with the protection of the environment and natural resources or with the maintenance of natural areas and the landscape. The programme only applied to those areas that had been designated for that purpose as being particularly sensitive by Member States. The latter were free to lay down rules and criteria to be observed in relation to the maintenance or decrease of production intensity and cattle density in the areas subject to the agreements.

An increasing number of Member States of the Community have adopted legislation to implement these provisions of the earlier Regulation.

In the United Kingdom, the country that took the initiative of introducing the scheme in the European Community, the Agriculture Act of 1986 provides for the designation of Environmentally Sensitive Areas and empowers the Ministry of Agriculture to conclude personal contracts with farmers. These specify authorised farming practices in keeping with the traditions of the region, in return for an annual payment per hectare. Project Officers of the Agricultural Development and Advisory Service ("ADAS") advise on detailed conservation prescriptions and monitor their implementation.

The contracts are for ten years but either party may take advantage of a "break clause" after five years. There are two tiers of payment depending on the amount of work involved, and rates vary according to the region. Tier 1 payments per hectare range from £10 in the Peak District to £350 in the Somerset Levels, though most are in the range of £100–200. By way of example, in the moorlands of the Peak District National Park, Tier One agreements prohibit cultivation, fertilisers, lime and pesticides and require the establishment of regular heather and grass burning. Tier Two agreements contain additional requirements for stock management and heather regeneration.

The scheme, which is open to all farmers in ESAs who volunteer, has so far been extremely successful. There are currently 10 ESAs in England, within which some 3,000 agreements have been concluded with farmers. A further 12 ESAs will be designated by the end of 1994, bringing the total to 12% of the land surface in England. Many of these new ESAs will be established on land within National Parks and AONBs: the Lake District National Park is shortly to become the largest ESA in England.

In the Netherlands, extensive areas have been designated as Nature Management Areas in which agricultural activities must be compatible with the conservation of nature and the landscape. To achieve this objective, a system of management agreements has been established whereby farmers commit themselves to take conservation objectives into account in return for quarterly payments. Each agreement lays down the specific obligations, negative or positive, of the farmer and the amounts payable by the Ministry of Agriculture. Agreements are made on the basis of management plans prepared by the Provincial Commissions for the Management of Agricultural Land. Members of the Commissions include farmers' organisations, conservation NGOs and local authorities. The plans are officially adopted by a National Commission. Agreements are concluded for a period of six years. During the first year which is considered as a trial period, the farmer may withdraw from the agreement at his discretion.

In Germany, a majority of the Länder have developed management agreement systems for the extensification of agriculture and the preservation of natural habitats and certain landscape features. In Schleswig-Holstein, for instance, agreements may include obligations to preserve hedgerows, spinneys, small ponds and watercourses. There are also special programmes for the protection of certain plants and the breeding ponds of species of amphibians. Agreements may also restrict the use of herbicides on the edge of cultivated fields to preserve weeds of cultivation which are now threatened with extinction almost everywhere in Europe. These schemes have met with considerable success. In Bavaria, for instance, 1,400,000 hectares have been designated as Environmentally Sensitive Areas under the EC Regulations.

The new Regulation of 30 June 1992 (2078/92) has considerably expanded the system. Its purpose is to promote the use of farming practices which reduce the polluting effects of agriculture; an environmentally favourable extensification of crop, sheep and cattle farming; the conversion of arable land into extensive grasslands; ways of using agricultural land which are compatible with the protection and improvement of the environment, including genetic diversity; the upkeep of abandoned farmland and woodlands; the long-term set-aside of agricultural land for purposes connected with the environment; and land management for public access and leisure activities.

For these purposes, aid may be provided to farmers who undertake to carry out farming practices in accordance with these objectives. Farmland set aside for environmental purposes should in particular be used for the establishment of biosphere reserves or natural parks or for the protection of hydrological systems. In addition, the scheme may include measures to improve the training of farmers with regard to farming or forestry practices compatible with the environment.

To implement the scheme, Member States must develop multi-annual zonal programmes reflecting the diversity of environmental situations, natural conditions and agricultural structures, as well as the main types of farming practices and Community environment priorities. Each zonal programme must cover an area which is homogenous in terms of the environment and the countryside, and be drawn up for a minimum of five years.

Member States may decide that one or more of the types of aid provided for by the Regulation may be applicable to the whole of their territory, in addition to the zonal programmes. An annual premium per hectare or livestock unit[46] is paid to farmers who have entered into such undertakings. The maximum premium varies according to the crops, type of livestock and other factors. An additional premium may be paid when several undertakings have been made in respect of the same land. Member States may require that a farmer's undertaking be given in the context of an overall plan for the entire holding or part thereof. The conditions for granting the aid are to be determined by the Member States. Aid may also be granted for attendance at training courses on agricultural and forestry production practices compatible with the requirements of

---

[46] A livestock unit refers to cattle more than 2 years old or horses more than 6 months old. Cattle between 6 months and 2 years, ewes and sheep, have lower values and several must be added together to make one unit.

environmental protection, natural resources and maintenance of the countryside and the landscape, and particular codes of good farming practice. The Community may contribute to demonstration projects in such matters.

All these measures are to be part-financed by the Community European Agricultural Guidance and Guarantee Fund, which will meet between 50% and 70% of the cost depending on the region. The Community may also contribute to the premiums granted by Member States to compensate for loss of income resulting from the mandatory application of restrictions imposed by Community legislation. This could for instance apply in the case of the establishment of Special Areas of Conservation under the new EC Habitats Directive. Member States are free to provide additional aid measures as long as these comply with the objectives of the Regulation and the provisions of the Treaty of Rome.

In consequence, the system is no longer limited to designated environmentally sensitive areas but can now be applied to whole ecological zones and, if a Member State so wishes, to its entire territory. The scheme remains purely voluntary, except in those cases where land-use restrictions are imposed by other Community provisions. In the latter case, compensation may be provided.

Outside the EC, Sweden adopted legislation in 1986 making it possible to conclude management agreements with farmers containing specific obligations adapted to each individual case. In exchange, farmers receive payments which are considered to be a remuneration for a public service.

Finally, the United Kingdom has recently developed two innovative types of management agreements which were first tested experimentally and have now been extended in scope or integrated into more permanent policies.

The first of these schemes, known as Countryside Premium, was designed to complement the European Community programme for the setting aside of agricultural land. Under this programme a premium was paid to farmers who withdraw land from cultivation. The scheme, which was launched in 1989 in seven counties in eastern England, consisted of paying an additional premium to those who agree to manage set-aside land in a way beneficial to wildlife and the landscape. Some 7,000 hectares were covered by the Scheme by 1992, but no more land will now be taken into the scheme in view of the fundamental changes to the agri-environment programme described above.

Another scheme begun in the United Kingdom in 1991 is called "Countryside Stewardship", which is intended to combine the conservation and public enjoyment of land with commercial land management through a national system of incentives. Administered by the Countryside Commission, the scheme is not area-specific nor is it restricted to farmers.

Countryside Stewardship provides for the payment of premiums for the management, restoration or recreation of certain habitats or landscape features. Farmers, landowners and other managers of land of any size may select options from a conservation "menu" drawn up for seven selected habitat types, and propose a ten-year agreement to the Countryside Commission. Acceptance into the scheme is not automatic, and is judged on the quality of the environmental improvements proposed by the landowner, as well as the opportunity for public benefit. Similarly, payment may be varied according to the work done. At the start of 1993, maximum annual payments per hectare for conservation measures within the different habitats were as follows: historic landscapes (£250), heathland (£250), waterside (£225), coastal land (£225), chalk and limestone grassland (£210), old meadow and pasture (£70), and uplands (£15, with additional payments per hectare for the regeneration and management of suppressed heather).

The Commission's target was to establish 3,000 to 5,000 contracts by 1994 covering between 60,000 and 90,000 hectares. The scheme has already been extremely successful: applications were 30% higher than expected, and 900 contracts covering 30,000 ha of land were taken into the scheme within its first year of operation. The original Government funding of £10 million for 1991–94 has consequently been increased to £25 million.

The particularly innovative aspect of the scheme is that it reverses the traditional procedure of management agreements. Instead of a government agency proposing to sometimes reluctant farmers to enter into such agreements, it is this time the farmers themselves who compete with one another to make the most attractive proposals from the point of view of conservation.

In conclusion, it can probably be said that management agreements as an incentive to voluntary conservation may well be one of the most effective instruments available for the preservation and management of natural and semi-natural areas and particular landscape features. However, there are still only a small number of countries where specific legislation has been enacted empowering government agencies to conclude such agreements and setting the legal framework that is necessary for the system to operate. Moreover, most of these laws are very recent.

Nevertheless, it is probable that this form of instrument will increasingly be used in the future, as it provides a more acceptable alternative both to acquisition (from the financial point of view) and to regulatory measures (from the political point of view).

The 1987 amendments to the Swiss federal Nature and Landscape Protection Act lay down the principle that the protection and management of those areas which have been designated as being of national, regional or local importance shall be secured as much as possible through agreements concluded with their owners or occupiers, as well as the adaptation of farming and forestry practices to specific conservation requirements. This is probably the first time that management agreements have been singled out by legislation as the preferred conservation instrument.

However, it must be emphasised that management agreements, useful as they may be, may not always be sufficient to secure important habitats if their owners do not accept voluntary restrictions. It follows that conservation agencies should always be empowered to take regulatory measures or to purchase land compulsorily to avoid irreversible damage to these areas.

B. Assistance to Conservation NGOs for the Acquisition, Preservation and Management of Land

NGOs which have acquired land for conservation are, like any other landowners, entitled to any particular benefit that may be available for the conservation of private lands. However, unlike most private landholders, they are in a peculiar situation in that it is not their will to conserve which must be encouraged by appropriate incentives but rather their legal and financial ability to do so. Nevertheless, since NGOs should be considered as performing a public service, often at a considerable cost for themselves, it would seem only fair that they be able to benefit from some assistance from the State.

Private reserves are generally owned by NGOs. However, the rights of these organisations to preserve the land they have acquired for conservation against trespassers are no different from those of any other private landowners. In addition, private reserves, like any other land, may be the subject of compulsory purchase orders should any Government agency need that land for its own development purposes.

These flaws can be remedied by special legislation conferring a legal status on private reserves.

1. Voluntary Reserves

In France, under the Nature Protection Act of 1976, voluntary reserves enjoy the same degree of protection as statutory reserves, when officially approved by an order of the Préfet. The approval order lists prohibited activities as well as the landowner's obligations. Violations of the prohibitions laid down by the order are criminal offences and are punished by the same penalties as the ones applicable in respect of official reserves. The only major difference between statutory and voluntary reserves is that approval for the latter is only given for a period of six years. It is, however, automatically renewed unless the owner disagrees. In addition, the approval order may be repealed at any time if the owner so requests or if he or she does not comply with the terms of the order.

With regard more particularly to the compulsory purchase of voluntary reserves by Government agencies for reasons of public interest, the law provides for a special procedure which requires consultation with the Nature Conservation Commission of the relevant "département", before the expropriation order can be adopted. However, the opinion of the Commission is not binding.

In Belgium, voluntary reserves are also legally protected when they have been approved by the Government, and protection rules are the same as for Government-owned reserves. The approval is given for a period of ten years and is automatically renewed. There do not seem to be any particular restrictions with regard to the compulsory purchase of land included in these reserves.

The protection of private reserves against expropriation by public authorities is rarely provided for by legislation in practice. Nevertheless, a few interesting examples of such laws do exist.

In a certain number of States of the United States, land may be dedicated to conservation by its owner, as seen above. This requires a formal procedure involving the approval of "Articles of Dedication" by the appropriate authority. Once these "Articles" have been officially registered, they become binding upon both the owner and the State. The State cannot thereafter expropriate dedicated land, except in the case of imperative and unavoidable necessity and after a cumbersome public procedure, involving public hearings, has been followed. Depending upon the State, expropriation will require the prior approval of the legislature, the Governor or a court of law. Owners of dedicated land are usually Government departments, universities or conservation organisations, such as The Nature Conservancy, for which dedication represents an important guarantee against Government interference.

Still in the United States, the National Fish and Wildlife Foundation is a private body which was established by a Special Act of Congress of 1984 for the purpose, inter alia, of acquiring land for the conservation of fish and wildlife. The property of the Foundation cannot be compulsorily purchased by a State or its political subdivisions where such property has been determined as being valuable for the conservation or management of wildlife by the Director of either the federal Fish and Wildlife Service or the Migratory Birds Conservation Commission, as appropriate.

2. National Trusts

In certain common law countries, special bodies have been created by law for the purpose of preserving historical monuments and buildings, land of scientific interest or natural beauty and other valuable property. These bodies are generally called National Trusts as they hold land in trust for future generations. The first such body was established in 1895 in the United Kingdom as a charitable association, and as a statutory body by an Act of Parliament of 1907. Under this Act as amended, the Trust enjoys a number of privileges and exemptions which place it in a unique position among land-holding private organisations.

The National Trust is not only a major landowner but has also been empowered by special legislation to declare any of its landholdings inalienable. Where this has been done, not only does this prevent the Trust from disposing of its property, but it also makes expropriation legally impossible unless Parliament itself adopts an Act to the opposite effect.

The example of the National Trust was followed by a certain number of British Commonwealth jurisdictions, such as the States of South Australia and Victoria in Australia, New Zealand and the Bahamas.

However, the right to declare real property inalienable does not seem to have been recognised in law for these other Trusts, except in the Bahamas. Other privileges include the right to make by-laws or regulations, to enter into covenants with other landowners, and to benefit from tax exemptions.

The right to make regulations, which has been granted to the National Trusts in the United Kingdom, the Bahamas and South Australia, is particularly unusual since these bodies are private organisations. Since their property is open to the public, the main purpose of this right is of course to enable them to control the behaviour of visitors and to prevent trespassing. For example, the Council of the National Trust of South Australia may make regulations to prevent damage to property vested in the Trust, to regulate traffic on its property, to prevent persons from injuring, destroying or taking animals, birds, trees and plants and to establish admission fees.

Most of the laws establishing National Trusts empower these bodies to enter into restrictive covenants with other landowners over their property and to enforce such covenants in perpetuity and against all successors in title. There is no requirement for contiguity of the land concerned. Covenants, as seen above, may only provide for negative obligations. However, they can be extremely useful to National Trusts as an inexpensive way of securing important habitats, since no payment is due to the owner. For the landowner concerned, this is a means to ensure that his land will be preserved after his death, although he continues to own it. In addition, a parcel of land burdened with a covenant will lose a part of its market value and the land taxes payable by the owner may be reduced accordingly. National Trusts often assist the owners of covenanted property to manage their land in an environmentally friendly way. In Great Britain, the National Trust generally tries to negotiate covenants in respect of land surrounding the property which it has declared inalienable, to provide it with a sort of buffer zone.

National Trusts also benefit from important tax concessions. In particular, they are generally exempted from land or property tax, estate or succession duties and capital gains taxes on gifts and bequests. This provides a considerable incentive to private landowners to transfer their property to the Trusts.

In the United Kingdom, Land of Outstanding Scenic or Scientific Interest (LOSI), together with buildings and works of art, may be given inter vivos or bequeathed to the nation in lieu of taxation. The State reimburses the Inland Revenue according to the value of the property

253

transferred to the State in lieu of tax, and then frequently hands over the property to the National Trust. However, taking estates into public ownership or passing them to the protective ownership of the National Trust can be relatively expensive. In 1992, for example, the State lost some £80 million by accepting just six properties in lieu of taxation, which were then passed on to the Trust.

Furthermore, no tax is payable on gifts or legacies made directly to the National Trust in the United Kingdom.

To illustrate the role that National Trusts may play in the safeguarding of natural areas, the National Trust in England protects nearly 300,000 hectares of land by ownership, including one eighth of the coastline of England, Wales and Northern Ireland. It also protects a further 35,000 hectares by means of restrictive covenants, which bind the owners or occupiers of the land to preserve its natural amenity. About one third of this land is included in SSSIs. The National Trust also owns at least 57 national nature reserves.

National Trusts enjoy exceptional privileges as a result of their special status. However, a certain number of countries now grant tax exemptions to ordinary conservation NGOs and sometimes also assist them by paying them subsidies for the acquisition or management of natural areas.

3. Tax Exemptions and Subsidies

The great majority of the States in the United States exempt charities, including conservation organisations, from property taxes for the land they hold or use for public purposes. Of particular interest in the United States is the possibility of deducting from taxable income gifts made to Government agencies or non-profit-making charitable organisations, including conservation NGOs. The federal Tax Code explicitly provides that gifts made for the protection of natural environmental systems are deductible in this way. Such gifts may include conservation easements.

The rules applicable to these deductions are fairly complex and cannot be considered in detail here. Summarising them briefly, it can be said that the donor may deduct as much as 30% of the value of the gift in a year. However, since natural persons may spread the deduction over a period of six years, the full value of the donated land becomes totally deductible in practice. Although the Tax Reform Act of 1986 has eliminated certain kinds of deductions and has imposed certain limitations on others, the system still continues to provide considerable incentives for the donation of land or easements for conservation. There are similar provisions in some of the State tax codes.

In rare cases, NGOs are allowed by law to acquire easements from private landowners without the usual requirement of the existence of a contiguous dominant tenement. This is possible, in particular, in a certain number of American States where legislation specifically authorises non-profit-making organisations to hold easements for the purpose of conserving natural resources. In Switzerland, where the Civil Code is exceptionally liberal in this regard, conservation NGOs have entered into many contracts establishing easements on individual landowners' land.

Perhaps one of the best examples of an organisation which has specialised in the acquisition of land for conservation and, to that end, uses all the possibilities provided by legislation with regard to tax exemptions and the creation of easements is The Nature Conservancy in the United States. This organisation has now become a major owner of land, half of which has been obtained by donations. It concentrates on the acquisition of areas which are important for the preservation of biological diversity.

Subsidies are another important means of assisting NGOs to purchase and manage land for conservation. The laws of several countries expressly provide that public conservation agencies are empowered to make grants to non-profit-making private organisations.

In the United Kingdom, for instance, the Countryside Commission and English Nature and its sister bodies are authorised to make grants to the National Trust and various NGOs to assist them in the acquisition of natural areas. The land thus purchased must be managed according to approved management plans. When making grants, English Nature and its sister bodies give priority to the acquisition of SSSIs.

In Denmark, the Nature Protection Act of 1992 allows the Ministry of the Environment to pay grants and make loans to NGOs for the acquisition, management and restoration of natural areas.

It is perhaps in the Netherlands that the grant system has reached its maximum degree of development, as it is established Government policy to rely heavily on NGOs for the preservation and management of natural areas. The major land-holding NGO is called "Natuurmonumenten" which, together with the Dutch provincial conservation organisations, owns and manages most of the privately-owned conservation areas. The total area covered by such conservation areas is almost equal to that of State-owned reserves.

Provided sufficient money is available, NGOs obviously enjoy a large degree of flexibility in their decisions to buy land for conservation, in contrast with Government departments and public agencies which are often hampered by budgetary constraints and administrative procedures. It is therefore much easier for a Government to subsidise private organisations to buy land rather than to purchase the land itself. If necessary, the public conservation agency will generally be able to acquire the land from the NGO subsequently: with the proceeds of such sales, the NGO will then buy new land. This is, of course, also possible when an NGO is not subsidised.

The acquisition and resale of land for conversation is a common practice, as mentioned above, for certain large conservation organisations, such as The Nature Conservancy in the United States. The organisation buys land when it comes on the market and sells it back to federal, State or local Government departments when the latter have the funds to enable them to do so. Important opportunities to acquire areas of value for conservation would otherwise often be lost.

CHAPTER IX
THE CONSERVATION OF MARINE AND COASTAL AREAS

A. General Legal Considerations

The jurisdictional rules of the Law of the Sea Convention over different parts of the marine environment have been already been explained in section (D)(4) of the Introduction, but will be briefly summarised here to illustrate the legal problems surrounding the establishment by States of protected areas in the sea.

The marine zone closest to the shore is called the "internal waters" and encompasses all waters on the landward side of the baseline, generally up to the high water mark or the salinity limit of the river mouth. The normal baseline is the low water mark along the coast. Internal waters and land areas have the same legal status. Coastal States may therefore establish protected areas in their internal waters, without any restrictions derived from international law.

The next zone is the "territorial sea" which may extend to a distance of twelve nautical miles from the baseline. This has a very similar legal status to that of internal waters and is also considered to form part of the territory of the State concerned. Coastal States may therefore establish protected areas in this zone as they please, provided that they do not prohibit the innocent passage of foreign ships through these waters.

The Exclusive Economic Zone (EEZ) may extend to a maximum distance of 200 nautical miles from the baseline. In their EEZ, States only have jurisdiction with regard to fisheries and the preservation of the marine environment. It is therefore not possible to establish protected areas in this zone if this would entail imposing restrictions on the navigation of foreign vessels. However, the competent State may of course close any part of the EEZ to fishing.

Notwithstanding, the Convention does provide for the possibility of establishing special areas in the EEZ for the purpose of preventing pollution. In these areas, mandatory rules for the prevention of pollution from vessels may be adopted by coastal States, but only where this is justified by oceanographical and ecological conditions, the protection of living resources and the special character of ship traffic. However, the designation of such areas must first follow a long and protracted procedure involving the approval of the designation by the competent international organisation, namely the International Maritime Organisation which is a United Nations agency.

As mentioned earlier, coastal States enjoy sovereign rights over the continental shelf for the purpose of exploiting its natural resources, including the sedentary species found there. The shelf is deemed to extend to the 200 mile limit, but may in practice extend much further. Sovereign coastal States are perfectly entitled to establish protected areas on the shelf, even where it extends under the high sea, provided that the rights of other States in or on the superjacent waters are not affected.

Comprehensive as it may be in its treatment of most matters, the Law of the Sea Convention is remarkably silent on the subject of protected areas. As mentioned in Part II, Chapter I, article 194.5[47] is the only provision to address the problem of habitat conservation. Not a single one of the many provisions dealing with fisheries mentions the need to protect important habitats, such as spawning or nursery areas, for species of commercial importance.

From the point of view of national law, the sea-bed and sub-soil as far as the edge of the continental shelf belong to what is commonly called the public maritime domain. In other words, these are both owned by the State and no problem of private ownership can therefore arise. In contrast, the water column usually has an uncertain status, often being considered as common property, although in certain countries, such as Spain, it too belongs to the State.

The public maritime domain generally extends landward as far as the high water mark, although there are differences between countries in that respect, but rarely goes beyond that limit. This means that beaches, dunes, cliffs, coastal wetlands and other near-shore habitats do not usually belong to the public maritime domain. Instead, they may be owned by various public departments, municipalities or private persons.

Reflecting the difference in ownership on either side of the upper boundary of the public maritime domain, different agencies have jurisdiction over the areas concerned. To the seaward side, there is generally a single agency which is competent in all matters relating to the use of the public domain and its superjacent waters. There is also often a special police force to enforce maritime legislation. However, fisheries sometimes come under a separate ministry or under a different department within the same ministry. Landwards, the administrative system is the same as in any other part of the country, with no specific agency usually being given responsibility for the coastal area.

The upper limit of the public maritime domain thus constitutes a most formidable legal barrier, separating not only the land from the sea but also the different public administrations.

This makes it constantly difficult and often impossible to provide for the integrated management of the coastal ecotone and especially to establish protected areas straddling both land and sea. As a result, most of the national parks and reserves which have been created on the landward side stop at the sea edge, whereas marine nature reserves do not go beyond the high water mark. In the relatively rare cases where legislation allows for the setting up of mixed areas covering both land and sea, conflicts between agencies having jurisdiction over different parts of that area may render management difficult.

These problems are of course exacerbated in federal or regionalised States, where some powers have been vested in the federated entities whilst others are retained by the federal or central Government. In Italy and Spain, for instance, jurisdiction over the establishment of terrestrial protected areas is, with minor exceptions, vested in the Regions, whereas the public maritime domain belongs to the national Government. This made it impossible to establish single or unitary reserves extending over both land and sea in Italy, until the Protected Areas Act of 6 December 1991, discussed below, and it is still impossible in Spain.

---

[47] Which was used as the basis for developing the Protocols on Marine Protected Areas to the Regional Seas Conventions.

It is therefore necessary to examine separately the terrestrial and marine parts of the coastal area, as far as legal conservation instruments are concerned, and then to assess to what extent innovative legislation could be developed to overcome the problems posed by the existence of the jurisdictional barrier between the two environments.

B. The Sea

As the State is the owner of the public maritime domain, no problem of private ownership arises. The State is the legal master of these areas and, in the exercise of its police powers, may regulate therein any human activities as it pleases, subject to its obligations under international law. The State can, of course, also exploit the resources of these areas directly or through concessions that it grants to public or private bodies or individuals. There are therefore no legal difficulties in the establishment of marine areas in which certain activities are prohibited or otherwise regulated.

Despite the foregoing, only a relatively small number of countries have so far enacted legislation enabling them to create national parks or other protected areas in the sea.

In contrast, most if not all coastal States have enacted legislation empowering them to regulate fishing in their waters and their EEZ. These laws always empower the authority responsible for fisheries to establish closed areas where fishing is prohibited or strictly controlled. When these areas are set up on a permanent basis, they are often considered to be marine reserves. Where the only threat to an area is overfishing or the use of certain types of fishing gear, such as bottom trawls which may cause considerable damage to sea-bed habitats, this method may be sufficient to protect the area concerned. There are many marine areas in the world which are protected in that way.

Nevertheless, there are many other activities which may be harmful to the marine environment and which, in many cases, should also be controlled if certain areas are to be effectively protected. These include the extraction of sand and minerals and other disturbances of the sea-bed and sub-soil; the anchoring of ships which may, taken cumulatively, be highly destructive to sea-bed communities such as coral reefs, sea-grass and sea-weed beds; and of course pollution, whether originating from ships or land-based sources.

Fisheries legislation is generally unable to cope with these threats as that is not its purpose. Other laws applicable to activities in the sea can sometimes be used to solve particular problems. Some countries, such as France and Italy, have for instance granted concessions to scientific or conservation organisations for the exclusive use of certain marine areas. However, such concessions can only ever be of a temporary nature, as the law never allows the public domain to be permanently assigned to a third party.

Different instruments are sometimes used in combination. In France, there is a marine reserve at Carry le Rouet on the Mediterranean coast, where no less than three distinct legal instruments were used to ensure its protection: a concession for scientific research to a private body, an order prohibiting fishing and another order prohibiting dredging and the movement of ships.

It follows that none of these legal instruments is sufficiently well adapted to the conservation needs of marine areas. Special legislation would therefore seem to be widely required to achieve that purpose, for which two important elements should be taken into consideration. Legislation should first provide for the possibility of establishing protected areas in the sea, and special rules should be adopted to meet the particular conditions, both ecological and legal, that prevail in the marine environment.

In many countries, protected area legislation contains no specific provisions for the creation of marine parks or reserves, and the department or agency responsible for the implementation of the law, which is often the Forest Department, has no jurisdiction over the public maritime domain. In a few other countries, the law has expressly been made applicable to both land and marine areas, but still does not contain specific provisions relating to the sea. As the authority in charge of applying the law usually has no scientific competence in marine matters and its jurisdiction over marine areas is moreover often doubtful, generally having to be shared with at least one other department, it is not surprising that the authority in question usually shows little inclination to establish marine parks or reserves.

In consequence, there seems to be no other alternative than to enact special legislation dealing specifically with marine protected areas, to be implemented by an authority or agency that has effective jurisdiction over the marine environment. The countries that have done so are still relatively few, but their number is increasing as the need for such special legislation becomes more widely recognised.

Examples of modern laws providing for the establishment of marine protected areas are those of the United States, several Caribbean island nations, Malaysia, Australia, New Zealand, Italy and the United Kingdom. Their characteristics will now be briefly reviewed.

These laws generally provide for the designation of marine protected areas by ministerial orders or other form of regulations. Prohibitions generally apply to fishing, the destruction of fauna and flora, particularly on the sea-bed, dredging, mining, the extraction of sand and gravel, the release or dumping of pollutants and waste and, more generally, to any activity which may cause harm to the natural environment. There are often some exceptions, particularly with regard to fishing, which may remain authorised when compatible with conservation objectives. This is normally the case for line fishing by amateurs or for traditional fishing by local fishermen. Spear-gun fishing, with or even without aqualung equipment, is almost universally prohibited and diving often regulated.

The main problem would seem to be the regulation of navigation, especially the anchoring or mooring of ships when this may damage vulnerable habitats. The only way to ensure that ships do not release or dump pollutants or waste may be to impose restrictions on navigation in certain areas, as the no-dumping rule is extremely difficult to enforce except when the culprits are caught red-handed which is of course infrequent.

However, there are often strong objections to the placing of prohibitions or restrictions on navigation and mooring, on the grounds that these run counter to the freedom of navigation enshrined in the new Convention on the Law of the Sea. Nevertheless, a few laws do restrict navigation in marine protected areas which proves that it can be done, presumably in the exercise of the State's jurisdiction under the Convention to protect the marine environment. Malaysia, for instance, in its Fisheries Act of 1985 prohibits the anchoring of ships on rock or coral reefs. The Fish and Fisheries Act of the Northern Territory of Australia provides that regulations may prohibit or restrict access to any aquatic park. In Italy, strict nature reserves may be established within marine protected areas, in which access is prohibited not only to ships but also to swimmers.

It would seem, therefore, that at least in internal waters and the territorial sea where the coastal State exercises sovereign rights, restrictions on navigation which are specifically enacted to preserve a marine protected area do not conflict with rules of international law, unless of course, these result in the denial of innocent passage. Coastal States find it quite natural to close certain areas to navigation for national defence reasons. There is nothing that prevents them from doing so for conservation purposes.

The situation is obviously different in the EEZ where, in principle, there should be no navigation restrictions. The only exception is, as mentioned above, where a special procedure has first been followed in order to establish special areas for the control of pollution. However, nothing prevents coastal States from designating protected areas in which fishing may be controlled as well as any activity affecting the sea-bed, since the States have sovereign rights over all living resources and all the resources of the continental shelf, whether living or mineral. For the time being, there seem to be few, if any, marine protected areas established in the EEZ of any State. Notwithstanding, the laws of certain countries, for instance Saint Lucia and Malaysia, do allow for this possibility.

Most of the specific laws that provide for marine parks or reserves settle the jurisdictional matter by appointing the fishery department or the administration specifically in charge of the sea as the authority responsible for the establishment and management of such areas. An unusual system exists in the Bahamas, where the land and waters of the public maritime domain may be leased to the Bahamas National Trust. The Trust then has the power to make by-laws prohibiting or restricting damaging activities, and also to manage the areas concerned in the interests of conservation. However, it cannot regulate navigation.

Some recent laws have developed management systems for marine reserves which are based on corresponding practices in terrestrial protected areas, particularly the practice of zoning. In Italy, marine reserves are established pursuant to an Act of 1982 on the Protection of the Sea and the Protected Areas Act of 1991. The former law simply provides that in such reserves, any activity may be prohibited or made subject to a permit, and that these requirements may be varied according to the different zones created within the reserve.

For example, the Ustica reserve in Sicily has three zones. In Zone A, which is a strict nature reserve, access by ships, boats and swimmers is totally prohibited, together with any modification of the environment, whether by direct or indirect means, and the release of any substances that may result in changes in the character of the area, even temporarily. In Zone B, called the "general reserve", spear-gun fishing is prohibited and the use of other fishing methods is subject to a permit. Fishing with hooks and lines remains authorised. In Zone C, the "partial reserve", commercial fishing is subject to a permit but sport fishing may be freely carried out, subject to any restrictions which may be required under general fishery legislation. Stricter rules may, however, be imposed if the situation so requires.

The problem of damage caused to a marine reserve by activities exercised outside its boundaries may be particularly difficult to resolve, as this entails the need to impose prohibitions or restrictions on such activities, although they may occur a long distance away. Few laws provide for this possibility. An example is the Port Launay Marine National Park Regulations of the Seychelles, which prohibit the discharge of pollutants outside the park where these may cause pollution in the park.

The overwhelming majority of marine protected areas are parks and reserves of the conventional type, although there are a few exceptions. The zoning system applicable in the Italian reserves, for example, is clearly representative of a trend towards the better adaptation of regulatory measures to the specific requirements of each particular zone, but Italian reserves are small. There would nonetheless seem to be a need to extend the nature park concept to the sea, which has already become increasingly popular and, it appears, effective on land. This has already been done in the Tuscan Archipelago National Park, discussed below.

There is at least one major example of the use of the nature park instrument in the marine environment: the Great Barrier Reef Park in Australia. The Park extends over a very large area along almost the whole length of the coral barrier in eastern Australia. There are hardly any prohibitions applicable to the entire area, except those relating to mineral exploration and exploitation. On the other hand, there is a sophisticated zoning system with zones ranging from strict nature reserves, where access is forbidden except for scientific research, to zones with very few use restrictions, if any. The whole area is managed by a special authority, the Great Barrier Reef Marine Park Authority, which is empowered to grant permits for activities subject to authorisation and also to enforce the regulations applicable to the Park. Just like a nature park on land, the Barrier Reef Park is therefore not closed to human activities, except in small reserve zones. Instead, the Park constitutes a marine space in which activities may be regulated and controlled so that its values remain preserved and the public may continue to enjoy it.

Conflicts between different uses of the marine environment, especially near the shore, are bound to increase with time and the development of water-based activities, particularly recreation. The need for zoning systems similar to terrestrial zoning plans is therefore likely to become more and more apparent. Such systems may be instituted within the marine equivalent of nature parks or may even stand on their own as specific sea-use plans.

In France, a law provides for the development of plans of this kind which are called "sea development schemes" (schémas pour la mise en valeur de la mer). These schemes apply to particular areas which are divided into zones, and specified uses are laid down for each zone, together with measures for the protection of the marine environment. The schemes are adopted by the national Government and are binding on the local authorities concerned.

Another method, also used in France, which can easily be combined with a zoning system consists of protecting certain particular marine habitat types. These are listed in the Coastal Act of 1986 and its implementing regulations, and include all areas with natural concentrations of animal and plant species, such as sea-weed and sea-grass beds, spawning areas and nurseries, natural live shellfish beds, as well as lagoons and coral reefs. The presence of these protected habitats must be taken into consideration when developing the sea-use schemes mentioned earlier. In addition, any extraction of materials from the sea, such as sand and gravel, must be prohibited or restricted where this is liable to affect the integrity of these protected habitat types, whether directly or indirectly.

Coral reefs are often the subject of specific conservation measures. In the United States, the Fishery Management Plan for Coral and Coral Reefs of the Gulf of Mexico and the South Atlantic was developed pursuant to federal fishery legislation. The Plan calls for the designation of coral habitats as areas of particular concern in which specific restrictions apply. In particular, it is prohibited to fish with bottom long lines, traps, pots and bottom trawls in those areas. The taking of coral is also forbidden.

In Bermuda, the extraction of coral is prohibited throughout the fishing zone of the islands. In the Australian State of Queensland, a permit is required to take coral, coral limestone and coral sand. In the Philippines, the collection of coral is prohibited, except under a permit.

Mention should also be made of the increasing trend to prohibit or restrict trawling in near-shore areas. This is a matter which is exclusively regulated by fishery departments. In Spain, for instance, regulations adopted in 1988 completely forbid trawling in areas shallower than 50 metres. The objective of this type of measure is twofold : to increase the survival rate of young fish, which are often caught indiscriminately by these nets, and to preserve sea-bed habitats from the destruction caused by the dragging of these nets on the bottom.

C. The Shore

The establishment on coastal land of national parks, nature reserves or nature parks is governed by the legislation which is generally applicable to these types of protected areas, and there are no specific rules deriving from their particular location near the sea.

The other instruments available for the protection of terrestrial, freshwater or brackish water habitats along the coast have already been examined in the earlier chapters of this presentation. It may, however, be useful to review them briefly again here in the specific context of the protection of the coastal zone.

It is necessary to identify the potential afforded by the public maritime domain, or at least the natural components of the domain, as it also generally includes ports and harbours, dykes, jetties, lighthouses and other manmade structures.

The first possibility is to extend the public domain landward to cover all natural ecosystems under the influence of the sea, including beaches, dunes and dune slacks, salt marshes, coastal wetlands and cliffs. As will be recalled, this is what has been done in Spain on the basis of an article of the national Constitution, which provides that the whole of the "maritime-terrestrial" zone and the beaches belong to the State.

This concept was further elaborated by the Coastal Act of 1988 which incorporates dunes, coastal wetlands, cliffs, islets and other features into the public domain. The Act provides, in addition, that the protection of the maritime-terrestrial domain must be understood as comprising the preservation of its natural elements and characteristics. For that purpose, the Act establishes a one hundred metre-wide protection zone measured from the sea-shore, within which the construction of dwellings and roads and other activities are, in principle, prohibited. Exceptions to this rule may be made in the public interest if they do not affect the protection strip lying behind beaches, wetlands and other areas of special interest.

Another possibility is to enact special rules for the protection of certain coastal habitat types. Examples are the laws of many States of the United States which establish permit requirements for activities that destroy or alter coastal wetlands, as well as the Danish Nature Conservation Act which similarly protects salt marshes and other natural habitats. The laws protecting mangroves and coral reefs in certain tropical countries belong to the same category.

Restrictions on the destruction of particular habitat types may be combined with land-use planning rules containing special provisions applicable to coastal areas. The French Coastal Act of 1986, for example, requires that areas containing natural or semi-natural habitat types along the coast, including dunes, wetlands, forests, cliffs and many others, be protected by their inclusion in the zones of highest protection in municipal zoning plans, and that the grant of permits for activities in these areas be compatible with the conservation of habitats. In Portugal, most of these coastal habitats must be included in the "National Ecological Reserve", which means that most human activities in these areas will be severely restricted. In certain Spanish Autonomous Communities, such as the Balearic islands, some coastal areas have been designated as special land-use units where severe restrictions apply, particularly with regard to certain habitat types such as wetlands or cliffs.

In an increasing number of countries, although no specific provisions are made for the conservation of particular habitat types, new legislation has been enacted to prevent excessive development in the coastal zone.

Examples of this approach include the Coastal Zone Act of 1990 in Turkey, and the legislation of many Latin American countries. In Brazil, the constitutions of several federated States consider coastal areas as zones of particular concern, in respect of which special planning legislation will need to be enacted. In Costa Rica, an Act of 1977 establishes a special regime for the maritime-terrestrial zone. Within the first 50 metres of that zone, measured from the mean high tide line, development is generally prohibited. In the next 150 metres, development is subject to a permit system based on local land-use plans. Mangroves are not covered by the Act and are managed by the Forest Administration. However, no development permits can be granted in respect of mangrove areas.

In France, in addition to the provisions relating to habitat types, the Coastal Act severely regulates development in those parts of the coasts which have not yet been urbanised. In Spain, the purpose of the establishment of the protection strip along the coastline is also to limit construction. In Denmark, construction prohibitions in the coastal zone have recently been extended further inland. In Turkey, the Coastal Zone Act of 1990 also provides for specific construction restrictions in the coastal area.

In the United States, the federal Coastal Zone Management Act of 1972 provides for a system of incentives for States that establish particular land-use controls along their coasts. To benefit from federal assistance, States must designate areas of particular concern including exceptional, rare, fragile or vulnerable natural habitats; areas of high natural productivity; essential habitats for wild species; coast protection areas, particularly dunes, reefs, beaches, sand-banks and mangroves; and zones necessary for the maintenance or supply of coastal resources, such as coastal flood plains and zones essential to the recharge of the water table. The areas in question must be preserved through land-use regulations adopted by the State or by local authorities made in accordance with State regulations or with the State's approval. Most of the coastal States of the United States have availed themselves of this opportunity and have now developed their own coastal zone programmes approved by the federal Government.

The California Coastal Act of 1976 establishes a permit system for most activities that may be carried out in the coastal zone. Land-use plans and regulations are prepared by the local authorities, under the control of a Californian Coastal Commission. The plans must be approved by the Commission.

In the Australian State of South Australia, a Coast Protection Act was enacted in 1972. The Act establishes a Coast Protection Board whose functions are, *inter alia*, to protect the coast against erosion, damage, deterioration or misuse, and to restore damaged areas. A management plan must be drawn up and adopted by regulations for each Coast Protection District. The Board has the status of a Planning Authority. It may acquire land and carry out works to implement the management plan. Construction, excavations and mining, as well as works that will change the nature, configuration or use of the coast, are subject to permits from the Board. Restricted areas, in which access may be prohibited or restricted, may be designated by the Minister in charge. The Board is financed by a Coast Protection Fund which is supplied by budgetary allocations.

The laws of certain countries establish special acquisition programmes by Government agencies to preserve coastal land from potential development. In France, this is the sole function of the Conservatoire du littoral et des espaces lacustres which is, as seen above, an autonomous Government body financed by the general budget. In the United States, the Coastal Zone Management Act provides for the conservation of remaining natural estuaries through grants made to the States for the acquisition of the land necessary for this purpose. Once secured, the estuaries concerned and adjacent associated habitats must be managed as Research Natural Areas.

Specific incentives or disincentives may also be used in coastal areas. One of the best examples is the removal in the United States of all federal subsidies for activities relating to the development of certain barrier islands, including a prohibition against federal agencies carrying out, authorising or financing any public works in these areas.

Finally, voluntary measures may also be used for the protection of the coast. The National Trust in the United Kingdom has acquired more than 800 km of coastline, largely after a special fund-raising campaign called "Operation Neptune".

Also in the United Kingdom, the Countryside Commission has defined a certain number of Heritage Coasts in England and Wales. This definition has no legal effects, although it must be endorsed by the local authorities concerned. Once a coast has been defined, a Heritage Coast Officer is appointed and a management plan prepared. There are no specific development restrictions, but local authorities are encouraged not to grant planning permission for operations incompatible with the management plan. All other conservation and management measures are voluntary and therefore depend on persuasion. The role of the Heritage Coast officer is of considerable importance in that respect. In 1992, there were 43 Heritage Coasts covering over 1400 km of coastline in England and Wales: there are no Heritage Coasts in Scotland or Northern Ireland. Some of the National Trust coastal land is of course included in Heritage Coasts.

D. Bringing Land and Sea Together

The land-sea jurisdictional split, widespread as it may be, is not the result of a universally applicable rule of law. The existence of this quasi-insurmountable legal barrier is due more to history and administrative convenience or routine than to substantial legal principles. Solutions can be found, therefore, if the administrations concerned agree to relinquish some of their powers for the sake of the unitary management of the land-sea interface.

The first possibility consists in providing for a marine protected area to extend to adjoining land areas beyond the high water mark or upper limit of the public maritime domain, where this is necessary for the management of the interface. Provisions along these lines appear in the fishery legislation of several Caribbean island States, such as Antigua and Barbuda, Saint Lucia, Saint Vincent and the Grenadines, and Trinidad and Tobago. Pursuant to these laws, marine reserves may include adjacent or surrounding land.

The Law of Trinidad and Tobago of 1970 on the Preservation of Marine Areas allows for the designation of restricted areas in the territorial sea and on any adjoining land or swamp areas forming a single ecological entity with certain submarine areas. Clearly, however, the extension of a marine protected area onto adjacent land will generally not be practicable, as it is impossible to bring the control of many land-based activities under the jurisdiction of marine or fishery agencies. This means that landward extensions may most often only concern limited areas of adjoining land.

Secondly, it is possible to extend a coastal or island terrestrial park or reserve into its adjacent waters. This may be achieved by the creation of a marine protected area which is legally distinct from the terrestrial one and is, as a result, managed by the department in charge of the sea. In Turkey, for example, coastal national parks which are administered by the Forest Department are bordered in the sea by a 200 metre strip in which commercial fishing is prohibited under fishery legislation. Another example is that of the Tunisian National Park of Zembra and Zembretta, in which the island of Zembra is surrounded by a protection zone where all fishing is prohibited, again under fishery legislation.

In both the above cases, the instruments establishing the terrestrial and marine protected areas were adopted quite independently by each of the agencies concerned. The management of the areas concerned as units therefore depends on the willingness of these agencies to co-operate with each other on a purely voluntary basis.

A more integrated method appears in the Act creating the Doñana National Park in Spain, which establishes a one nautical mile buffer zone along its shores. The Park management plan, as approved by royal decree, provides that fishing and the use of boats will be jointly regulated in that zone by the park authority and the department in charge of navigation and fisheries.

The laws of certain countries go further and broadly provide that terrestrial national parks and other protected areas may extend to the adjoining public maritime domain. However, there are often legal restrictions that make this difficult to achieve in practice.

In Italy, for instance, State-owned reserves may extend to the adjacent sea. This was formerly impossible for those reserves established by the Regions, as most Regions have no jurisdiction over the public maritime domain. However, the new framework Protected Areas Act of 6 December 1991 now makes it possible for regional parks to be established not only on land, rivers and lakes but also in adjacent marine areas.

In France, the only park which encompasses both land and marine areas is the Port Cros National Park in the Mediterranean. The Park is a small island with a 600 metre-wide surrounding marine area. The decree establishing the Park only prohibits spear-gun fishing and the use of certain gear, including trawls, in the Park waters. All other regulatory measures are to be taken by the competent agencies in charge of marine matters. These include the regulation of other fishing methods and the movement and mooring of ships, particularly pleasure crafts which visit the area in very large numbers and cause considerable damage to sea-grass beds when they anchor in the Park. The Park authorities therefore have no control over these practices.

However, there are now a few examples of laws that provide for the establishment of protected areas covering both land and sea areas as single units. Such examples include the law of Israel on Nature Reserves, the Turkish and Greek laws on the Protection of the Environment of 1983 and 1986 respectively, and the Spanish Act on the Conservation of Natural Areas and of Wild Flora and Fauna of 27 March 1989.

Taking the latter first, the Spanish national Government is now empowered to establish protected areas in the sea. It may also create national parks on land to preserve areas which are representative of the main Spanish natural systems, including coastal areas and the continental shelf. All other forms of protected areas can only be established by the Autonomous Communities. The first land and sea national park to be established pursuant to these new provisions was created in 1991 to protect the small archipelago of Cabrera and its surrounding waters in the Balearic islands. The whole of the park is to be administered by ICONA, the central Government administrative body in charge of national parks, and all activities within the park are to be regulated under a single management plan.

In Turkey, a number of Special Protection Areas covering both land and marine environments have now been established under the 1983 legislation on the Protection of the Environment.

In Greece, the Act of 1986 empowers the Government to establish similar areas. A national park has been established to protect the Northern Sporades islands and their adjacent waters in the northern Aegean Sea, by means of a presidential decree providing for permanent conservation measures. The park is organised like a nature park, with different zones on land and in the sea which are subject to different regulations according to the specific conservation requirements of each zone. There will clearly be a need for a single body to manage, or at least to co-ordinate, the management of the park as a single unit, but this does not seem to be envisaged in the immediate future. This creates a risk that the various departments or agencies that have jurisdiction over the many different activities to be carried out in the park will continue to act independently, without having enough regard for the global objectives of the protected area.

Another example of a terrestrial and marine park based on the nature park concept is the new Italian Tuscan Archipelago National Park, which was established in the Tyrrhenian Sea in 1989. The park comprises some of the islands of the archipelago and the surrounding terrestrial sea down to a depth of one hundred metres. The land and sea areas are zoned and different rules apply to each zone. There are strict nature reserves on some of the islands and in the adjoining sea. A park authority is to be appointed to manage the park as a single unit.

Where for legal reasons, especially in federal States where the federated entities have jurisdiction over land areas and the federal Government over the sea, it is impossible to establish and manage a land and sea park in a unitary way, it should always be made possible to create juxtaposed terrestrial and marine protected areas and to set up a joint agency to co-ordinate the management of the whole.

In Spain, for instance, the Columbretes Archipelago and the surrounding sea area are protected by a law of the Valencia Autonomous Community for the islands, as well as by another law enacted by the central Government in respect of the adjacent waters. There is a joint management commission composed of State and Regional representatives to manage the marine area, and there is a central Government representative on the managing body of the land area.

In view of the difficulties faced by most countries in overcoming the jurisdictional barrier that separates the land from the sea, even in the relatively limited case of protected areas, it is not surprising that there are very few jurisdictions where the whole of the coastal zone on either side of the coastline is managed by a single authority.

Nevertheless, there are a few cases where this has been achieved. One example is that of the American State of California, where the California Coastal Commission has jurisdiction over the whole of the territorial sea of the State, all islands and a 1000 yard-wide strip of land, measured from the mean high water mark, except in the cities. In some cases, the width of that strip can extend as far as 5000 yards inland. The carrying out of any works in the coastal zone as thus defined requires a coastal development permit.

In South Australia, the Coast Protection Board has jurisdiction over a distance of three nautical miles seaward of the mean high tide mark, and one hundred metres on land on the other side of that limit. All works within that area are subject to a permit from the Board.

CHAPTER X
CONCLUSION

Recent years have been characterised by a considerable increase in the use of new legal instruments for area conservation by the developed world. These include special regulatory measures for the protection of certain habitat types, particularly wetlands; the gradual integration of conservation into land-use planning in certain countries; the expansion of the nature park concept; and an increased reliance on voluntary instruments, especially management agreements, to promote conservation on private land.

The main conservation objective of many developed countries is now the safeguarding of remaining important natural habitats and the habitats of endangered species. The new Habitats Directive, which was adopted by the European Community in May 1992, constitutes a major step in that direction.

The greater emphasis which is now placed on the need to preserve habitats is matched by the new trend towards greater extensification of farming practices. Extensification is increasingly considered necessary to restore a certain degree of naturalness to the general countryside, to abate pollution by fertilisers and pesticides and to resolve, at least partially, the unwieldy problem of agricultural surpluses.

The strategies adopted by the countries concerned all have now the same objectives, although the conservation methods used may vary. The United Kingdom mostly relies on the SSSI system and the conclusion of management agreements, but has developed a successful extensification scheme in recent years which is heavily based on financial incentives. Denmark now protects all natural habitat types and uses conservation orders extensively to preserve natural areas. Switzerland is drawing up inventories of all remaining important natural sites and will use different methods to protect them, albeit with an explicit preference for voluntary conservation through the conclusion of contracts.

The case of the Netherlands is perhaps unique in that it is apparently the only country at present where nature conservation has become a major component of government policy through the development of an overall National Nature Conservation Plan which sets out a number of important objectives for the next thirty years or so. The goal of the Plan is to achieve the sustainable conservation, rehabilitation and development of nature and the landscapes in the Netherlands. More than 250.000 hectares of land will be designated as Nature Conservation or Nature Management Areas. That land will be acquired by the State or by conservation NGOs with government subsidies, or be the subject of management agreements with the farmers. In addition land will also be acquired for "nature development" that is to say for the re-creation of natural or semi-natural habitats. Finally a system of ecological corridors will connect the different elements of the system.

The protection of areas in industrialised countries will increasingly depend on a combination of instruments, including public ownership, regulatory measures, planning, contracts and financial incentives. All these should contribute to the building up of a coherent framework within which the main instruments may be tuned to maximum effectiveness.

For the time being, this integrated approach is still applied unevenly, with a few countries leading the way. Nevertheless, the trend is clear. As conventional instruments such as national parks and nature reserves are seen to be insufficient, legislative changes are gradually being introduced to develop new instruments. The nature park instrument is developing fast.

The role of public property remains important. Those areas important for conservation which are already in public ownership should remain so.

The role of private property needs to be developed considerably through appropriate incentives and contractual agreements with particular emphasis on positive management action. Conservation advice to farmers is essential in this respect. This requires an ecological survey of individual farms and the identification of practices best able to maintain, enhance or restore natural environments and biological diversity, as well as advice on how best carry out these practices. To do this, farm advisory conservation services are needed. However, these are still in their infancy almost everywhere.

In the United Kingdom, model demonstration farms have been established by the Countryside Commission and advice for the development of farm conservation plans is now becoming gradually available, particularly in national parks. In Switzerland, a small number of individual farm conservation plans are now in operation. France is developing a similar mechanism for the preparation of 'sustainable development plans'.

The new EC Regulation 2078/92 of 1992, on agricultural production methods compatible with requirements for the protection of the environment and the maintenance of the countryside, now provides the legal and financial basis for the development of farmers' training, farm demonstration projects and providing payments to farmers as remuneration for a service, namely the environmentally friendly management of farm land. Member States may also make aid subject to ecological farm plans. This is clearly a step in the right direction towards the integration of farm production and environmental and landscape conservation.

For such measures to work, there is a need for departments of agriculture and their farm advisory services to accept what amounts to a revolution after decades of promoting intensification, and to become technically trained to implement the new systems. It is also important that a sufficient number of field ecologists should be trained to make the necessary surveys and evaluation of farmland, to identify the best-adapted practices and to monitor their implementation.

However, most of these considerations only concern the developed world. Many of the innovative conservation instruments that have emerged in the past few years may not necessarily be suitable in many tropical countries, where conservation problems and the causes of environmental destruction may be very different.

In developing countries, a lot will of course continue to depend on public ownership. Large traditional national parks should continue to play a major role, but can be supplemented by wilderness areas which do not require considerable wardening expenses and infrastructures.

Wilderness areas can only be established on public lands which are uninhabited, with no roads and to which access is difficult. The designation of such areas implies that the State prohibits itself from developing the area or building roads. The effectiveness of wilderness areas depends greatly on the type of habitat: the instrument clearly works better in dense forest, mountains or marshes than in desert or dry forest, where it may be difficult to control access. In addition, the instrument is not designed to control such activities as shifting cultivation or over-grazing.

A further possibility is the protection forest, or a permanent protection order prohibiting certain activities. These may generally be used on private land. For example, Brazil makes wide use of both instruments, protection orders being known as *tombamento*. Once again, the main problem relates to difficulties of enforcement.

The best suited instrument would seem to be the use of the biosphere reserve concept with a legal basis similar to that of the Nature Park, provided that it is adapted to the specific circumstances of each country. It is not always necessary to adopt the complex structure that Nature Parks have in western countries and Japan. However, it is essential to ensure that local populations are consulted over the establishment of both the park and the zones therein and that they participate in its management. Moreover, these populations should benefit directly from the establishment of such areas. Systems of incentives should be developed as counterparts to the necesssary restrictions.

In some cases, the intensification of certain agricultural practices and the development of village woodlands as alternatives to shifting cultivation or firewood collection should be encouraged and subsidised. In return, villagers would have to commit themselves to maintain specific areas in their natural state and to limit certain uses in other zones.

Such parks should ideally be managed by bodies made up of representatives of local authorities, with a minimum degree of control being exercised by central Government.

Bearing these ideas in mind, it is important to translate into practice the biosphere reserve concept as it has evolved in recent years, and to try to establish successful multi-purpose parks which could serve as field models, particularly in tropical environments.

If this is not done, and for as long as protected areas continue to be considered as a liability rather than an asset by local populations and sometimes even by Governments as well, we may perhaps succeed only in protecting high security areas with barbed wire and armed guards in the middle of an ecological desert.

If, on the other hand, the nature park concept becomes generally accepted because of the economic and social benefits it procures to the local economy, this may well give rise to an increased demand for new parks. With some assistance from the world community, which hopefully will be made available under the new Convention on Biological Diversity, this may not be an unattainable goal.

GENERAL CONCLUSION

The new Convention on Biological Diversity mainly lays down performance obligations and leaves the Contracting Parties free as to the choice of means to discharge their obligations. They must therefore enact or further develop their national legislation for that purpose.

For a long time, the basic instruments have almost universally concentrated on the protection of species and areas. These are two sides of the same coin and are inseparable. However, a new type of basic instrument is now emerging, which seeks to control or manage those processes which are potentially damaging to biodiversity. There is now an obligation under the Convention on Biological Diversity to identify processes and categories of activities which have or are likely to have significant adverse impacts on the conservation and sustainable use of biological diversity and to regulate or manage such processes or categories of activities.

This novel approach is a further step towards integrated conservation and should considerably assist in the resolution of many current problems, such as external threats to protected areas, as well as problems which affect species and natural habitats generally and which are not related to specific areas.

The Australian State of Victoria is probably the first jurisdiction which has developed legislation to cope with the matter of processes in a systematic way. However, many other countries have obviously used legislation to deal with certain processes, mostly by means of prohibitions or restrictions rather than in terms of management. Examples include prohibitions on the use of lead shot in waterfowl hunting, the ban on the use of lead fishing weights in the United Kingdom, bans on certain pesticides such as DDT, and restrictions on the use of off-road vehicles.

Of particular importance are the prohibitions or restrictions on the release and sometimes the import of exotic species which may harm indigenous species and ecosystems. A review of that legislation is outside the scope of this paper and would require considerable research, as applicable measures are often found in separate texts not directly connected with conservation law. The purpose of such texts is often the protection of human interests, such as agricultural production, against introduced pests and diseases.

A. The Legitimacy of Conservation Legislation

The enactment of conservation legislation, and especially the effectiveness and enforcement of the legislation in force, depends to a great extent on the recognition by both the administration and the general public of the legitimacy of conservation itself. This is because legislation imposes restrictions on public freedoms and private property, and limits certain activities or projects which have consequences for the economy or for employment. Moreover, public expenditure on conservation, including the payment of incentives for the safeguarding of biological diversity, may not be considered acceptable if the sums go beyond the nominal amounts which are presently paid almost everywhere.

The reasons for the difficulty in securing such acceptance is that damage to biological diversity does not affect human interests recognised by law.

This is illustrated by the difficulties encountered in respect of the automatic protection of endangered species' habitats or endangered habitat types. In practice, either the protection rule is riddled with exceptions (for example, for farming, forestry, construction and public works) or the conservation authorities do not really dare to implement the law.

As regards species' habitats, France has a legislative provision which protects the habitat of protected species, but hardly uses it. The State of Victoria has been very slow in designating critical habitats. Spain provides for the listing of species vulnerable to alteration of their habitat, but has so far not listed any. In the United States, the designation of critical habitats met with no serious problems as long as only very small areas were concerned. However, the critical habitat of the spotted owl covers many much larger areas and the very legitimacy of the Endangered Species Act has increasingly been put in doubt as a result of the associated controversy.

The fate of biological diversity depends on the recognition of the legitimacy of measures for its conservation, in spite of the restrictions that these may entail. The signature of the Convention on Biological Diversity is a reflection of the consensus on the need for its conservation that has gradually evolved.

However, this consensus exists principally at State level and needs to be supported by public opinion in each individual country. Such legitimacy can only be gained through the demands of the general public, rather than those of a minority of conservationists. Society must recognise that the conservation of biological diversity is at least as important as other interests. Countries in which there is already a strong conservation-minded segment of the public are also the most advanced in conservation legislation and achievements. In contrast, many countries lag behind. It will often take a long time before public opinion is sufficiently mature everywhere, yet we cannot afford to wait.

Measures must therefore be taken to facilitate the acceptance of conservation needs. To disarm opposition to conservation, attempts should be made to eliminate attitudes which maintain that there is no legitimate reason why persons should suffer economically or otherwise because they happen to own or occupy areas important for biological diversity. Once again, such areas should be regarded as an asset rather than a liability. "Compensation" is not an appropriate term in the context of use restrictions. Fianancial payments should instead be viewed as a premium or reward for conservation. In the same way, municipalities and other local authorities should receive benefits if they preserve the biological diversity on land that they own or over which they have powers, including powers to regulate land-use.

Another important step towards achieving such legitimacy is to lay down a clear duty for the State and its citizens to preserve biological diversity. This is best achieved through a constitutional provision, as has already been done in a certain number of countries. Recent examples include the new Constitution of Namibia of 1990, article 95 of which reads:

"The State shall actively promote and maintain the welfare of the people by adopting, *inter alia*, policies aimed at ... [the] maintenance of ecosystems, essential ecological processes and biological diversity of Namibia and utilisation of living natural resources on a sustainable basis for the benefits of all Namibians, both present and future."

The influence of the World Conservation Strategy can be clearly seen in the above provisions.

The Brazilian Constitution of 1988 provides that

"the State must preserve and restore essential ecological processes,... preserve diversity and integrity of the country's genetic heritage, establish protected areas, require environmental impact assessments, promote ecological education and protect fauna and flora."

There is a further provision declaring that the Amazonian forest, the coastal Atlantic forest, the Serra do Mar, the Pantanal and the coastal zone constitute a national heritage, the utilisation of which must be governed by Act of Parliament and under conditions guaranteeing the conservation of the environment.

Most of the Brazilian States have adopted sometimes detailed provisions in their own constitutions on the conservation of natural areas and habitat types.

The value of such constitutional provisions is that conservation is thereby elevated to the same level as human interests of acknowledged legitimacy. This creates an obligation for all Government agencies to weigh carefully all interests at stake before taking any decision significantly affecting the natural environment. If a permit is denied on that basis, the legitimacy of the decision is clearly much greater when based on a constitutional provision.

In addition, some basic laws may be of importance for promoting the legitimacy of conservation where these set out State policy and lay down the principle that conservation is in the general interest. The laws of the State of Victoria and of Spain are particularly interesting examples of this approach, having the stated purpose of 'guaranteeing' the survival of native species of fauna and flora.

Both the implementation and enforcement of laws remain difficult, as there is a natural trend on the part of the part of public authorities to disregard conservation legislation when making regulations or issuing permits. As the law stands in the great majority of countries, no human interest is at stake in such cases and nobody has, as a matter of principle, the standing to challenge the legality of such administrative decisions in the courts. As a result, Government agencies may violate conservation legislation with impunity.

This situation is slowly changing as, often against considerable opposition, an increasing number of countries have now given standing to conservation NGOs to institute judicial proceedings to strike down administrative decisions which they believe have been taken in violation of the law.

Allowing for judicial review of Government action may not, however, be enough. Firstly, court judgments are often pronounced long after the challenged decision has been implemented and works completed which are subsequently declared illegal by the judges. Secondly, even where this is not the case, Government agencies may tend simply to ignore a judgment quashing an illegal administrative order, if they believe that this is in the public interest.

The remedies to these problems are procedural. Courts should be empowered, and in the case of a risk of serious damage to biological diversity, obliged to stay administrative decisions and to enjoin Government agencies to execute court judgments. However, there are still some countries where although NGOs have been given standing to sue, these procedural guarantees do not exist or are not sufficiently used.

Finally, where the law is violated by private persons and the State does not prosecute, as is still frequently the case, certain procedures should be developed to allow NGOs to bring the matter to court.

B.   The Precautionary Principle

Reflecting the emergence of the precautionary principle as a principle of law, the Convention on Biological Diversity states in the Preamble that it is vital to anticipate, prevent and attack the causes of significant reduction or loss of biological diversity at source.

The translation of the precautionary principle into binding rules of law is particularly difficult. The logical approach would be for no activities to be undertaken unless it has first been proved that no significant harm to biological diversity would result. This would require a generalised permit system applicable to all activities and processes which may potentially affect wild fauna and flora and ecosystems. Such a system would entail the transformation of a society in which the accepted rule is that everything which is not specifically prohibited or subject to a permit may be freely undertaken into the opposite form of society, in which everything which is not specifically recognised as harmless to biological diversity would have to be regulated. As this would clearly not be acceptable in democratic societies, some form of compromise is necessary if the legitimacy of biodiversity conservation is ever to be secured.

The essence of such a compromise should be to limit the permit system to those activities or processes which have been clearly identified as presenting unacceptable risks for biological diversity because of their nature, magnitude or location. This may of course vary from one country to another, as a result of local circumstances and the degree of support from public opinion.

One of the best examples of this approach is the permit requirement in the State of South Australia for the clearing of native vegetation. This legislation is a remarkable illustration of the implementation of the precautionary principle and seems to have been well-accepted, as sufficient incentives are apparently provided for native vegetation to be perceived as an asset.

The consequence of this approach is that since the permit system cannot be extended *ad infinitum*, a way must be found to make the promoters of activities and the users of potentially damaging processes more responsible for their actions regarding biological diversity, by encouraging them to implement the precautionary principle on their own initiative.

The above-mentioned compromise should be based upon a clear statement of the obligations of both State and citizens with regard to the maintenance of biological diversity, preferably on the basis of a constitutional provision. The State should be bound not to carry out, authorise or finance any programmes, projects, activities or processes which may significantly affect biological diversity. There should be a corresponding obligation on the part of citizens to avoid causing significant damage to biological diversity, even or especially where no permit is required, that is to say, where the State has no powers to control the activity. This would reflect the position in civil law, whereby persons are liable for substantial damage caused to others, whether or not the action that has been taken is prohibited or allowed by the law.

Such an approach would to a great extent substitute the principles of responsibility and civil liability for those of coercion and criminal liability, which would represent a considerable change in comparison with the present situation.

However, for this system to work, a certain number of legal problems would have to be resolved.

C. A Proposed New System of Liability for Damage to Biological Diversity caused by Public or Private Persons

1. To whom should the system apply?

The system should ideally apply to all persons, whether public or private, which would include all Government departments, local authorities and Government agencies or other bodies. These should all be liable as any other person if they have carried out or authorised projects or activities harmful to biological diversity.

The system should also apply to landowners in respect of damage to biological diversity caused by them on their own land. This should of course be counterbalanced by a right for landowners to enter into conservation agreements and to receive payments accordingly.

2. What would be the nature of the liability?

Liability should be "strict", which means that any person who damages biological diversity may be held legally responsible without it being necessary to prove that the person was at fault. On such a basis, liability would attach to all cases of damage to biological diversity, whether the harm was wilful, caused by negligence or otherwise, except in cases of force majeure. Liability should also arise in cases where the activity has been authorised by a permit, unless the permit-issuing authority has explicitly and specifically authorised the damage. In the latter case, it would be the authority itself which could be held liable.

3. How can significant harm to biological diversity be defined in an objective way?

The concept of "significant harm" is rarely defined by texts. Nevertheless, it is necessary to provide guidance to permit-issuing authorities, the authors of environmental impact assessments, the public in general and the courts. Some criteria are therefore essential and should be laid down by the law.

It is evident that the extinction or endangerment of a species, a habitat type or an area of great biological diversity would constitute "significant harm", whereas the destruction of common plants would generally not do so. A line has to be drawn somewhere between these two extremes. In the case of uncertainty, it would of course be for the courts to decide.

4. How can it be ascertained that significant harm may occur?

First and foremost, this should be ascertained by environmental impact assessments in respect of activities for which permits are required. However, as mentioned in Part II, Chapter VII, there is a need to improve requirements as to the content, scope and review of EIAs which regard to potential harm to biological diversity. In particular, the scope of EIAs should be broadened for proposed activities in or near areas of particular value for biological diversity, so as to cover potentially harmful activities which are normally exempted from the EIA requirement elsewhere.

Moreover, the scope of EIAs should be expanded to cover policies, plans and programmes and potentially harmful processes, even where no permit is required. The objective should be to assess the effects on biodiversity of future activities well before projects are ready, as by that time it may be too late to stop them on the grounds of harm to biodiversity alone, in view of the expenditure incurred and the expectations raised.

With regard to those activities for which no permit is usually required, a simple habitat evaluation methodology should be developed so as to assist landowners or project promoters in ascertaining that proposed activities will not make them liable for significant harm to biological diversity. This could be provided as a public service free of charge or at a nominal charge. The availability of such a service could also be useful to assist landowners to assess biological diversity on their land with a view to the possible conclusion of conservation agreements.

5. How can damage to biological diversity be assessed in monetary terms?

As no human, namely monetary interests are generally affected by damage to biological diversity,[1] the monetary evaluation of such damage is difficult. In the still relatively few cases in which courts have granted damages for harm caused to nature, judges have been trying various alternatives, none of which is really satisfactory. Tables of values assigned to individuals of certain species for the purpose of assessing damages in civil law do exist in certain countries and may help. This system could be expanded to cover natural or semi-natural habitats, on the basis of the amount of destroyed or altered surface area and according to such factors as the species richness and rarity of the habitat concerned.

6. Who would be entitled to claim damages for harm to biological diversity?

The claimant could be the State as the trustee or custodian of biological diversity, as is the position in the United States,[2] or as under the Italian Act of 1986 which established the Ministry of the Environment and laid down rules relating to compensation for environmental damage.

There is of course a risk that the State will not exercise this right for various reasons, including political expediency. The right of action could therefore be vested in an independent body which would be the Defensor of the natural environment. Cases could be brought before

---

[1] If such interests are affected, they can of course be compensated separately.

[2] Comprehensive Environmental Response, Compensation and Liability Act of 1980 (CERCLA, also called the Superfund Act).

the Defensor by the State or approved NGOs. Defensors should also have powers to investigate the facts.

7. Who would make the evaluation?

Although in the last resort this is down to the courts, the evaluation could be made by the Defensor of the natural environment before any proceedings are instituted. The Defensor would draw up its own table of values and develop evaluation criteria which could be endorsed by the State by means of regulations. If the author of the damage agreed with the evaluation, the matter could thus be settled without the need for judicial proceedings. If the author disagreed with the evaluation, the matter would be referred to the civil court for determination. In this case, the evaluation by the Defensor could still be considered as *prima facie* evidence of the value of the damage. It would therefore be for the author of the damage to rebut the presumption and prove to the satisfaction of the court that the evaluation was wrong.

There are some precedents for this system in legislation. Most notorious is that provided by the United States under CERCLA. However, the scope of that Act is limited to damage caused by hazardous substances.

8. What would be the destination of the sums paid in compensation?

The sums thus raised should be used exclusively for conservation, with priority being given to the restoration of the area in respect of which damages have been paid. The best solution is for the moneys to be paid into a special fund managed by the Defensor under the control of the State.

The above system is not completely without precedent, following the adoption of the Convention on Civil Liability for Damage Resulting from Activities Dangerous to the Environment under the auspices of the Council of Europe on 9 March 1993. The object of the Convention is to ensure adequate compensation for damage resulting from activities dangerous to the environment and also to provide for means of prevention and reinstatement.

The Convention covers damage resulting from "dangerous activities" which include the production, handling, storage, use or discharge of "dangerous substances", namely those substances which "have properties which consistute a significant risk for man, the environment or property" and which are to be identified by means of criteria set out in Annex I. The Convention also covers damage resulting from specified operations involving genetically modified organisms and micro-organisms, as well as the operation of certain types of waste incineration or other treatment plant. The Convention excludes damage caused by nuclear substances as well as damage arising from carriage. No reference is made to damage resulting from the introduction of exotic species.

The Convention provides that "the operator of a dangerous activity ... shall be liable for damage caused by the activity" during the time when he was exercising the control over it. When determining the causal link between the activity and the damage, the court "shall take due account of the increased danger of causing the damage inherent in the dangerous activity."

Liability is strict, which means that the victim of any damage is not required to prove that the operator was at fault. However, the operator is exempted from liability where he or she can prove, *inter alia*, that the damage resulted from an act of war or *force majeure*, from an act deliberately carried out by a third party despite appropriate safety precautions, or through compliance with a compulsory measure of a public authority. Where a person suffers damage through his own fault, compensation may be reduced or disallowed, having regard to all the circumstances.

Damage is defined to include not only harm to persons or property but also, more innovatively, "loss or damage by impairment of the environment", to the extent that this loss or damage results from the hazardous properties of the dangerous substances, genetically-modified organisms or micro-organisms or arises from waste. "Damage" also includes the costs of preventive measures and any loss or damage caused by preventive measures.

Other than any loss of profit resulting from impairment of the environment, compensation for such impairment is limited to "the costs of measures of reinstatement actually undertaken or to be undertaken. "Measures of reinstatement" are defined as

"any reasonable measures aiming to reinstate or restore damaged or destroyed components of the environment, or to introduce, where reasonable, the equivalent of these components into the environment. Internal law may indicate who will be entitled to take such measures."

Parties are also required, where appropriate in view of the risks of the activity, to ensure that operators on its territory participate in a financial security scheme or that they have or maintain a financial guarantee, as specified under internal law, sufficient to cover liability under the Convention.

Finally, the Convention is silent as to who may issue proceedings for compensation in the event of damage. Where the relevant damage is to persons or property, the claimant is obviously the person or legal entity which has suffered the damage. However, where it is purely a question of loss or damage by impairment of the environment, each Party is apparently free to determine the persons or bodies who have standing in such circumstances.

The Convention does provide at article 18 that those environmental protection organisations which comply with any further conditions of internal law of the Party concerned may request

- the prohibition of a dangerous activity which is unlawful and poses a grave threat to the environment;
- that the operator be ordered to take measures to prevent an incident or damage;
- that the operator be ordered to take measures, after an incident, to prevent damage; or
- that the operator be ordered to take measures of reinstatement.

It remains to be seen whether Parties will enact or amend internal law loosely or restrictively, so as to grant or withhold standing for environmental NGOs to bring legal proceedings 'on behalf of' the environment which has been damaged in this way.

D. The Position of Developing Countries

Many of the instruments reviewed or proposed in this paper may not necessarily be suitable for developing countries, in which conservation problems and the causes of environmental destruction may be very different. It would be absurd, for instance, to claim civil damages from shifting cultivators who destroy tropical rain forest. On the other hand, there is nothing to prevent such damages being claimed from a contractor who has logged a forest without the required permit.

In developing countries, it is not only a matter of political will but also of financial and human needs. The signature of the Convention on Biological Diversity by almost all countries of the world is indeed a remarkable achievement and a sure sign of the global consensus on the legitimacy of conserving biological diversity. The Preamble recognises that the conservation of biological diversity is a common concern of humankind and specifies that States are responsible for conserving their biological diversity and for using their biological resources in a sustained manner.

To translate these words into practice, the same principles that apply in domestic law should be applied in international law. The presence of biological diversity, especially at high level, should be an asset rather than a liability.

However, until the Convention on Biological Diversity, there has been pressure from the developed world on developing countries to conserve biological diversity, but without providing any incentives to this end. All international conservation instruments laid down obligations but provided no means to assist the poorest Parties to discharge their obligations. The only exception to this rule is the EC Habitats Directive which provides for the possibility of co-financing the conservation of certain species and habitat types.

Moreover, no benefits have traditionally accrued to the countries of origin from the use of genetic resources from their territory, as these were considered to be free goods which anyone could use as he or she pleased. This was a further reason for the lack of any incentive to conserve biological diversity.

The new Convention completely changes the situation by regulating access to biological material and providing that such access may be made subject to payments on mutually agreed terms. Above all, Parties are required to share in a fair and equitable way the results of research and development and the benefits arising from the commercial and other utilisation of genetic resources with the Contracting Party providing such resources.

For the first time in international law, biological diversity is accordingly becoming seen as an asset.

Moreover, the Convention commits developed country Parties to provide new and additional financial resources to enable developing country Parties to meet the incremental costs of discharging their obligations under the Convention. The text makes it clear that the extent to which the developing countries will effectively implement their conservation commitments will depend on the effective implementation by the developed countries of their commitment to provide the necessary resources. In consequence, the fate of biological diversity in the world will depend to a considerable degree on the political will of the industrialised Parties.

Nevertheless, conservation is of course not only a matter of money but also, *inter alia*, a question of adequate legislation and institutions. Most developing countries only use conventional instruments of species protection, national parks and nature reserves. The species-based approach, centred on taking and trade prohibitions or restrictions, is probably essential for large animals, especially where these are subject to trade. Import controls by importing countries under CITES are essential as an enforcement mechanism. Integrated species conservation with critical habitat designations and recovery plans will probably only work in exceptional cases.

However, except for species of exceptional value, such as certain wild relatives of domestic animals or cultivated plants, or well-identified endangered species for which specific conservation measures, including habitat protection, are necessary, the major problem continues to be the destruction of natural forests, coral reefs or other species-rich environments. The conservation of areas is therefore essential, through conventional protected areas, protection forests, wilderness areas and nature parks based on the biosphere reserve concept. Enforcement will often continue to be difficult unless alternatives to the clearing of natural vegetation can be found which provide sufficiently attractive benefits to local populations.

Land-use planning instruments in many developing countries are generally limited to urban planning. Consideration could be given to the development of country-wide or regional plans identifying important areas for conservation, which should be binding upon the State and local authorities and to which any lower-level planning planning document developed in the future should conform. Environmental impact assessment procedures should also be developed which place particular emphasis on activities threatening biological diversity, including policies, programmes and plans. Processes constituting potential threats to biological diversity should be identified and made subject to EIAs. Their development or continuation should be prohibited or limited wherever these would result in unacceptable damage.

These are of course very general recommendations, which should be adapted in each individual case to local conditions. The sheer variety of situations and differing state of development within developing countries is such that it is difficult to be more specific.

What is essential, however, is to find measures and procedures which will make individual farmers and local communities benefit from conservation action: these should therefore provide incentives for conservation or at least attenuate opposition to the associated restrictions. The example of CAMPFIRE in Zimbabwe shows that this is possible, at least in the context of big game found in relatively open country rather than in dense forest. Other methods will have to be found elsewhere.

It is clear that the almost universal system of prohibitions and regulatory measures without the provision of corresponding incentives is not viable in the long term. Incentives will obviously cost money, although not much in comparison with other types of expenditure, such as for military purposes.

With the relaxation of tension in today's world and in spite of the economic crisis, more money could be made available for conservation. This is a firm requirement under the new Convention. It remains to be seen if industrialised countries are prepared to recognise the legitimacy of conservation spending, especially for the benefit of the developing world, and whether developing countries are ready to re-orientate development to conserve biological diversity for the world community. It is to be hoped that the mechanisms established under the Convention on Biological Diversity will assist in reaching these objectives.

BIBLIOGRAPHY

Bean, Michael J. (1983): *The Evolution of National Wildlife Law*, 2nd edition. Praeger Publishers, New York. 449pp.

Bilderbeek, Simone (ed.)(1992): *Biodiversity and International Law*. IOS Press, Oxford, UK. 213pp.

Burhenne-Guilmin, Françoise and Susan Casey-Lefkowitz (1992): "The Convention on Biological Diversity: A Hard-Won Global Achievement" *in* Yearbook of International Environmental Law, Vol. 3 (ed. Handl, Günther). Graham & Trotman/Martinus Nijhoff Publishers, London. pp. 43–59

Chauvet, Michel and Louis Olivier (1993): *La Biodiversité, Enjeu Planétaire: Préserver notre Patrimoine Génétique*. Le Sang de la Terre, Paris. 415pp.

Edelman, Bernard and Marie-Angèle Hermitte (eds.)(1988): *L'Homme, la Nature et le Droit*. Christian Bourgois Editeur, Paris. 392pp.

Experts Group on Environmental Law, World Commission on Environment and Development (1986): *Environmental Protection and Sustainable Development— Legal Principles and Recommendations*. Graham & Trotman/Martinus Nijhoff, London. 196pp.

Favre, David S. (1989): *International Trade in Endangered Species: A Guide to CITES*. Martinus Nijhoff, Dordrecht, The Netherlands. 410pp.

Holt, Sidney J. and Lee M. Talbot (eds.)(1975): *The Conservation of Wild Living Resources*. American Appeal/WWF, Washington DC. 51pp.

IUCN Conservation Monitoring Centre (1987): *Protected Landscapes: Experience Around the World*. IUCN, Gland, Switzerland. 404pp.

IUCN (1980): *World Conservation Strategy*. IUCN/UNEP/WWF, Gland, Switzerland. 66pp.

Kiss, Alexandre (1989): *Droit International de l'Environnement*. Editions A. Pedone, Paris. 349pp.

Kiss, Alexandre (ed.)(1989): *L'Ecologie et la Loi—Le Statut Juridique de l'Environnement*. Editions l'Harmattan, Paris. 391pp.

Klemm, Cyrille de (1993): *Guidelines for Legislation to Implement CITES*; IUCN Environmental Policy and Law Paper No. 26. IUCN, Gland, Switzerland. 107pp.

Klemm, Cyrille de (1991): *Conservation of Biological Diversity and International Law*. IUCN Environmental Law Centre, Bonn. 82pp.

Klemm, Cyrille de (1990): *Wild Plant Conservation and the Law*; IUCN Environmental Policy and Law Paper No. 24. IUCN, Gland, Switzerland. 215pp.

Klemm, Cyrille de (1989): "Migratory Species in International Law" *in* Natural Resources Journal, Vol. 29 No. 4. University of New Mexico, Albuquerque, NM. pp. 935–987.

Klemm, Cyrille de (1989): "The Conservation of Biological Diversity: State Obligations and Citizens' Duties" *in* Environmental Policy and Law, Vol. 19 No. 2. ICEL, Bonn. pp. 50–57.

Klemm, Cyrille de (1985): "Le Patrimoine Naturel de l'Humanité" *in* Colloque 1984: l'Avenir du Droit International de l'Environnement. Martinus Nijhoff, Dordrecht, The Netherlands. pp. 117–150.

Klemm, Cyrille de (1982): "Conservation of Species: The Need for a New Approach" *in* Environmental Policy and Law, Vol. 9

No. 4. North Holland Publishing, Amsterdam. pp. 118–128.

Lyster, Simon (1985): *International Wildlife Law—An Analysis of International Treaties Concerned with the Conservation of Wildlife*. Grotius Publications, Cambridge, UK. 470pp.

McNeely, Jeffrey A. and Kenton R. Miller, Walter V. Reid, Russell A. Mittermeier, Timothy B. Werner (1990): *Conserving the World's Biological Diversity*. WRI Publications, Baltimore, MD. 193pp.

Overseas Development Administration (1991): *Biological Diversity and Developing Countries: Issues and Options*. ODA Foreign and Commonwealth Office, London. 50pp.

Prieur, Michel (1991): *Droit de l'Environnement*, 2nd Edition. Editions Dalloz, Paris. 775pp.

Reid, Walter V. and Kenton R. Miller (1989): *Keeping Options Alive: The Scientific Basis for the Conservation of Biological Diversity*. WRI Publications, Baltimore, MD. 128pp.

Rémond-Gouilloud, Martine (1989): *Du Droit de Détruire—Essai sur le Droit de l'Environnement*. Presses Universitaires de France, Paris. 304pp.

Sand, Peter H. (1990): *Lessons Learned in Global Environmental Governance*. WRI Publications, Washington, DC. 60pp.

Shine, Clare and Palitha T.B. Kohona (1992): "The Convention on Biological Diversity: Bridging the Gap between Conservation and Development" *in* Review of European Community and International Environmental Law, Vol. 1 No. 3. Basil Blackwell, London. pp. 278–288.

Trombetti, Oriana and Kenneth W. Cox (1990): *Land, Law and Wildlife Conservation*. Wildlife Habitat Canada, Ottawa. 51pp.

Wijnstekers, Willem (1992): *The Evolution of CITES: A Reference to the Convention on International Trade in Endangered Species of Wild Fauna and Flora*, 3rd revised edition. CITES Secretariat, Lausanne, Switzerland. 387pp.

The World Bank (1990): *The World Bank and the Environment: First Annual Report Fiscal 1990*. The World Bank, Washington, DC. 102pp.

World Commission on Environment and Development (1987): *Our Common Future*. Oxford University Press, Oxford, UK . 383pp.

Selective Index

K

L

M